PRINCIPLES OF ASYNCHRONOUS CIRCUIT DESIGN

Principles of Asynchronous Circuit Design

A Systems Perspective

Edited by

Jens Sparsø
Technical University of Denmark

and

Steve Furber
The University of Manchester, U.K.

Series Editors

René van Leuken
Alexander de Graaf
Reinder Nouta
TU Delft/DIMES, Delft, The Netherlands

KLUWER ACADEMIC PUBLISHERS
BOSTON / DORDRECHT / LONDON

A C.I.P. Catalogue record for this book is available from the Library of Congress.

ISBN 0-7923-7613-7

Published by Kluwer Academic Publishers,
P.O. Box 17, 3300 AA Dordrecht, The Netherlands.

Sold and distributed in North, Central and South America
by Kluwer Academic Publishers,
101 Philip Drive, Norwell, MA 02061, U.S.A.

In all other countries, sold and distributed
by Kluwer Academic Publishers,
P.O. Box 322, 3300 AH Dordrecht, The Netherlands.

Printed on acid-free paper

Printed in the Netherlands.

Contents

Preface

This book was compiled to address a perceived need for an introductory text on asynchronous design. There are several highly technical books on aspects of the subject, but no obvious starting point for a designer who wishes to become acquainted for the first time with asynchronous technology. We hope this book will serve as that starting point.

The reader is assumed to have some background in digital design. We assume that concepts such as logic gates, flip-flops and Boolean logic are familiar. Some of the latter sections also assume familiarity with the higher levels of digital design such as microprocessor architectures and systems-on-chip, but readers unfamiliar with these topics should still find the majority of the book accessible.

The intended audience for the book comprises the following groups:

- Industrial designers with a background in conventional (clocked) digital design who wish to gain an understanding of asynchronous design in order, for example, to establish whether or not it may be advantageous to use asynchronous techniques in their next design task.

- Students in Electronic and/or Computer Engineering who are taking a course that includes aspects of asynchronous design.

The book is structured in three parts. Part I is a tutorial in asynchronous design. It addresses the most important issue for the beginner, which is how to think about asynchronous systems. The first big hurdle to be cleared is that of mindset – asynchronous design requires a different mental approach from that normally employed in clocked design. Attempts to take an existing clocked system, strip out the clock and simply replace it with asynchronous handshakes are doomed to disappoint. Another hurdle is that of circuit design methodology – the existing body of literature presents an apparent plethora of disparate approaches. The aim of the tutorial is to get behind this and to present a single unified and coherent perspective which emphasizes the common ground. In this way the tutorial should enable the reader to begin to understand the characteristics of asynchronous systems in a way that will enable them to 'think outside the box' of conventional clocked design and to create radical new design solutions that fully exploit the potential of clockless systems.

Once the asynchronous design mindset has been mastered, the second hurdle is designer productivity. VLSI designers are used to working in a highly productive environment supported by powerful automatic tools. Asynchronous design lags in its tools environment, but things are improving. Part II of the book gives an introduction to Balsa, a high-level synthesis system for asynchronous circuits. It is written by Doug Edwards (who has managed the Balsa development at the University of Manchester since its inception) and Andrew Bardsley (who has written most of the software). Balsa is not the solution to all asynchronous design problems, but it is capable of synthesizing very complex systems (for example, the 32-channel DMA controller used on the DRACO chip described in Chapter 15) and it is a good way to develop an understanding of asynchronous design 'in the large'.

Knowing how to think about asynchronous design and having access to suitable tools leaves one question: what can be built in this way? In Part III we offer a number of examples of complex asynchronous systems as illustrations of the answer to this question. In each of these examples the designers have been asked to provide descriptions that will provide the reader with insights into the design process. The examples include a commercial smart card chip designed at Philips and a Viterbi decoder designed at the University of Manchester. Part III closes with a discussion of the issues that come up in the design of advanced asynchronous microprocessors, focusing on the Amulet processor series, again developed at the University of Manchester.

Although the book is a compilation of contributions from different authors, each of these has been specifically written with the goals of the book in mind – to provide answers to the sorts of questions that a newcomer to asynchronous design is likely to ask. In order to keep the book accessible and to avoid it becoming an intimidating size, much valuable work has had to be omitted. Our objective in introducing you to asynchronous design is that you might become acquainted with it. If your relationship develops further, perhaps even into the full-blown affair that has smitten a few, included among whose number are the contributors to this book, you will, of course, want to know more. The book includes an extensive bibliography that will provide food enough for even the most insatiable of appetites.

JENS SPARSØ AND STEVE FURBER, SEPTEMBER 2001

Acknowledgments

Many people have helped significantly in the creation of this book. In addition to writing their respective chapters, several of the authors have also read and commented on drafts of other parts of the book, and the quality of the work as a whole has been enhanced as a result.

The editors are also grateful to Alan Williams, Russell Hobson and Steve Temple, for their careful reading of drafts of this book and their constructive suggestions for improvement.

Part I of the book has been used as a course text and the quality and consistency of the content improved by feedback from the students on the spring 2001 course "49425 Design of Asynchronous Circuits" at DTU.

Any remaining errors or omissions are the responsibility of the editors.

The writing of this book was initiated as a dissemination effort within the European Low-Power Initiative for Electronic System Design (ESD-LPD), and this book is part of the book series from this initiative. As will become clear, the book goes far beyond the dissemination of results from projects within in the ESD-LPD cluster, and the editors would like to acknowledge the support of the working group on asynchronous circuit design, ACiD-WG, that has provided a fruitful forum for interaction and the exchange of ideas. The ACiD-WG has been funded by the European Commission since 1992 under several Framework Programmes: FP3 Basic Research (EP7225), FP4 Technologies for Components and Subsystems (EP21949), and FP5 Microelectronics (IST-1999-29119).

Foreword

This book is the third in a series on novel low-power design architectures, methods and design practices. It results from a large European project started in 1997, whose goal is to promote the further development and the faster and wider industrial use of advanced design methods for reducing the power consumption of electronic systems.

Low-power design became crucial with the widespread use of portable information and communication terminals, where a small battery has to last for a long period. High-performance electronics, in addition, suffers from a continuing increase in the dissipated power per square millimeter of silicon, due to increasing clock-rates, which causes cooling and reliability problems or otherwise limits performance.

The European Union's Information Technologies Programme 'Esprit' therefore launched a 'Pilot action for Low-Power Design', which eventually grew to 19 R&D projects and one coordination project, with an overall budget of 14 million EUROs. This action is known as the European Low-Power Initiative for Electronic System Design (ESD-LPD) and will be completed in the year 2002. It aims to develop or demonstrate new design methods for power reduction, while the coordination project ensures that the methods, experiences and results are properly documented and publicised.

The initiative addresses low-power design at various levels. These include system and algorithmic level, instruction set processor level, custom processor level, register transfer level, gate level, circuit level and layout level. It covers data-dominated, control-dominated and asynchronous architectures. 10 projects deal mainly with digital circuits, 7 with analog and mixed-signal circuits, and 2 with software-related aspects. The principal application areas are communication, medical equipment and e-commerce devices.

The following list describes the objectives of the 20 projects. It is sorted by decreasing funding budget.

CRAFT CMOS Radio Frequency Circuit Design for Wireless Application

- Advanced CMOS RF circuit design including blocks such as LNA, down converter mixers & phase shifters, oscillators and frequency synthesisers, integrated filters delta sigma conversion, power amplifiers
- Development of novel models for active and passive devices as well as fine-tuning and validation based on first silicon prototypes
- Analysis and specification of sophisticated architectures to meet, in particular, low-power single-chip implementation

PAPRICA Power and Part Count Reduction Innovative Communication Architecture

- Feasibility assessment of DQIF, through physical design and characterisation of the core blocks
- Low-power RF design techniques in standard CMOS digital processes
- RF design tools and framework; PAPRICA Design Kit
- Demonstration of a practical implementation of a specific application

MELOPAS Methodology for Low Power Asic design

- To develop a methodology to evaluate the power consumption of a complex ASIC early on in the design flow
- To develop a hardware/software co-simulation tool
- To quickly achieve a drastic reduction in the power consumption of electronic equipment

TARDIS Technical Coordination and Dissemination

- To organise the communication between design experiments and to exploit their potential synergy
- To guide the capturing of methods and experiences gained in the design experiments
- To organise and promote the wider dissemination and use of the gathered design know-how and experience

LUCS Low-Power Ultrasound Chip Set.

- Design methodology on low-power ADC, memory and circuit design
- Prototype demonstration of a hand-held medical ultrasound scanner

ALPINS Analog Low-Power Design for Communications Systems

- Low-voltage voice band smoothing filters and analog-to-digital and digital-to-analog converters for an analog front-end circuit for a DECT system
- High linear transconductor-capacitor (gm-C) filter for GSM Analog Interface Circuit operating at supply voltages as low as 2.5V
- Formal verification tools, which will be implemented in the industrial partner's design environment. These tools support the complete design process from system level down to transistor level

SALOMON System-level analog-digital trade-off analysis for low power

- A general top-down design flow for mixed-signal telecom ASICs
- High-level models of analog and digital blocks and power estimators for these blocks
- A prototype implementation of the design flow with particular software tools to demonstrate the general design flow

DESCALE Design Experiment on a Smart Card Application for Low Energy

- The application of highly innovative handshake technology
- Aiming at some 3 to 5 times less power and some 10 times smaller peak currents compared to synchronously operated solutions

SUPREGE A low-power SUPerREGEnerative transceiver for wireless data transmission at short distances

- Design trade-offs and optimisation of the micro power receiver/transmitter as a function of various parameters (power consumption, area, bandwidth, sensitivity, etc)
- Modulation/demodulation and interface with data transmission systems
- Realisation of the integrated micro power receiver/transmitter based on the super-regeneration principle

PREST Power REduction for System Technologies

- Survey of contemporary Low-Power Design techniques and commercial power analysis software tools
- Investigation of architectural and algorithmic design techniques with a power consumption comparison
- Investigation of Asynchronous design techniques and Arithmetic styles
- Set-up and assessment of a low-power design flow
- Fabrication and characterisation of a Viterbi demonstrator to assess the most promising power reduction techniques

DABLP Low-Power Exploration for Mapping DAB Applications to Multi-Processors

- A DAB channel decoder architecture with reduced power consumption
- Refined and extended ATOMIUM methodology and supporting tools

COSAFE Low-Power Hardware-Software Co-Design for Safety-Critical Applications

- The development of strategies for power-efficient assignment of safety critical mechanisms to hardware or software
- The design and implementation of a low-power, safety-critical ASIP, which realises the control unit of a portable infusion pump system

AMIED Asynchronous Low-Power Methodology and Implementation of an Encryption/Decryption System

- Implementation of the IDEA encryption/decryption method with drastically reduced power consumption
- Advanced low-power design flow with emphasis on algorithm and architecture optimisations
- Industrial demonstration of the asynchronous design methodology based on commercial tools

LPGD A Low-Power Design Methodology/Flow and its Application to the Implementation of a DCS1800-GSM/DECT Modulator/Demodulator

- To complete the development of a top-down, low-power design methodology/flow for DSP applications
- To demonstrate the methods on the example of an integrated GFSK/GMSK Modulator-Demodulator (MODEM) for DCS1800-GSM/DECT applications

SOFLOPO Low-Power Software Development for Embedded Applications

- Develop techniques and guidelines for mapping a specific algorithm code onto appropriate instruction subsets
- Integrate these techniques into software for the power-conscious ARM-RISC and DSP code optimisation

I-MODE Low-Power RF to Base Band Interface for Multi-Mode Portable Phone

- To raise the level of integration in a DECT/DCS1800 transceiver, by implementing the necessary analog base band low-pass filters and data converters in CMOS technology using low-power techniques

COOL-LOGOS Power Reduction through the Use of Local don't Care Conditions and Global Gate Resizing Techniques: An Experimental Evaluation.

- To apply the developed low-power design techniques to an existing 24-bit DSP, which is already fabricated
- To assess the merit of the new techniques using experimental silicon through comparisons of the projected power reduction (in simulation) and actually measured reduction of new DSP; assessment of the commercial impact

LOVO Low Output VOltage DC/DC converters for low-power applications

- Development of technical solutions for the power supplies of advanced low-power systems
- New methods for synchronous rectification for very low output voltage power converters

PCBIT Low-Power ISDN Interface for Portable PC's

- Design of a PC-Card board that implements the PCBIT interface
- Integrate levels 1 and 2 of the communication protocol in a single ASIC
- Incorporate power management techniques in the ASIC design:
  - system level: shutdown of idle modules in the circuit
  - gate level: precomputation, gated-clock FSMs

COLOPODS Design of a Cochlear Hearing Aid Low-Power DSP System

- Selection of a future oriented low-power technology enabling future power reduction through integration of analog modules
- Design of a speech processor IC yielding a power reduction of 90% compared to the 3.3 Volt implementation

The low power design projects have achieved the following results:

- Projects that have designed prototype chips can demonstrate power reductions of 10 to 30 percent.
- New low-power design libraries have been developed.
- New proven low-power RF architectures are now available.
- New smaller and lighter mobile equipment has been developed.

Instead of running a number of Esprit projects at the same time independently of each other, during this pilot action the projects have collaborated strongly. This is achieved mostly by the novel feature of this action, which is the presence and role of the coordinator: DIMES - the Delft Institute of Microelectronics and Submicron-technology, located in Delft, the Netherlands (http://www.dimes.tudelft.nl). The task of the coordinator is to co-ordinate, facilitate, and organize:

- the information exchange between projects;
- the systematic documentation of methods and experiences;
- the publication and the wider dissemination to the public.

The most important achievements, credited to the presence of the coordinator are:

- New personnel contacts have been made, and as a consequence the resulting synergy between partners resulted in better and faster developments.

- The organization of low-power design workshops, special sessions at conferences, and a low-power design web site:

  `http://www.esdlpd.dimes.tudelft.nl.`

  At this site all of the public reports from the projects can be found, as can all kinds of information about the initiative itself.

- The design methodology, design methods and/or design experience are disclosed, are well-documented and available.

  Based on the work of the projects, and in cooperation with the projects, the publication of a low-power design book series is planned. Written by members of the projects, this series of books on low-power design will disseminate novel design methodologies and design experiences that were obtained during the run-time of the European Low Power Initiative for Electronic System Design, to the general public.

In conclusion, the major contribution of this project cluster is, in addition to the technical achievements already mentioned, the acceleration of the introduction of novel knowledge on low-power design methods into mainstream development processes.

We would like to thank all project partners from all of the different companies and organizations who make the Low-Power Initiative a success.

Rene van Leuken, Reinder Nouta, Alexander de Graaf, Delft, June 2001

I

ASYNCHRONOUS CIRCUIT DESIGN – A TUTORIAL

Author: Jens Sparsø

Technical University of Denmark

jsp@imm.dtu.dk

Abstract Asynchronous circuits have characteristics that differ significantly from those of synchronous circuits and, as will be clear from some of the later chapters in this book, it is possible exploit these characteristics to design circuits with very interesting performance parameters in terms of their power, performance, electromagnetic emissions (EMI), etc.

Asynchronous design is not yet a well-established and widely-used design methodology. There are textbooks that provide comprehensive coverage of the underlying theories, but the field has not yet matured to a point where there is an established curriculum and university tradition for teaching courses on asynchronous circuit design to electrical engineering and computer engineering students.

As this author sees the situation there is a gap between understanding the fundamentals and being able to design useful circuits of some complexity. The aim of Part I of this book is to provide a tutorial on asynchronous circuit design that fills this gap.

More specifically the aims are: (i) to introduce readers with background in synchronous digital circuit design to the fundamentals of asynchronous circuit design such that they are able to read and understand the literature, *and* (ii) to provide readers with an understanding of the "nature" of asynchronous circuits such that they are to design non-trivial circuits with interesting performance parameters.

The material is based on experience from the design of several asynchronous chips, and it has evolved over the last decade from tutorials given at a number of European conferences and from a number of special topics courses taught at the Technical University of Denmark and elsewhere. In May 1999 I gave a one-week intensive course at Delft University of Technology and it was when preparing for this course I felt that the material was shaping up, and I set out to write the following text. Most of the material has recently been used and debugged in a course at the Technical University of Denmark in the spring 2001. Supplemented by a few journal articles and a small design project, the text may be used for a one semester course on asynchronous design.

Keywords: asynchronous circuits, tutorial

Chapter 1

INTRODUCTION

1.1. Why consider asynchronous circuits?

Most digital circuits designed and fabricated today are "synchronous". In essence, they are based on two fundamental assumptions that greatly simplify their design: (1) all signals are binary, and (2) all components share a common and discrete notion of time, as defined by a clock signal distributed throughout the circuit.

Asynchronous circuits are fundamentally different; they also assume binary signals, *but there is no common and discrete time.* Instead the circuits use handshaking between their components in order to perform the necessary synchronization, communication, and sequencing of operations. Expressed in 'synchronous terms' this results in a behaviour that is similar to systematic fine-grain clock gating and local clocks that are not in phase and whose period is determined by actual circuit delays – registers are only clocked where and when needed.

This difference gives asynchronous circuits inherent properties that can be (and have been) exploited to advantage in the areas listed and motivated below. The interested reader may find further introduction to the mechanisms behind the advantages mentioned below in [140].

- Low power consumption, [136, 138, 42, 45, 99, 102]
  ...due to fine-grain clock gating and zero standby power consumption.

- High operating speed, [156, 157, 88]
  ...operating speed is determined by actual local latencies rather than global worst-case latency.

- Less emission of electro-magnetic noise, [136, 109]
  ...the local clocks tend to tick at random points in time.

- Robustness towards variations in supply voltage, temperature, and fabrication process parameters, [87, 98, 100]
  ...timing is based on matched delays (and can even be insensitive to circuit and wire delays).

J. Sparsø and S. Furber (eds.), Principles of Asynchronous Circuit Design, 3–8.

- Better composability and modularity, [92, 80, 142, 128, 124]
  ...because of the simple handshake interfaces and the local timing.

- No clock distribution and clock skew problems,
  ...there is no global signal that needs to be distributed with minimal phase skew across the circuit.

On the other hand there are also some drawbacks. The asynchronous control logic that implements the handshaking normally represents an overhead in terms of silicon area, circuit speed, and power consumption. It is therefore pertinent to ask whether or not the investment pays off, i.e. whether the use of asynchronous techniques results in a substantial improvement in one or more of the above areas. Other obstacles are a lack of CAD tools and strategies and a lack of tools for testing and test vector generation.

Research in asynchronous design goes back to the mid 1950s [93, 92], but it was not until the late 1990s that projects in academia and industry demonstrated that it is possible to design asynchronous circuits which exhibit significant benefits in nontrivial real-life examples, and that commercialization of the technology began to take place. Recent examples are presented in [106] and in Part III of this book.

1.2. Aims and background

There are already several excellent articles and book chapters that introduce asynchronous design [54, 33, 34, 35, 140, 69, 124] as well as several monographs and textbooks devoted to asynchronous design including [106, 14, 25, 18, 95] – why then write yet another introduction to asynchronous design? There are several reasons:

- My experience from designing several asynchronous chips [123, 103], and from teaching asynchronous design to students and engineers over the past 10 years, is that it takes more than knowledge of the basic principles and theories to design efficient asynchronous circuits. In my experience there is a large gap between the introductory articles and book chapters mentioned above explaining the design methods and theories on the one side, and the papers describing actual designs and current research on the other side. It takes more than knowing the rules of a game to play and win the game. Bridging this gap involves experience and a good understanding of the nature of asynchronous circuits. An experience that I share with many other researchers is that "just going asynchronous" results in larger, slower and more power consuming circuits. *The crux is to use asynchronous techniques to exploit characteristics in the algorithm and architecture of the application in question.* This further implies that asynchronous techniques may not always be the right solution to the problem.

- Another issue is that asynchronous design is a rather young discipline. Different researchers have proposed different circuit structures and design methods. At a first glance they may seem different – an observation that is supported by different terminologies; but a closer look often reveals that the underlying principles and the resulting circuits are rather similar.

- Finally, most of the above-mentioned introductory articles and book chapters are comprehensive in nature. While being appreciated by those already working in the field, the multitude of different theories and approaches in existence represents an obstacle for the newcomer wishing to get started designing asynchronous circuits.

Compared to the introductory texts mentioned above, the aims of this tutorial are: (1) to provide an introduction to asynchronous design that is more selective, (2) to stress basic principles and similarities between the different approaches, and (3) to take the introduction further towards designing practical and useful circuits.

1.3. Clocking versus handshaking

Figure 1.1(a) shows a synchronous circuit. For simplicity the figure shows a pipeline, but it is intended to represent any synchronous circuit. When designing ASICs using hardware description languages and synthesis tools, designers focus mostly on the data processing and assume the existence of a global clock. For example, a designer would express the fact that data clocked into register $R3$ is a function $CL3$ of the data clocked into $R2$ at the previous clock as the following assignment of variables: $R3 := CL3(R2)$. Figure 1.1(a) represents this high-level view with a universal clock.

When it comes to physical design, reality is different. Todays ASICs use a structure of clock buffers resulting in a large number of (possibly gated) clock signals as shown in figure 1.1(b). It is well known that it takes CAD tools and engineering effort to design the clock gating circuitry and to minimize and control the skew between the many different clock signals. Guaranteeing the two-sided timing constraints – the setup to hold time window around the clock edge – in a world that is dominated by wire delays is not an easy task. The buffer-insertion-and-resynthesis process that is used in current commercial CAD tools may not converge and, even if it does, it relies on delay models that are often of questionable accuracy.

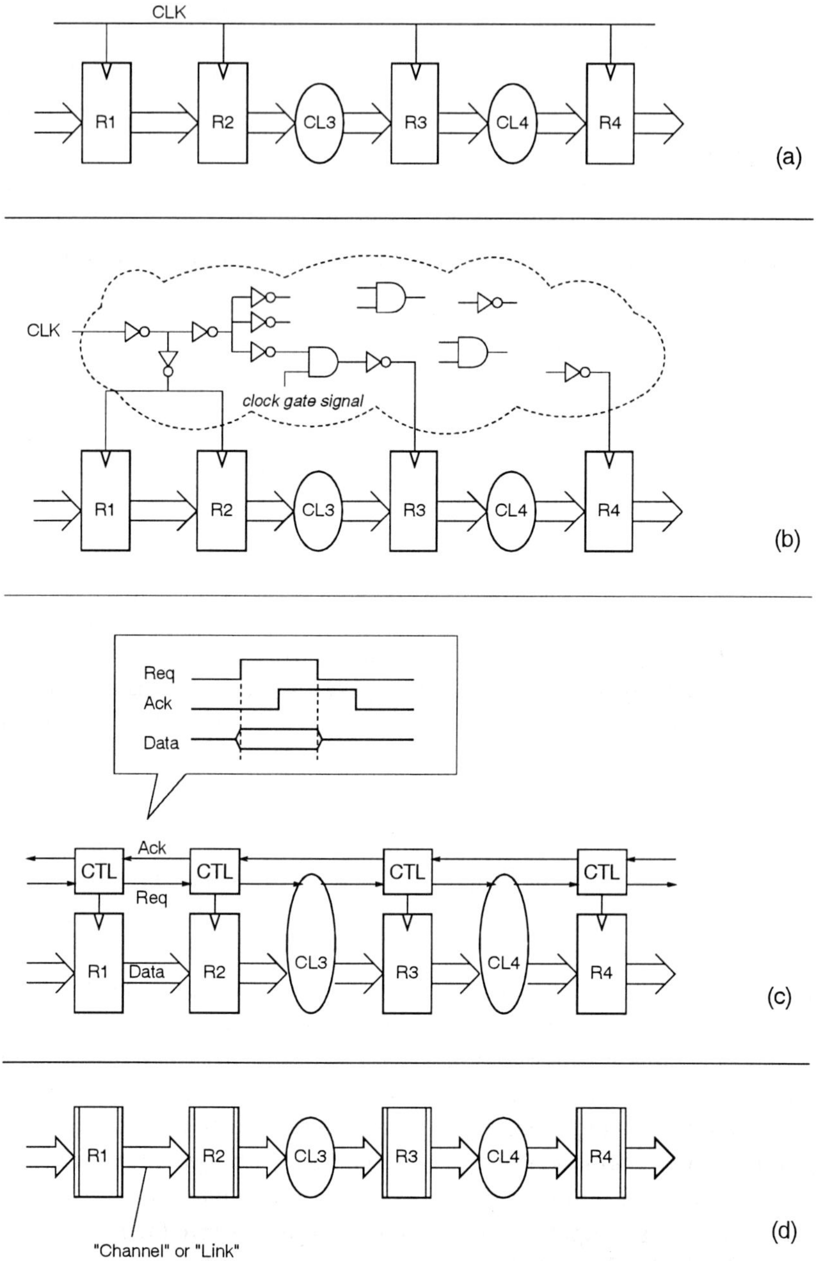

Figure 1.1. (a) A synchronous circuit, (b) a synchronous circuit with clock drivers and clock gating, (c) an equivalent asynchronous circuit, and (d) an abstract data-flow view of the asynchronous circuit. (The figure shows a pipeline, but it is intended to represent any circuit topology).

Asynchronous design represents an alternative to this. In an asynchronous circuit the clock signal is replaced by some form of handshaking between neighbouring registers; for example the simple request-acknowledge based handshake protocol shown in figure 1.1(c). In the following chapter we look at alternative handshake protocols and data encodings, but before departing into these implementation details it is useful to take a more abstract view as illustrated in figure 1.1(d):

- think of the data and handshake signals connecting one register to the next in figure 1.1(c) as a "handshake channel" or "link,"
- think of the data stored in the registers as tokens tagged with data values (that may be changed along the way as tokens flow through combinational circuits), and
- think of the combinational circuits as being transparent to the handshaking between registers; a combinatorial circuit simply absorbs a token on each of its input links, performs its computation, and then emits a token on each of its output links (much like a transition in a Petri net, c.f. section 6.2.1).

Viewed this way, an asynchronous circuit is simply a static data-flow structure [36]. Intuitively, correct operation requires that data tokens flowing in the circuit do not disappear, that one token does not overtake another, and that new tokens do not appear out of nowhere. A simple rule that can ensure this is the following:

> *A register may input and store a new data token from its predecessor if its successor has input and stored the data token that the register was previously holding.* [The states of the predecessor and successor registers are signaled by the incoming request and acknowledge signals respectively.]

Following this rule data is copied from one register to the next along the path through the circuit. In this process subsequent registers will often be holding copies of the same data value but the old duplicate data values will later be overwritten by new data values in a carefully ordered manner, and a handshake cycle on a link will always enclose the transfer of exactly one data-token. Understanding this "token flow game" is crucial to the design of efficient circuits, and we will address these issues later, extending the token-flow view to cover structures other than pipelines. Our aim here is just to give the reader an intuitive feel for the fundamentally different nature of asynchronous circuits.

An important message is that the "handshake-channel and data-token view" represents a very useful abstraction that is equivalent to the register transfer level (RTL) used in the design of synchronous circuits. This *data-flow abstraction*, as we will call it, separates the structure and function of the circuit from the implementation details of its components.

Another important message is that it is the handshaking between the registers that controls the flow of tokens, whereas the combinational circuit blocks must be fully transparent to this handshaking. Ensuring this transparency is not always trivial; it takes more than a traditional combinational circuit, so we will use the term 'function block' to denote a combinational circuit whose input and output ports are handshake-channels or links.

Finally, some more down-to-earth engineering comments may also be relevant. The synchronous circuit in figure 1.1(b) is "controlled" by clock pulses that are in phase with a periodic clock signal, whereas the asynchronous circuit in figure 1.1(c) is controlled by locally derived clock pulses that can occur at any time; the local handshaking ensures that clock pulses are generated where and when needed. This tends to randomize the clock pulses over time, and is likely to result in less electromagnetic emission and a smoother supply current without the large di/dt spikes that characterize a synchronous circuit.

1.4. Outline of Part I

Chapter 2 presents a number of fundamental concepts and circuits that are important for the understanding of the following material. Read through it but don't get stuck; you may want to revisit relevant parts later.

Chapters 3 and 4 address asynchronous design at the data-flow level: chapter 3 explains the operation of pipelines and rings, introduces a set of handshake components and explains how to design (larger) computing structures, and chapter 4 addresses performance analysis and optimization of such structures, both qualitatively and quantitatively.

Chapter 5 addresses the circuit level implementation of the handshake components introduced in chapter 3, and chapter 6 addresses the design of hazard-free sequential (control) circuits. The latter includes a general introduction to the topics and in-depth coverage of one specific method: the design of speed-independent control circuits from signal transition graph specifications. These techniques are illustrated by control circuits used in the implementation of some of the handshake components introduced in chapter 3.

All of the above chapters 2–6 aim to explain the basic techniques and methods in some depth. The last two chapters are briefer. Chapter 7 introduces more advanced topics related to the implementation of circuits using the 4-phase bundled-data protocol, and chapter 8 addresses hardware description languages and synthesis tools for asynchronous design. Chapter 8 is by no means comprehensive; it focuses on CSP-like languages and syntax-directed compilation, but also describes how asynchronous design can be supported by a standard language, VHDL.

Chapter 2

FUNDAMENTALS

This chapter provides explanations of a number of topics and concepts that are of fundamental importance for understanding the following chapters and for appreciating the similarities between the different asynchronous design styles. The presentation style will be somewhat informal and the aim is to provide the reader with intuition and insight.

2.1. Handshake protocols

The previous chapter showed one particular handshake protocol known as a return-to-zero handshake protocol, figure 1.1(c). In the asynchronous community it is given a more informative name: the 4-phase bundled-data protocol.

2.1.1 Bundled-data protocols

The term *bundled-data* refers to a situation where the data signals use normal Boolean levels to encode information, and where separate request and acknowledge wires are bundled with the data signals, figure 2.1(a). In the *4-phase* protocol illustrated in figure 2.1(b) the request and acknowledge wires also use normal Boolean levels to encode information, and the term 4-phase refers to the number of communication actions: (1) the sender issues data and sets request high, (2) the receiver absorbs the data and sets acknowledge high, (3) the sender responds by taking request low (at which point data is no longer guaranteed to be valid) and (4) the receiver acknowledges this by taking acknowledge low. At this point the sender may initiate the next communication cycle.

The 4-phase bundled data protocol is familiar to most digital designers, but it has a disadvantage in the superfluous return-to-zero transitions that cost unnecessary time and energy. The 2-phase bundled-data protocol shown in figure 2.1(c) avoids this. The information on the request and acknowledge wires is now encoded as signal transitions on the wires and there is no difference between a $0 \rightarrow 1$ and a $1 \rightarrow 0$ transition, they both represent a "signal event". Ideally the 2-phase bundled-data protocol should lead to faster circuits than the 4-phase bundled-data protocol, but often the implementation of circuits

J. Sparsø and S. Furber (eds.), Principles of Asynchronous Circuit Design, 9–28.

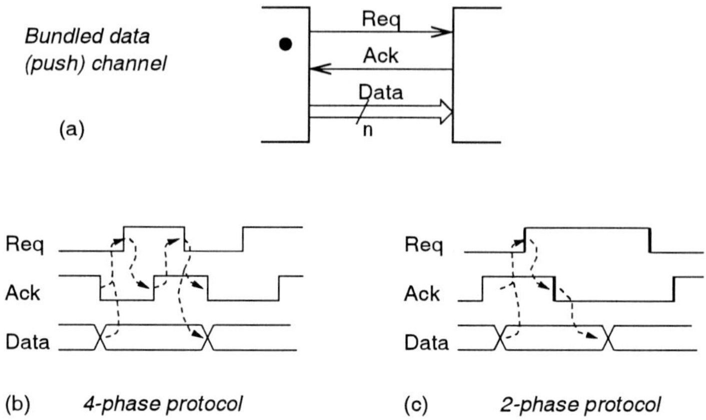

Figure 2.1. (a) A bundled-data channel. (b) A 4-phase bundled-data protocol. (c) A 2-phase bundled-data protocol.

responding to events is complex and there is no general answer as to which protocol is best.

At this point some discussion of terminology is appropriate. Instead of the term *bundled-data* that is used throughout this text, some texts use the term *single-rail*. The term 'bundled-data' hints at the timing relationship between the data signals and the handshake signals, whereas the term 'single-rail' hints at the use of one wire to carry one bit of data. Also, the term single-rail may be considered consistent with the dual-rail data representation discussed in the next section. Instead of the term *4-phase* handshaking (or signaling) some texts use the terms *return-to-zero (RTZ) signaling* or *level signaling*, and instead of the term *2-phase* handshaking (or signaling) some texts use the terms *non-return-to-zero (NRZ) signaling* or *transition signaling*. Consequently a return-to-zero single-track prococol is the same as a 4-phase bundled-data protocol, etc.

The protocols introduced above all assume that the sender is the active party that initiates the data transfer over the channel. This is known as a *push channel*. The opposite, the receiver asking for new data, is also possible and is called a *pull channel*. In this case the directions of the request and acknowledge signals are reversed, and the validity of data is indicated in the acknowledge signal going from the sender to the receiver. In abstract circuit diagrams showing links/channels as one symbol we will often mark the active end of a channel with a dot, as illustrated in figure 2.1(a).

To complete the picture we mention a number of variations: (1) a channel without data that can be used for synchronization, and (2) a channel where data is transmitted in both directions and where *req* and *ack* indicate validity of the data that is exchanged. The latter could be used to interface a read-only memory: the address would be bundled with *req* and the data would be bundled with *ack*. These alternatives are explained later in section 7.1.1. In the following sections we will restrict the discussion to push channels.

All the bundled-data protocols rely on delay matching, such that the order of signal events at the sender's end is preserved at the receiver's end. On a push channel, data is valid before request is set high, expressed formally as $Valid(Data) \prec Req$. This ordering should also be valid at the receiver's end, and it requires some care when physically implementing such circuits. Possible solutions are:

- To control the placement and routing of the wires, possibly by routing all signals in a channel as a bundle. This is trivial in a tile-based datapath structure.
- To have a safety margin at the sender's end.
- To insert and/or resize buffers after layout (much as is done in today's synthesis and layout CAD tools).

An alternative is to use a more sophisticated protocol that is robust to wire delays. In the following sections we introduce a number of such protocols that are completely insensitive to delays.

2.1.2 The 4-phase dual-rail protocol

The 4-phase dual-rail protocol encodes the request signal into the data signals using two wires per bit of information that has to be communicated, figure 2.2. In essence it is a 4-phase protocol using two request wires per bit of information *d*; one wire *d.t* is used for signaling a logic 1 (or true), and another wire *d.f* is used for signaling logic 0 (or false). When observing a 1-bit channel one will see a sequence of 4-phase handshakes where the participating

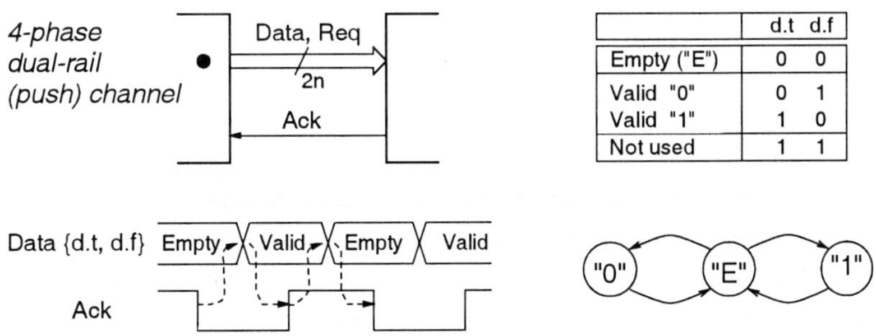

	d.t	d.f
Empty ("E")	0	0
Valid "0"	0	1
Valid "1"	1	0
Not used	1	1

Figure 2.2. A delay-insensitive channel using the 4-phase dual-rail protocol.

"request" signal in any handshake cycle can be either $d.t$ or $d.f$. This protocol is very robust; two parties can communicate reliably regardless of delays in the wires connecting the two parties – the protocol is *delay-insensitive*.

Viewed together the $\{x.f, x.t\}$ wire pair is a codeword; $\{x.f, x.t\} = \{1,0\}$ and $\{x.f, x.t\} = \{0,1\}$ represent "valid data" (logic 0 and logic 1 respectively) and $\{x.f, x.t\} = \{0,0\}$ represents "no data" (or "spacer" or "empty value" or "NULL"). The codeword $\{x.f, x.t\} = \{1,1\}$ is not used, and a transition from one valid codeword to another valid codeword is not allowed, as illustrated in figure 2.2.

This leads to a more abstract view of 4-phase handshaking: (1) the sender issues a valid codeword, (2) the receiver absorbs the codeword and sets acknowledge high, (3) the sender responds by issuing the empty codeword, and (4) the receiver acknowledges this by taking acknowledge low. At this point the sender may initiate the next communication cycle. An even more abstract view of what is seen on a channel is a data stream of valid codewords separated by empty codewords.

Let's now extend this approach to bit-parallel channels. An N-bit data channel is formed simply by concatenating N wire pairs, each using the encoding described above. A receiver is always able to detect when all bits are valid (to which it responds by taking acknowledge high), and when all bits are empty (to which it responds by taking acknowledge low). This is intuitive, but there is also some mathematical background – the dual-rail code is a particularly simple member of the family of delay-insensitive codes [147], and it has some nice properties:

- any concatenation of dual-rail codewords is itself a dual-rail codeword.
- for a given N (the number of bits to be communicated), the set of all possible codewords can be *disjointly* divided into 3 sets:
  - the *empty codeword* where all N wire pairs are $\{0,0\}$.
  - the *intermediate codewords* where some wire-pairs assume the empty state and some wire pairs assume valid data.
  - the 2^N different *valid codewords*.

Figure 2.3 illustrates the handshaking on an N-bit channel: a receiver will see the empty codeword, a sequence of intermediate codewords (as more and more bits/wire-pairs become valid) and eventually a valid codeword. After receiving and acknowledging the codeword, the receiver will see a sequence of intermediate codewords (as more and more bits become empty), and eventually the empty codeword to which the receiver responds by driving acknowledge low again.

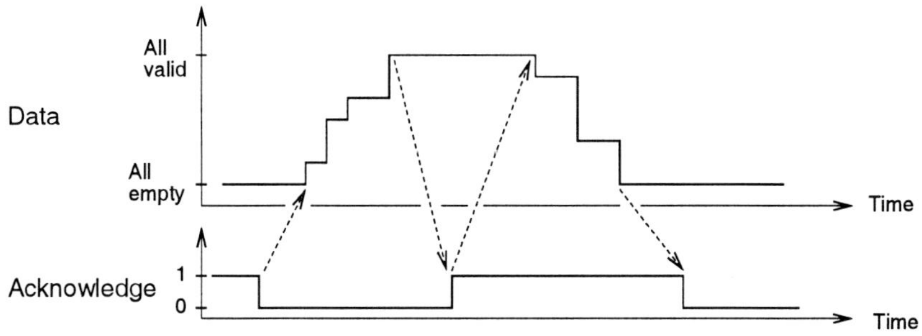

Figure 2.3. Illustration of the handshaking on a 4-phase dual-rail channel.

2.1.3 The 2-phase dual-rail protocol

The 2-phase dual-rail protocol also uses 2 wires $\{d.t, d.f\}$ per bit, but the information is encoded as transitions (events) as explained previously. On an N-bit channel a new codeword is received when exactly one wire in each of the N wire pairs has made a transition. There is no empty value; a valid message is acknowledged and followed by another message that is acknowledged. Figure 2.4 shows the signal waveforms on a 2-bit channel using the 2-phase dual-rail protocol.

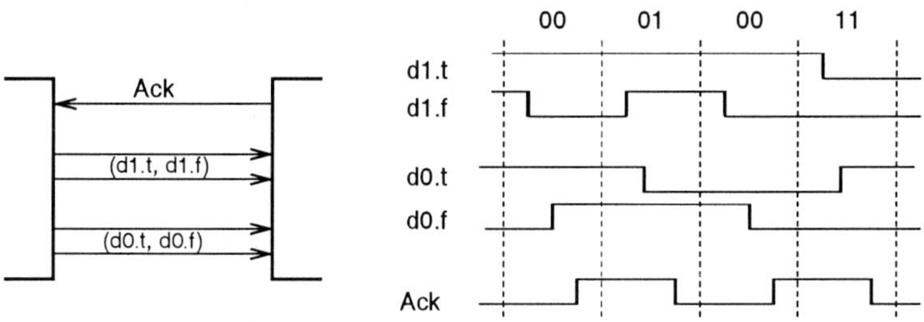

Figure 2.4. Illustration of the handshaking on a 2-phase dual-rail channel.

2.1.4 Other protocols

The previous sections introduced the four most common channel protocols: the 4-phase bundled-data push channel, the 2-phase bundled-data push channel, the 4-phase dual-rail push channel and the 2-phase dual-rail push channel; but there are many other possibilities. The two wires per bit used in the dual-rail protocol can be seen as a one-hot encoding of that bit and often it is useful to extend to 1-of-n encodings in control logic and higher-radix data encodings.

If the focus is on communication rather than computation, m-of-n encodings may be of relevance. The solution space can be expressed as the cross product of a number of options including:

$$\{2\text{-phase}, 4\text{-phase}\} \times \{\text{bundled-data}, \text{dual-rail}, 1\text{-of-n}, \ldots\} \times \{\text{push}, \text{pull}\}$$

The choice of protocol affects the circuit implementation characteristics (area, speed, power, robustness, etc.). Before continuing with these implementation issues it is necessary to introduce the concept of indication or acknowledgement, as well as a new component, the Muller C-element.

2.2. The Muller C-element and the indication principle

In a synchronous circuit the role of the clock is to define points in time where signals are stable and valid. In between the clock-ticks, the signals may exhibit hazards and may make multiple transitions as the combinational circuits stabilize. This does not matter from a functional point of view. In asynchronous (control) circuits the situation is different. The absence of a clock means that, in many circumstances, signals are required to be valid all the time, that every signal transition has a meaning and, consequently, that hazards and races must be avoided.

Intuitively a circuit is a collection of gates (normally including some feedback loops), and when the output of a gate changes it is seen by other gates that in turn may decide to change their outputs accordingly. As an example figure 2.5 shows one possible implementation of the CTL circuit in figure 1.1(c). The intention here is not to explain its function, just to give an impression of the type of circuit we are discussing. It is obvious that hazards on the *Ro*, *Ai*, and *Lt* signals would be disastrous if the circuit is used in the pipeline of figure 1.1(c).

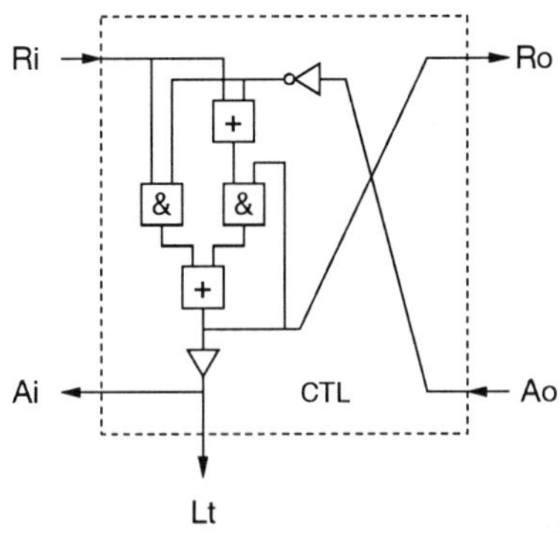

Figure 2.5. An example of an asynchronous control circuit. *Lt* is a "local" clock that is intended to control a latch.

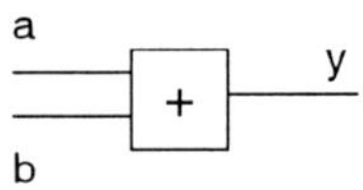

a	b	y
0	0	0
0	1	1
1	0	1
1	1	1

Figure 2.6. A normal OR gate

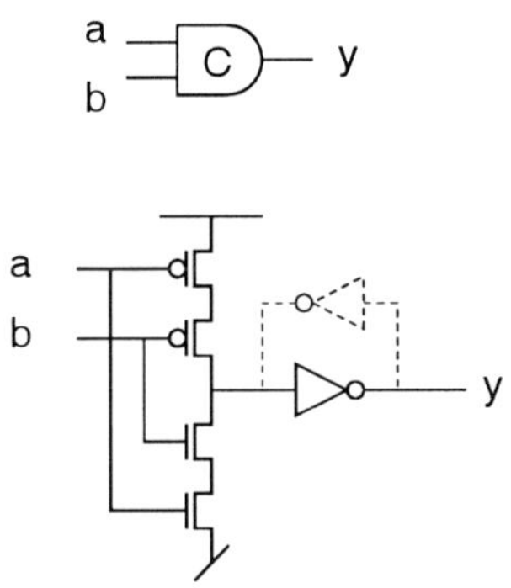

Some specifications:

1: if $a = b$ then $y := a$

2: $a = b \mapsto y := a$

3: $y = ab + y(a+b)$

4:

a	b	y
0	0	0
0	1	no change
1	0	no change
1	1	1

Figure 2.7. The Muller C-element: symbol, possible implementation, and some alternative specifications.

The concept of *indication* or *acknowledgement* plays an important role in the design of such circuits. Consider the simple 2-input OR gate in figure 2.6. An observer seeing the output change from 1 to 0 may conclude that *both* inputs are now at 0. However, when seeing the output change from 0 to 1 the observer is not able to make conclusions about *both* inputs. The observer only knows that at least one input is 1, but it does not know which. We say that the OR gate only indicates or acknowledges when both inputs are 0. Through similar arguments it can be seen that an AND gate only indicates when both inputs are 1.

Signal transitions that are not indicated or acknowledged in other signal transitions are the source of hazards and should be avoided. We will address this issue in greater detail later in section 2.5.1 and in chapter 6.

A circuit that is better in this respect is the Muller C-element shown in figure 2.7. It is a state-holding element much like an asynchronous set-reset latch. When both inputs are 0 the output is set to 0, and when both inputs are 1 the output is set to 1. For other input combinations the output does not change. Consequently, an observer seeing the output change from 0 to 1 may conclude that *both* inputs are now at 1; and similarly, an observer seeing the output change from 1 to 0 may conclude that *both* inputs are now 0.

Combining this with the observation that all asynchronous circuits rely on handshaking that involves cyclic transitions between 0 and 1, it should be clear that the Muller C-element is indeed a fundamental component that is extensively used in asynchronous circuits.

2.3. The Muller pipeline

Figure 2.8 shows a circuit that is built from C-elements and inverters. The circuit is known as a Muller pipeline or a Muller distributor. Variations and extensions of this circuit form the (control) backbone of almost all asynchronous circuits. It may not always be obvious at a first glance, but if one strips off the cluttering details, the Muller pipeline is always there as the crux of the matter. The circuit has a beautiful and symmetric behaviour, and once you understand its behaviour, you have a very good basis for understanding most asynchronous circuits.

The Muller pipeline in figure 2.8 is a mechanism that relays handshakes. After all of the C-elements have been initialized to 0 the left environment may start handshaking. To understand what happens let's consider the ith C-element, $C[i]$: It will propagate (i.e. input and store) a 1 from its predecessor, $C[i-1]$, only if its successor, $C[i+1]$, is 0. In a similar way it will propagate (i.e. input and store) a 0 from its predecessor if its successor is 1. It is often useful to think of the signals propagating in an asynchronous circuit as a sequence of waves, as illustrated at the bottom of figure 2.8. Viewed this way, the role of a C-element stage in the pipeline is to propagate crests and troughs of waves in a carefully controlled way that maintains the integrity of each wave.

On any interface between C-element pipeline stages an observer will see correct handshaking, but the timing may differ from the timing of the handshaking on the left hand environment; once a wave has been injected into the Muller pipeline it will propagate with a speed that is determined by actual delays in the circuit.

Eventually the first handshake (request) injected by the left hand environment will reach the right hand environment. If the right hand environment does not respond to the handshake, the pipeline will eventually fill. If this happens the pipeline will stop handshaking with the left hand environment – the Muller pipeline behaves like a ripple through FIFO!

In addition to this elegant behaviour, the pipeline has a number of beautiful symmetries. Firstly, it does not matter if you use 2-phase or 4-phase handshaking. It is the same circuit. The difference is in how you interpret the signals and use the circuit. Secondly, the circuit operates equally well from right to left. You may reverse the definition of signal polarities, reverse the role of the request and acknowledge signals, and operate the circuit from right to left. It is analogous to electrons and holes in a semiconductor; when current flows in

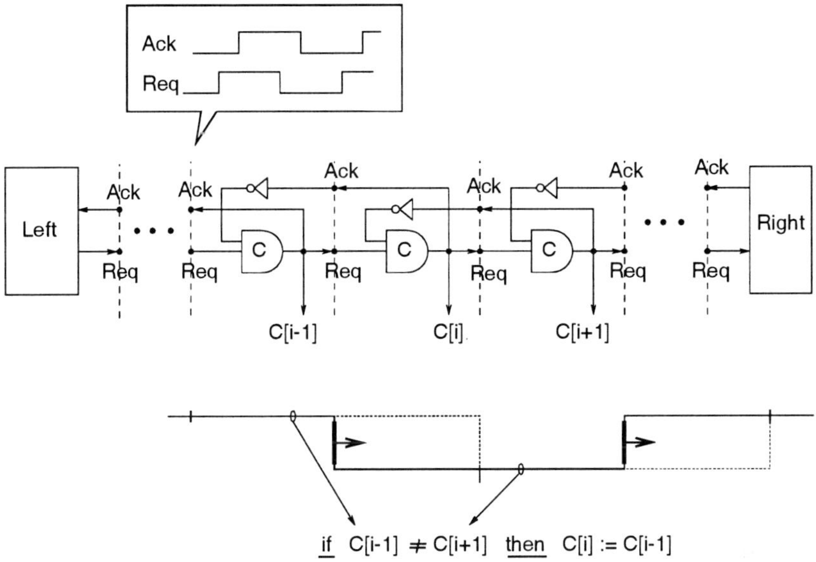

Figure 2.8. The Muller pipeline or Muller distributor.

one direction it may be carried by electrons flowing in one direction or by holes flowing in the opposite direction.

Finally, the circuit has the interesting property that it works correctly regardless of delays in gates and wires – the Muller-pipeline is delay-insensitive.

2.4. Circuit implementation styles

As mentioned previously, the choice of handshake protocol affects the circuit implementation (area, speed, power, robustness, etc.). Most practical circuits use one of the following protocols introduced in section 2.1:

4-phase bundled-data – which most closely resembles the design of synchronous circuits and which normally leads to the most efficient circuits, due to the extensive use of timing assumptions.

2-phase bundled-data – introduced under the name *Micropipelines* by Ivan Sutherland in his 1988 Turing Award lecture.

4-phase dual-rail – the classic approach rooted in David Muller's pioneering work in the 1950s.

Common to all protocols is the fact that the corresponding circuit implementations all use variations of the Muller pipeline for controlling the storage elements. Below we explain the basics of pipelines built using simple transpar-

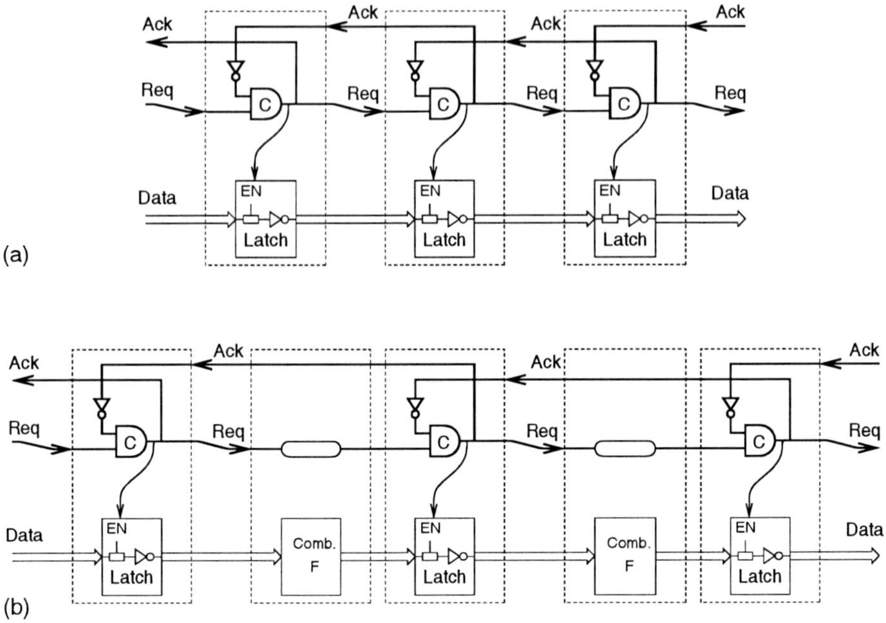

Figure 2.9. A simple 4-phase bundled-data pipeline.

ent latches as storage elements. More optimized and elaborate circuit implementations and more complex circuit structures are the topics of later chapters.

2.4.1 4-phase bundled-data

A 4-phase bundled-data pipeline is particularly simple. A Muller pipeline is used to generate local clock pulses. The clock pulse generated in one stage overlaps with the pulses generated in the neighbouring stages in a carefully controlled interlocked manner. Figure 2.9(a) shows a FIFO, i.e. a pipeline without data processing, and figure 2.9(b) shows how combinational circuits (also called function blocks) can be added between the latches. To maintain correct behaviour matching delays have to be inserted in the request signal paths.

You may view this circuit as a traditional "synchronous" data-path, consisting of latches and combinational circuits that are clocked by a distributed gated-clock driver, or you may view it as an asynchronous data-flow structure composed of two types of handshake components: latches and function blocks, as indicated by the dashed boxes.

The pipeline implementation shown in figure 2.9 is particularly simple but it has some drawbacks: when it fills the state of the C-elements is (0, 1, 0, 1, etc.), and as a consequence only every other latch is storing data. This

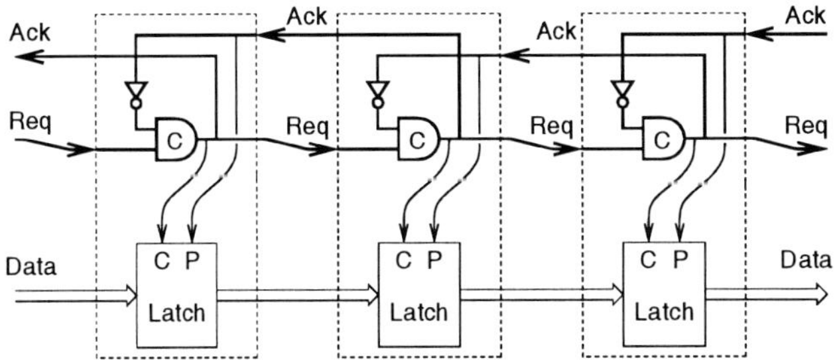

Figure 2.10. A simple 2-phase bundled-data pipeline.

is no worse than in a synchronous circuit using master-slave flip-flops, but it is possible to design asynchronous pipelines and FIFOs that are better in this respect. Another disadvantage is speed. The throughput of a pipeline or FIFO depends on the time it takes to complete a handshake cycle and for the above implementation this involves communication with both neighbours. Chapter 7 addresses alternative implementations that are both faster and have better occupancy when full.

2.4.2 2-phase bundled data (Micropipelines)

A 2-phase bundled-data pipeline also uses a Muller pipeline as the backbone control circuit, but the control signals are interpreted as events or transitions, figure 2.10. For this reason special capture-pass latches are needed: events on the C and P inputs alternate, causing the latch to alternate between capture mode and pass mode. This calls for a special latch design as shown in figure 2.11 and explained below. The switch symbol in figure 2.11 is a multiplexer, and the event controlled latch can be understood as two ordinary level sensitive latches (operating in an alternating fashion) followed by a multiplexer and a buffer.

Figure 2.10 shows a pipeline without data processing. Combinational circuits with matching delay elements can be inserted between latches in a similar way to the 4-phase bundled-data approach in figure 2.9.

The 2-phase bundled-data approach was pioneered by Ivan Sutherland in the late 1980s and an excellent introduction is given in his 1988 Turing Award Lecture [128]. The title *Micropipelines* is often used synonymously with the use of the 2-phase bundled-data protocol, but it also refers to the use of a particular set of components that are based on event signalling. In addition to the latch in figure 2.11 these are: AND, OR, Select, Toggle, Call and Arbiter. The above figures 2.10 and 2.11 are similar to figures 15 and 12 in [128], but they emphasise stronger the fact that the control structure is a Muller-pipeline. Some alter-

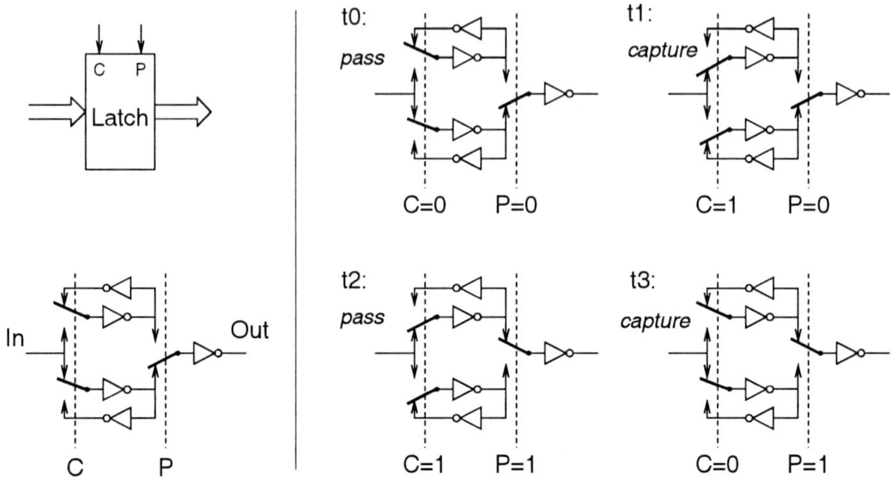

Figure 2.11. Implementation and operation of a capture-pass event controlled latch. At time t_0 the latch is transparent (i.e. in pass mode) and signals C and P are both low. An event on the C input turns the latch into capture mode, etc.

native latch designs that are (significantly) smaller and (significantly) slower are also presented in [128].

At the conceptual level the 2-phase bundled-data approach is elegant and efficient; compared to the 4-phase bundled-data approach it avoids the power and performance loss that is incurred by the return-to-zero part of the handshaking. However, as illustrated by the latch design, the implementation of components that respond to signal transitions is often more complex than the implementation of components that respond to normal level signals. In addition to the storage elements explained above, conditional control logic that responds to signal transitions tends to be complex as well. This has been experienced by this author [123], by the University of Manchester [42, 45] and by many others.

Having said this, the 2-phase bundled-data approach may be the preferred solution in systems with unconditional data-flows and very high speed requirements. But as just mentioned, the higher speed comes at a price: larger silicon area and higher power consumption. In this respect asynchronous design is no different from synchronous design.

2.4.3 4-phase dual-rail

A 4-phase dual-rail pipeline is also based on the Muller pipeline, but in a more elaborate way that has to do with the combined encoding of data and request. Figure 2.12 shows the implementation of a 1-bit wide and three stage deep pipeline without data processing. It can be understood as two Muller

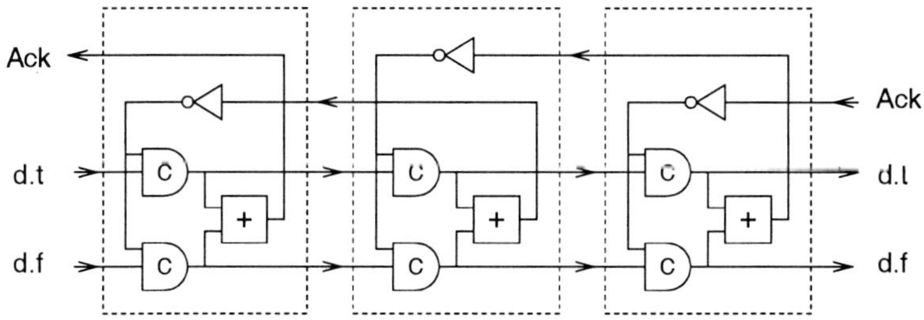

Figure 2.12. A simple 3-stage 1-bit wide 4-phase dual-rail pipeline.

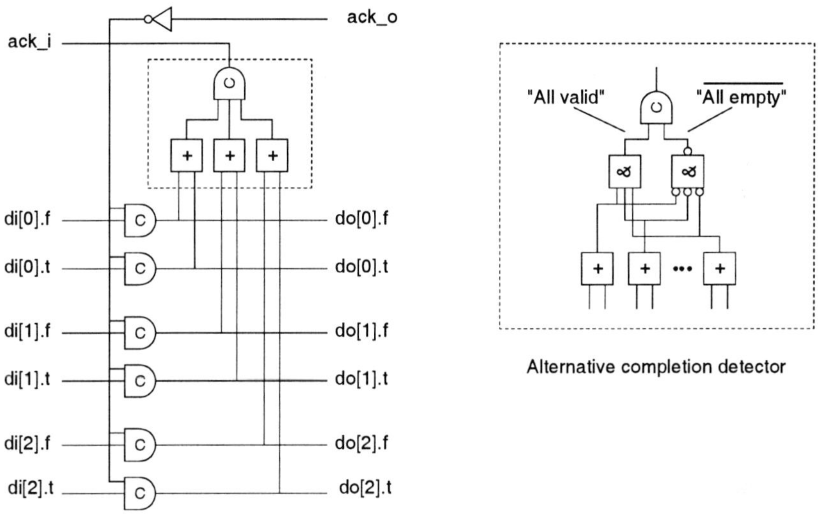

Figure 2.13. An N-bit latch with completion detection.

pipelines connected in parallel, using a common acknowledge signal per stage to synchronize operation. The pair of C-elements in a pipeline stage can store the empty codeword $\{d.t, d.f\} = \{0,0\}$, causing the acknowledge signal out of that stage to be 0, or it can store one of the two valid codewords $\{0,1\}$ and $\{1,0\}$, causing the acknowledge signal out of that stage to be logic 1. At this point, and referring back to section 2.2, the reader should notice that because the codeword $\{1,1\}$ is illegal and does not occur, the acknowledge signal generated by the OR gate safely indicates the state of the pipeline stage as being "valid" or "empty."

An N-bit wide pipeline can be implemented by using a number of 1-bit pipelines in parallel. This does not guarantee to a receiver that all bits in a word arrive at the same time, but often the necessary synchronization is done

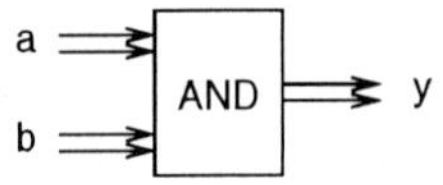

a	b	y.f	y.t
E	E	0	0
		NO CHANGE	
F	F	1	0
F	T	1	0
T	F	1	0
T	T	0	1

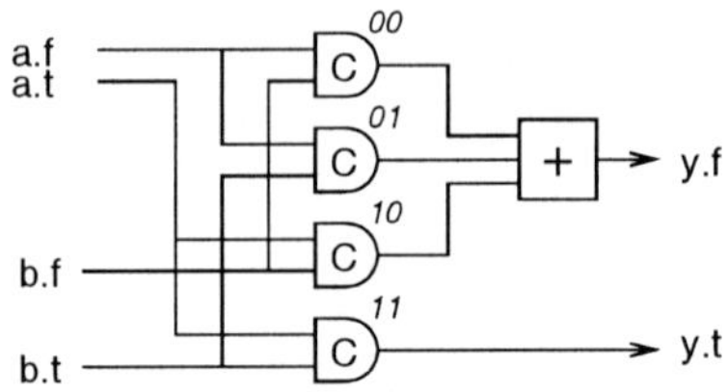

Figure 2.14. A 4-phase dual-rail AND gate: symbol, truth table, and implementation.

in the function blocks. In [124, 125] we describe a design of this style using the DIMS combinational circuits explained below.

If bit-parallel synchronization is needed, the individual acknowledge signals can be combined into one global acknowledge using a C-element. Figure 2.13 shows an N-bit wide latch. The OR gates and the C-element in the dashed box form a *completion detector* that indicates whether the N-bit dual-rail codeword stored in the latch is empty or valid. The figure also shows an implementation of a completion detector using only a 2-input C-element.

Let us now look at how combinational circuits for 4-phase dual-rail circuits are implemented. As mentioned in chapter 1 combinational circuits must be transparent to the handshaking between latches. Therefore, all outputs of a combinational circuit must not become valid until after all inputs have become valid. Otherwise the receiving latch may prematurely set acknowledge high (before all signals from the sending latch have become valid). In a similar way all outputs of a combinational circuit must not become empty until after all inputs have become empty. Otherwise the receiving latch may prematurely set acknowledge low (before all signals from the sending latch have become empty). Consequently a combinational circuit for the 4-phase dual-rail approach involves state holding elements and it exhibits a hysteresis-like behaviour in the empty-to-valid and valid-to-empty transitions.

A particularly simple approach, using only C-elements and OR gates, is illustrated in figure 2.14, which shows the implementation of a dual-rail AND gate.The circuit can be understood as a direct mapping from sum-of-minterms expressions for each of the two output wires into hardware. The circuit waits for all its inputs to become valid. When this happens exactly one of the four C-elements goes high. This again causes the relevant output wire to go high corresponding to the gate producing the desired valid output. When all inputs become empty the C-elements are all set low, and the output of the dual-rail AND gate becomes empty again. Note that the C-elements provide both the necessary 'and' operator and the hysteresis in the empty-to-valid and valid-to-empty transitions that is required for transparent handshaking. Note also that (again) the OR gate is never exposed to more than one input signal being high.

Other dual-rail gates such as OR and EXOR can be implemented in a similar fashion, and a dual-rail inverter involves just a swap of the true and false wires. The transistor count in these basic dual-rail gates is obviously rather high, and in chapter 5 we explore more efficient circuit implementations. Here our interest is in the fundamental principles.

Given a set of basic dual-rail gates one can construct dual-rail combinational circuits for arbitrary Boolean expressions using normal combinational circuit synthesis techniques. The transparency to handshaking that is a property of the basic gates is preserved when composing gates into larger combinational circuits.

The fundamental ideas explained above all go back to David Muller's work in the late 1950s and early 1960s [93, 92]. While [93] develops the fundamental theory for the design of speed-independent circuits, [92] is a more practical introduction including a design example: a bit-serial multiplier using latches and gates as explained above.

2.5. Theory

Asynchronous circuits can be classified, as we will see below, as being *self-timed*, *speed-independent* or *delay-insensitive* depending on the delay assumptions that are made. In this section we introduce some important theoretical concepts that relate to this classification. The goal is to communicate the basic ideas and provide some intuition on the problems and solutions, and a reader who wishes to dig deeper into the theory is referred to the literature. Some recent starting points are [95, 54, 69, 35, 18].

2.5.1 The basics of speed-independence

We will start by reviewing the basics of David Muller's model of a circuit and the conditions for it being speed-independent [93]. A circuit is modeled along with its (dummy) environment as a closed network of gates, closed meaning that all inputs are connected to outputs and vice versa. Gates are modeled as Boolean operators with arbitrary non-zero delays, and wires are assumed to be ideal. In this context the circuit can be described as a set of concurrent Boolean functions, one for each gate output. The state of the circuit is the set of all gate outputs. Figure 2.15 illustrates this for a stage of a Muller pipeline with an inverter and a buffer mimicing the handshaking behaviour of the left and right hand environments.

A gate whose output is consistent with its inputs is said to be stable; its "next output" is the same as its "current output", $z_i\prime = z_i$. A gate whose inputs

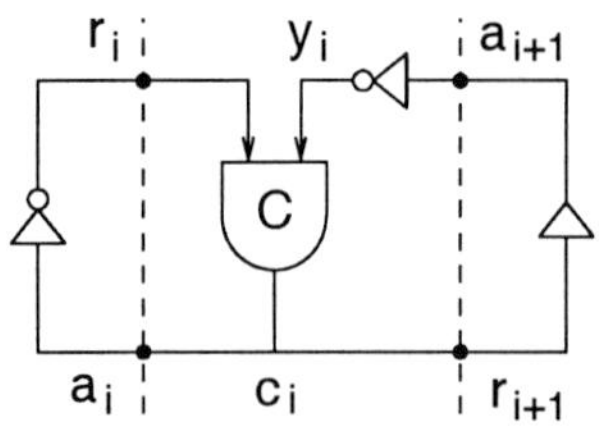

$$
\begin{aligned}
r_i\prime &= not(c_i) \\
c_i\prime &= r_i y_i + (r_i + y_i) c_i \\
y_i\prime &= not(a_{i+1}) \\
a_{i+1}\prime &= c_i
\end{aligned}
$$

Figure 2.15. Muller model of a Muller pipeline stage with "dummy" gates modeling the environment behaviour.

have changed in such a way that an output change is called for is said to be excited; its "next output" is different from its "current output", i.e. $z_i\prime \neq z_i$. After an arbitrary delay an excited gate may spontaneously change its output and become stable. We say that the gate fires, and as excited gates fire and become stable with new output values, other gates in turn become excited, etc.

To illustrate this, suppose that the circuit in figure 2.15 is in state $(r_i, y_i, c_i, a_{i+1}) = (0,1,0,0)$. In this state (the inverter) r_i is excited corresponding to the left environment being about to take request high. After the firing of $r_i \uparrow$ the circuit reaches state $(r_i, y_i, c_i, a_{i+1}) = (1,1,0,0)$ and c_i now becomes excited. For synthesis and analysis purposes one can construct the complete state graph representing all possible sequences of gate firings. This is addressed in detail in chapter 6. Here we will restrict the discussion to an explanation of the fundamental ideas.

In the general case it is possible that several gates are excited at the same time (i.e. in a given state). If one of these gates, say z_i, fires the interesting thing is what happens to the other excited gates which may have z_i as one of their inputs: they may remain excited, or they may find themselves with a different set of input signals that no longer calls for an output change. *A circuit is* speed-independent *if the latter never happens.* The practical implication of an excited gate becoming stable without firing is a potential hazard. Since delays are unknown the gate may or may not have changed its output, or it may be in the middle of doing so when the 'counter-order' comes calling for the gate output to remain unchanged.

Since the model involves a Boolean state variable for each gate (and for each wire segment in the case of delay-insensitive circuits) the state space becomes very large even for very simple circuits. In chapter 6 we introduce signal transition graphs as a more abstract representation from which circuits can be synthesized.

Now that we have a model for describing and reasoning about the behaviour of gate-level circuits let's address the classification of asynchronous circuits.

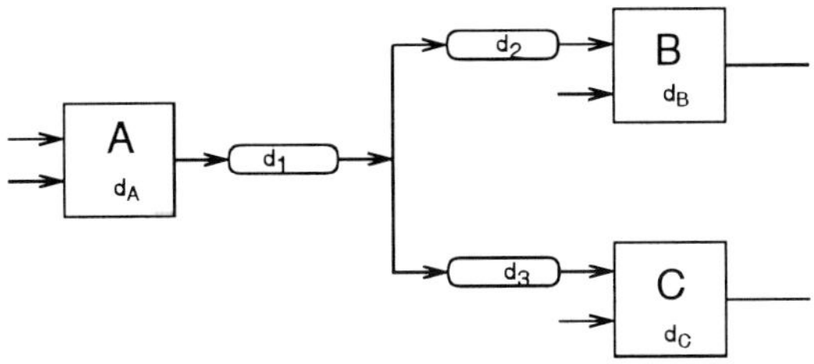

Figure 2.16. A circuit fragment with gate and wire delays. The output of gate A forks to inputs of gates B and C.

2.5.2 Classification of asynchronous circuits

At the gate level, asynchronous circuits can be classified as being self-timed, speed-independent or delay-insensitive depending on the delay assumptions that are made. Figure 2.16 serves to illustrate the following discussion. The figure shows three gates: A, B, and C, where the output signal from gate A is connected to inputs on gates B and C.

A *speed-independent* (SI) circuit as introduced above is a circuit that operates "correctly" assuming positive, bounded but unknown delays in gates and ideal zero-delay wires. Referring to figure 2.16 this means arbitrary d_A, d_B, and d_C, but $d_1 = d_2 = d_3 = 0$. Assuming ideal zero-delay wires is not very realistic in today's semiconductor processes. By allowing arbitrary d_1 and d_2 and by requiring $d_2 = d_3$ the wire delays can be lumped into the gates, and from a theoretical point of view the circuit is still speed-independent.

A circuit that operates "correctly" with positive, bounded but unknown delays in wires as well as in gates is *delay-insensitive* (DI). Referring to figure 2.16 this means arbitrary d_A, d_B, d_C, d_1, d_2, and d_3. Such circuits are obviously extremely robust. One way to show that a circuit is delay-insensitive is to use a Muller model of the circuit where wire segments (after forks) are modeled as buffer components. If this equivalent circuit model is speed-independent, then the circuit is delay-insensitive.

Unfortunately the class of delay-insensitive circuits is rather small. Only circuits composed of C-elements and inverters can be delay-insensitive [82], and the Muller pipeline in figures 2.5, 2.8, and 2.15 is one important example. Circuits that are delay-insensitive with the exception of some carefully identified wire forks where $d_2 = d_3$ are called *quasi-delay-insensitive* (QDI). Such wire forks, where signal transitions occur at the same time at all end-points, are called isochronic (and discussed in more detail in the next section). Typically these isochronic forks are found in gate-level implementations of basic building blocks where the designer can control the wire delays. At the higher levels of abstraction the composition of building blocks would typically be delay-insensitive. After these comments it is obvious that a distinction between DI, QDI and SI makes good sense.

Because the class of delay-insensitive circuits is so small, basically excluding all circuits that compute, most circuits that are referred to in the literature as delay-insensitive are only quasi-delay-insensitive.

Finally a word about *self-timed* circuits: speed-independence and delay-insensitivity as introduced above are (mathematically) well defined properties under the unbounded gate and wire delay model. Circuits whose correct operation relies on more elaborate and/or engineering timing assumptions are simply called self-timed.

2.5.3 Isochronic forks

From the above it is clear that the distinction between speed-independent circuits and delay-insensitive circuits relates to wire forks and, more specifically, to whether the delays to all end-points of a forking wire are identical or not. If the delays are identical, the wire-fork is called *isochronic.*

The need for isochronic forks is related to the concept of indication introduced in section 2.2. Consider a situation in figure 2.16 where gate A has changed its output. Eventually this change is observed on the inputs of gates B and C, and after some time gates B and C may respond to the new input by producing a new output. If this happens we say that the output change on gate A is indicated by output changes on gates B and C. If, on the other hand, only gate B responds to the new input, it is not possible to establish whether gate C has seen the input change as well. In this case it is necessary to strengthen the assumptions to $d_2 = d_3$ (i.e. that the fork is isochronic) and conclude that since the input signal change was indicated by the output of B, gate C has also seen the change.

2.5.4 Relation to circuits

In the 2-phase and 4-phase bundled-data approaches the control circuits are normally speed-independent (or in some cases even delay-insensitive), but the data-path circuits with their matched delays are self-timed. Circuits designed following the 4-phase dual-rail approach are generally quasi-delay-insensitive. In the circuits shown in figures 2.12 and 2.14 the forks that connect to the inputs of several C-elements must be isochronic, whereas the forks that connect to the inputs of several OR gates are delay-insensitive.

The different circuit classes, DI, QDI, SI and self-timed, are not mutually-exclusive ways to build complete systems, but useful abstractions that can be used at different levels of design. In most practical designs they are mixed. For example, in the Amulet processors [44, 43, 48] SI design is used for local asynchronous controllers, bundled-data for local data processing, and DI is used for high-level composition. Another example is the hearing-aid filter bank design presented in [103]. It uses the DI dual-rail 4-phase protocol inside

RAM-modules and arithmetic circuits to provide robust completion indication, and 4-phase bundled-data with SI control at the top levels of design, i.e. somewhat different from the Amulet designs. This emphasizes that the choice of handshake protocol and circuit implementation style is among the factors to consider when optimizing an asynchronous digital system.

It is important to stress that speed-independence and delay-insensitivity are mathematical properties that can be verified for a given implementation. If an abstract component – such as a C-element or a complex And-Or-Invert gate – is replaced by its implementation using simple gates and possibly some wire-forks, then the circuit may no longer be speed-independent or delay-insensitive. As an illustrative example we mention that the simple Muller pipeline stage in figures 2.8 and 2.15 is no longer delay-insensitive if the C-element is replaced by the gate-level implementation shown in figure 2.5 that uses simple AND and OR gates. Furthermore, even simple gates are abstractions; in CMOS the primitives are N and P transistors, and even the simplest gates include forks.

In chapter 6 we will explore the design of SI control circuits in great detail (because theory and synthesis tools are well developed). As SI circuits ignore wire delays completely some care is needed when physically implementing these circuits. In general one might think that the zero wire-delay assumption is trivially satisfied in small circuits involving 10-20 gates, but this need not be the case: a normal place and route CAD tool might spread the gates of a small controller all over the chip. Even if the gates are placed next to each other they may have different logic thresholds on their inputs which in combination with slowly rising or falling signals can cause (and have caused!) circuits to malfunction. For static CMOS and for circuits operating with low supply voltages (e.g. $V_{DD} \sim Vt_N + |Vt_P|$) this is less of a problem, but for dynamic circuits using a larger V_{DD} (e.g. 3.3 V or 5.0 V) the logic thresholds can be very different. This often overlooked problem is addressed in detail in [134].

2.6. Test

When it comes to the commercial exploitation of asynchronous circuits the problem of test comes to the fore. Test is a major topic in its own right, and it is beyond the scope of this tutorial to do anything more than mention a few issues and challenges. Although the following text is brief it assumes some knowledge of testing. The material does not constitute a foundation for the following chapters and it may be skipped.

The previous discussion about Muller circuits (excited gates and the firing of gates), the principle of indication, and the discussion of isochronoic forks ties in nicely with a discussion of testing for stuck at faults. In the stuck-at fault model defects are modeled at the gate level as (individual) inputs and outputs being stuck-at-1 or stuck-at-0. The principle of indication says that all input signal transitions on a gate must be indicated by an output signal transition on the gate. Furthermore, asynchronous circuits make extensive use of handshaking and this causes signals to exhibit cyclic transitions between 0 and 1. In this scenario, the presence of a stuck-at fault is likely to cause the circuit to halt; if one component stops handshaking the stall tends to "propagate" to neighbouring components, and eventually the entire circuit halts. Consequently, the development of a set of test patterns that exhaustively tests for all stuck-at faults is simply a matter of developing a set of test patterns that toggle all nodes, and this is generally a comparatively simple task.

Since isochronic forks are forks where a signal transition in one or more branches is not indicated in the gates that take these signals as inputs, it follows that isochronic forks imply untestable stuck-at faults.

Testing asynchronous circuits incurs additional problems. As we will see in the following chapters, asynchronous circuits tend to implement registers using latches rather than flip-flops. In combination with the absence of a global clock, this makes it less straightforward to connect registers into scan-paths. Another consequence of the distributed self-timed control (i.e. the lack of a global clock) is that it is less straightforward to single-step the circuit through a sequence of well-defined states. This makes it less straightforward to steer the circuit into particular quiescent states, which is necessary for I_{DDQ} testing, – the technique that is used to test for shorts and opens which are faults that are typical in today's CMOS processes.

The extensive use of state-holding elements (such as the Muller C-element), together with the self-timed behaviour, makes it difficult to test the feed-back circuitry that implements the state holding behaviour. Delay-fault testing represents yet another challenge.

The above discussion may leave the impression that the problem of testing asynchronous circuits is largely unsolved. This is not correct. The truth is rather that the techniques for testing synchronous circuits are not directly applicable. The situation is quite similar to the design of asynchronous circuits that we will address in detail in the following chapters. Here a mix of new and well-known techniques are also needed. A good starting point for reading about the testing of asynchronous circuits is [120]. Finally, we mention that testing is also touched upon in chapters 13 and 15.

2.7. Summary

This chapter introduced a number of fundamental concepts. We will now return to the main track of designing circuits. The reader will probably want to revisit some of the material in this chapter again while reading the following chapters.

Chapter 3

STATIC DATA-FLOW STRUCTURES

In this chapter we will develop a high-level view of asynchronous design that is equivalent to RTL (register transfer level) in synchronous design. At this level the circuits may be viewed as static data-flow structures. The aim is to focus on the behaviour of the circuits, and to abstract away the details of the handshake signaling which can be considered an orthogonal implementation issue.

3.1. Introduction

The various handshake protocols and the associated circuit implementation styles presented in the previous chapters are rather different. However, when looking at the circuits at a more abstract level – the data-flow handshake-channel level introduced in chapter 1 – these differences diminish, and it makes good sense to view the choice of handshake protocol and circuit implementation style as low level implementation decisions that can be made largely independently from the more abstract design decisions that establish the overall structure and operation of the circuit.

Throughout this chapter we will assume a 4-phase protocol since this is most common. From a data-flow point of view this means that the we will be dealing with data streams composed of alternating valid and empty values – in a two-phase protocol we would see only a sequence of valid values, but apart from that everything else would be the same. Furthermore we will be dealing with simple latches as storage elements. The latches are controlled according to the simple rule stated in chapter 1:

> *A latch may input and store a new token (valid or empty) from its predecessor if its successor latch has input and stored the token that it was previously holding.*

Latches are the only components that initiate and take an active part in handshaking; all other components are "transparent" to the handshaking. To ease the distinction between latches and combinational circuits and to emphasize the token flow in circuit diagrams, we will use a box symbol with double vertical lines to represent latches throughout the rest of this tutorial (see figure 3.1).

J. Sparsø and S. Furber (eds.), Principles of Asynchronous Circuit Design, 29–40.

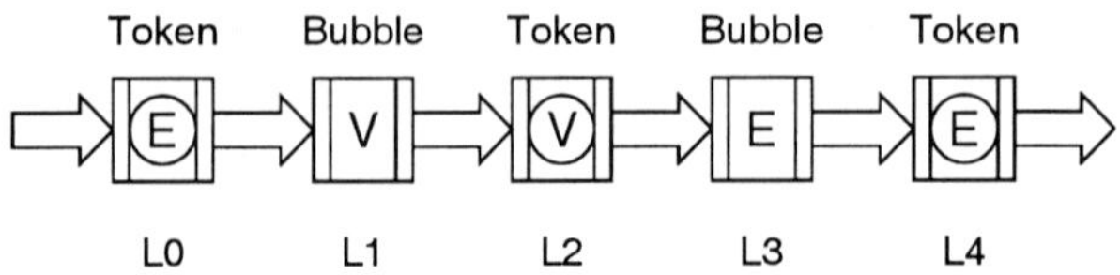

Figure 3.1. A possible state of a five stage pipeline.

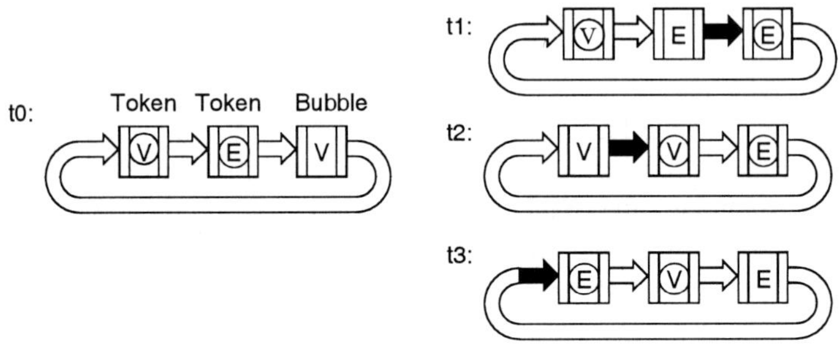

Figure 3.2. Ring: (a) a possible state; and (b) a sequence of data transfers.

3.2. Pipelines and rings

Figure 3.1 shows a snapshot of a pipeline composed of five latches. The "box arrows" represent channels or links consisting of request, acknowledge and data signals (as explained on page 5). The valid value in L1 has just been copied into L2 and the empty value in L3 has just been copied into L4. This means that L1 and L3 are now holding old duplicates of the values now stored in L2 and L4. Such old duplicates are called "bubbles", and the newest/rightmost valid and empty values are called "tokens". To distinguish tokens from bubbles, tokens are represented with a circle around the value. In this way a latch may hold a valid token, an empty token or a bubble. Bubbles can be viewed as catalysts: a bubble allows a token to move forward, and in supporting this the bubble moves backwards one step.

Any circuit should have one or more bubbles, otherwise it will be in a deadlock state. This is a matter of initializing the circuit properly, and we will elaborate on this shortly. Furthermore, as we will see later, the number of bubbles also has a significant impact on performance.

In a pipeline with at least three latches, it is possible to connect the output of the last stage to the input of the first, forming a ring in which data tokens can circulate autonomously. Assuming the ring is initialized as shown in figure 3.2(a) at time t_0 with a valid token, an empty token and a bubble, the first steps of the circulation process are shown in figure 3.2(b), at times t_1, t_2 and

t_3. Rings are the backbone structures of circuits that perform iterative computations. The cycle time of the ring in figure 3.2 is 6 "steps" (the state at t_6 will be identical to the state at t_0). Both the valid token and the empty token have to make one round trip. A round trip involves 3 "steps" and as there is only one bubble to support this the cycle time is 6 "steps". It is interesting to note that a 4-stage ring initialized to hold a valid token, an empty token and two bubbles can iterate in 4 "steps". It is also interesting to note that the addition of one more latch does not re-time the circuit or alter its function (as would be the case in a synchronous circuit); it is still a ring in which a single data token is circulating.

3.3. Building blocks

Figure 3.3 shows a minimum set of components that is sufficient to implement asynchronous circuits (static data-flow structures with deterministic behaviour, i.e. without arbiters). The components can be grouped in four categories as explained below. In the next section we will see examples of the token-flow behaviour in structures composed of these components. Components for mutual exclusion and arbitration are covered in section 5.8.

Latches provide storage for variables and implement the handshaking that supports the token flow. In addition to the normal latch a number of degenerate latches are often needed: a latch with only an output channel is a source that produces tokens (with the same constant value), and a latch with only an input channel is a sink that consumes tokens. Figure 2.9 shows the implementation of a 4-phase bundled-data latch, figure 2.11 shows the implementation of a 2-phase bundled-data latch, and figures 2.12 – 2.13 show the implementation of a 4-phase dual-rail latch.

Function blocks are the asynchronous equivalent of combinatorial circuits. They are transparent/passive from a handshaking point of view. A function block will: (1) wait for tokens on its inputs (an implicit join), (2) perform the required combinatorial function, and (3) issue tokens on its outputs. Both empty and valid tokens are handled in this way. Some implementations assume that the inputs have been synchronized. In this case it may be necessary to use an explicit join component. The implementation of function blocks is addressed in detail in chapter 5.

Unconditional flow control: Fork and join components are used to handle parallel threads of computation. In engineering terms, forks are used when the output from one component is input to more components, and joins are used when data from several independent channels needs to be synchronized – typically because they are (independent) inputs to a circuit. In the following we will often omit joins and forks from cir-

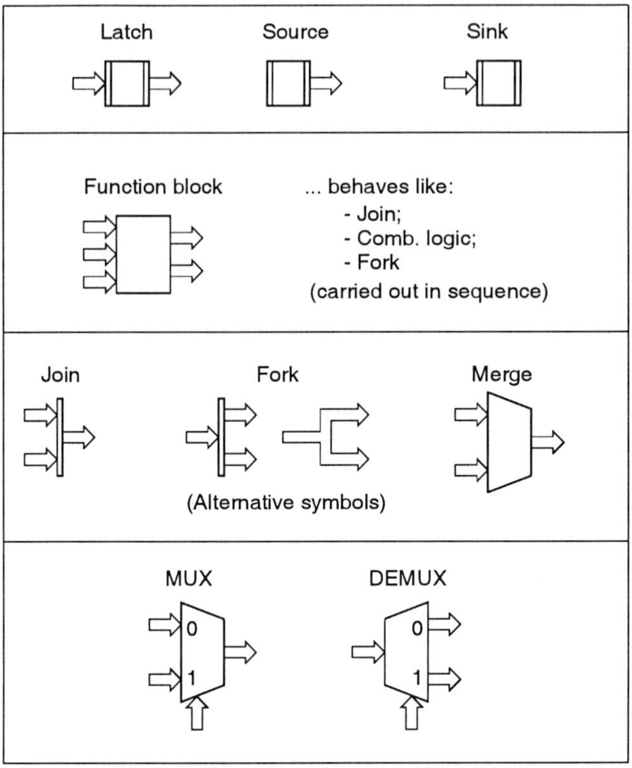

Figure 3.3. A minimum and, for most cases, sufficient set of asynchronous components.

cuit diagrams: the fan-out of a channel implies a fork, and the fan-in of several channels implies a join.

A merge component has two or more input channels and one output channel. Handshakes on the input channels are assumed to be mutually exclusive and the merge relays input tokens/handshakes to the output.

Conditional flow control: MUX and DEMUX components perform the usual functions of selecting among several inputs or steering the input to one of several outputs. The control input is a channel just like the data inputs and outputs. A MUX will synchronize the control channel and the relevant input channel and send the input data to the data output. The other input channel is ignored. Similarly a DEMUX will synchronize the control and data input channels and steer the input to the selected output channel.

As mentioned before the latches implement the handshaking and thereby the token flow in a circuit. All other components must be transparent to the handshaking. This has significant implications for the implementation of these components!

3.4. A simple example

Figure 3.4 shows an example of a circuit composed of latches, forks and joins that we will use to illustrate the token-flow behaviour of an asynchronous circuit. The structure can be described as pipeline segments and a ring connected into a larger structure using fork and join components.

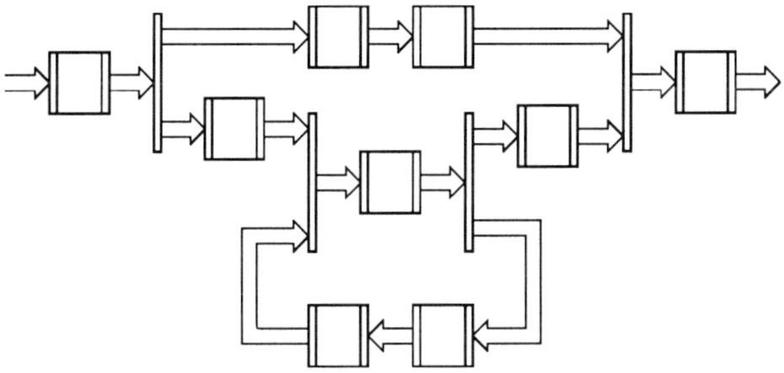

Figure 3.4. An example asynchronous circuit composed of latches, forks and joins.

Assume that the circuit is initialized as shown in figure 3.5 at time t_0: all latches are initialized to the empty value except for the bottom two latches in the ring that are initialized to contain a valid value and an empty value. Values enclosed in circles are tokens and the rest are bubbles. Assume further that the left and right hand environments (not shown) take part in the handshakes that the circuit is prepared to perform. Under these conditions the operation of the circuit (i.e. the flow of tokens) is as illustrated in the snapshots labeled $t_0 - t_{11}$. The left hand environment performs one handshake cycle inputting a valid value followed by an empty value. In a similar way the right environment takes part in one handshake cycle and consumes a valid value and an empty value.

Because the flow of tokens is controlled by local handshaking the circuit could exhibit many other behaviours. For example, at time t_5 the circuit is ready to accept a new valid value from its left environment. Notice also that if the initial state had no tokens in the ring, then the circuit would deadlock after a few steps. It is highly recommended that the reader tries to play the token-bubble data-flow game; perhaps using the same circuit but with different initial states.

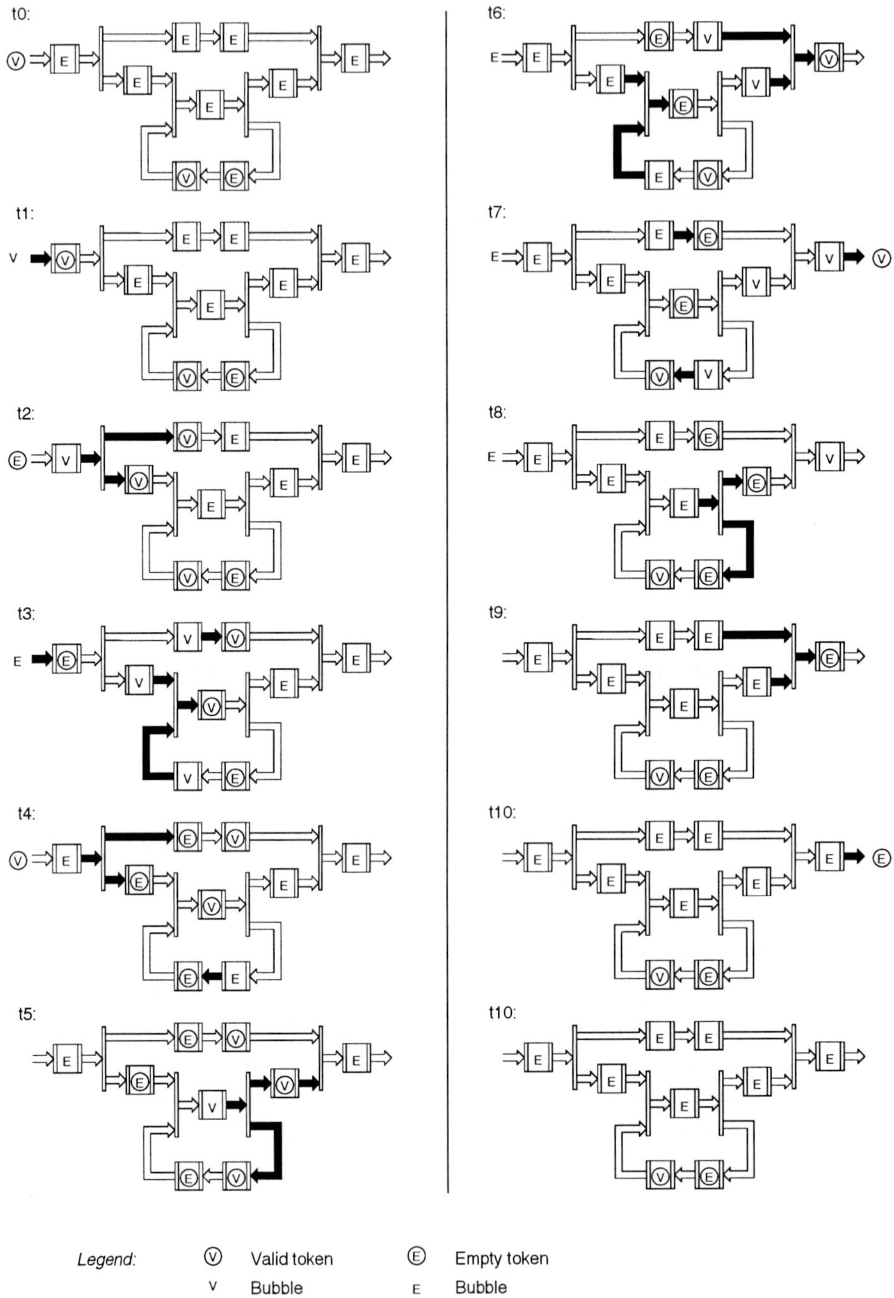

Figure 3.5. A possible operation sequence of the example circuit from figure 3.4.

3.5. Simple applications of rings

This section presents a few simple and obvious circuits based on a single ring.

3.5.1 Sequential circuits

Figure 3.6 shows a straightforward implementation of a finite state machine. Its structure is similar to a synchronous finite state machine; it consists of a function block and a ring that holds the current state. The machine accepts an "input token" that is joined with the "current state token". Then the function block computes the output and the next state, and finally the fork splits these into an "output token" and a "next state token."

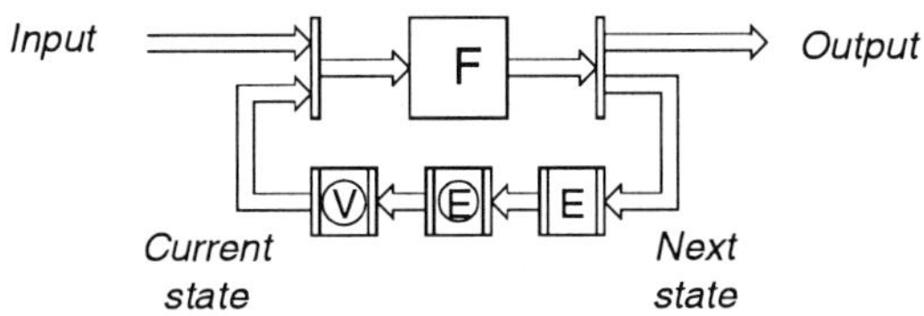

Figure 3.6. Implementation of an asynchronous finite state machine using a ring.

3.5.2 Iterative computations

A ring can also be used to build circuits that implement iterative computations. Figure 3.7 shows a template circuit. The idea is that the circuit will: (1) accept an operand, (2) sequence through the same operation a number of times until the computation terminates and (3) output the result. The necessary control is not shown. The figure shows one particular implementation. Pos-

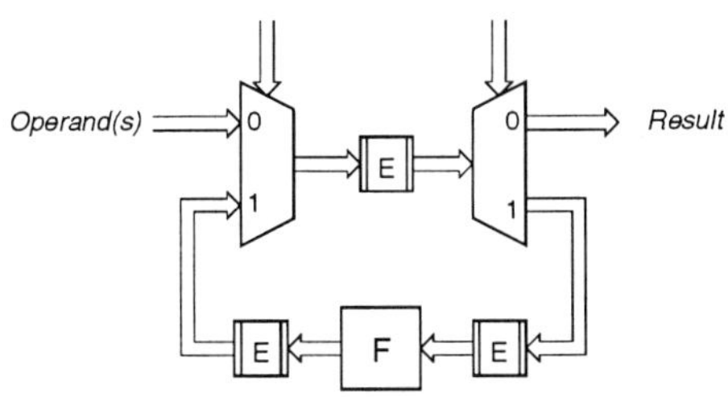

Figure 3.7. Implementation of an iterative computation using a ring.

sible variations involve locating the latches and the function block differently in the ring as well as decomposing the function block and putting these (simpler) function blocks between more latches. In [156] Ted Williams presents a circuit that performs division using a self-timed 5-stage ring. This design was later used in a floating point coprocessor in a commercial microprocessor [157].

3.6. FOR, IF, and WHILE constructs

Very often the desired function of a circuit is expressed using a programming language (C, C++, VHDL, Verilog, etc.). In this section we will show implementation templates for a number of typical conditional structures and loop structures. A reader who is familiar with control-data-flow graphs, perhaps from high-level synthesis, will recognize the great similarities between asynchronous circuits and control-data-flow graphs [36, 127].

if <cond> then <body1> else <body2> An asynchronous circuit template for implementing an *if* statement is shown in figure 3.8(a). The data-type of the input and output channels to the *if* circuit is a record containing all variables in the `<cond>` expression and the variables manipulated by `<body1>` and `<body2>`. The data-type of the output channel from the cond block is a Boolean that controls the DEMUX and MUX components. The FORK associated with this channel is not shown.

Since the execution of `<body1>` and `<body2>` is mutually exclusive it is possible to replace the controlled MUX in the bottom of the circuit with a simpler MERGE as shown in figure 3.8(b). The circuit in figure 3.8 contains

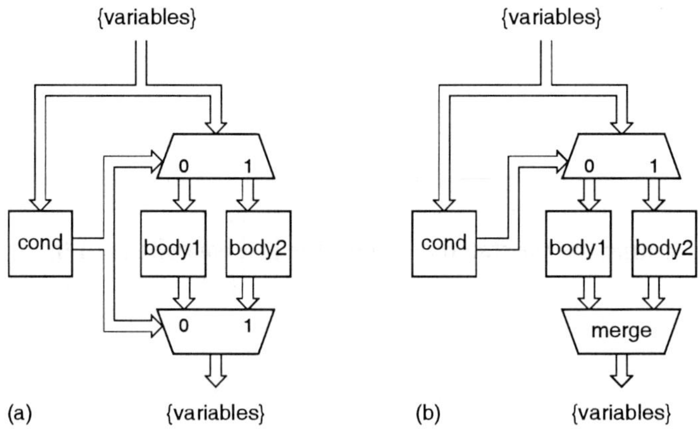

Figure 3.8. A template for implementing *if* statements.

no feedback loops and no latches – it can be considered a large function block. The circuit can be pipelined for improved performance by inserting latches.

for <count> do <body> An asynchronous circuit template for implementing a *for* statement is shown in figure 3.9. The data-type of the input channel to the *for* circuit is a record containing all variables manipulated in the `<body>` and the loop count, `<count>`, that is assumed to be a non-negative integer. The data-type of the output channel is a record containing all variables manipulated in the `<body>`.

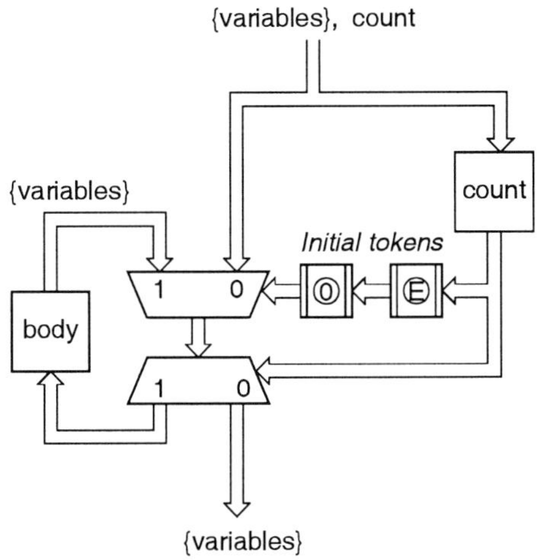

Figure 3.9. A template for implementing *for* statements.

The data-type of the output channel from the count block is a Boolean, and one handshake on the input channel of the count block encloses `<count>` handshakes on the output channel: `<count> - 1` handshakes providing the Boolean value "1" and one (final) handshake providing the Boolean value "0". Notice the two latches on the control input to the MUX. They must be initialized to contain a data token with the value "0" and an empty token in order to enable the *for* circuit to read the variables into the loop.

After executing the *for* statement once, the last handshake of the count block will steer the variables in the loop onto the output channel and put a "0" token and an empty token into the two latches, thereby preparing the *for* circuit for a subsequent activation. The FORK in the input and the FORK on the output of the count block are not shown. Similarly a number of latches are omitted. Remember: (1) all rings must contain at least 3 latches and (2) for each latch initialized to hold a data token there must also be a latch initialized to hold an empty token (when using 4-phase handshaking).

while <cond> do <body> An asynchronous circuit template for implementing a *while* statement is shown in figure 3.10. Inputs to (and outputs from) the circuit are the variables in the <cond> expression and the variables manipulated by <body>. As before in the *for* circuit, it is necessary to put two latches initialized to contain a data token with the value "0" and an empty token on the control input of the MUX. And as before a number of latches are omitted in the two rings that constitute the *while* circuit. When the *while* circuit terminates (after zero or more iterations) data is steered out of the loop and this also causes the latches on the MUX control input to become initialized properly for the subsequent activation of the circuit.

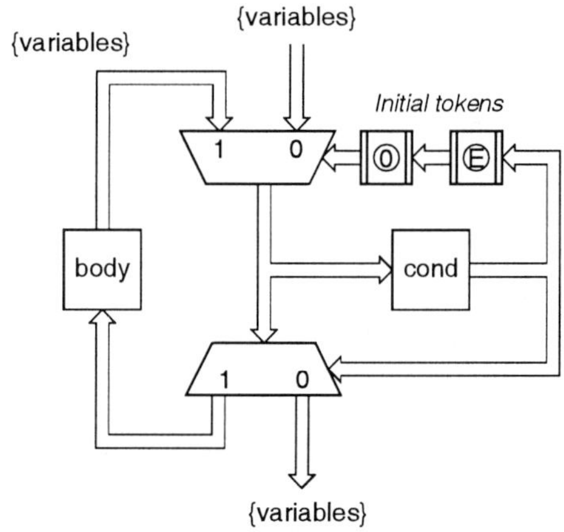

Figure 3.10. A template for implementing *while* statements.

3.7. A more complex example: GCD

Using the templates just introduced we will now design a small example circuit, GCD, that computes the greatest common divisor of two integers. GCD is often used as an introductory example, and figure 3.11 shows a programming language specification of the algorithm.

In addition to its role as a design example in the current context, GCD can also serve to illustrate the similarities and differences between different design techniques. In chapter 8 we will use the same example to illustrate the Tangram language and the associated syntax-directed compilation process (section 8.3.3 on pages 127–128).

The implementation of GCD is shown in figure 3.12. It consists of a *while* template whose body is an *if* template. Figure 3.12 shows the circuit including all the necessary latches (with their initial states). The implementation makes no attempt at sharing resources – it is a direct mapping following the implementation templates presented in the previous section.

```
input (a,b);
while a ≠ b do
    if a > b   then a ← a − b;
               else b ← b − a;
output (a);
```

Figure 3.11. A programming language specification of GCD.

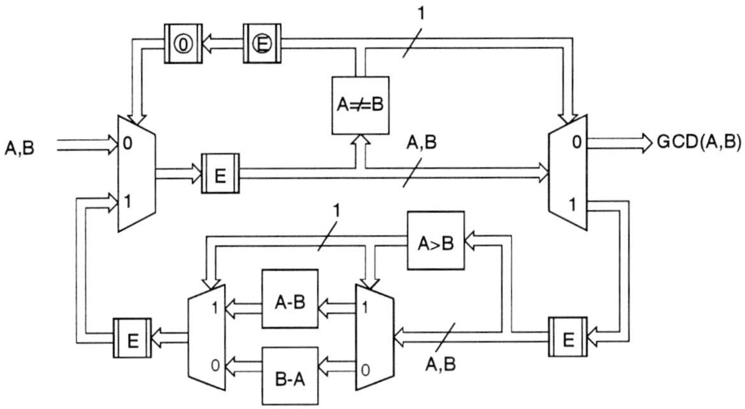

Figure 3.12. An asynchronous circuit implementation of GCD.

3.8. Pointers to additional examples

3.8.1 A low-power filter bank

In [103] we reported on the design of a low-power IFIR filter bank for a digital hearing aid. It is a circuit that was designed following the approach presented in this chapter. The paper also provides some insight into the design of low power circuits as well as the circuit level implementation of memory structures and datapath units.

3.8.2 An asynchronous microprocessor

In [23] we reported on the design of a MIPS microprocessor, called ARISC. Although there are many details to be understood in a large-scale design like a microprocessor, the basic architecture shown in figure 3.13 can be understood as a simple data-flow structure. The solid-black rectangles represent latches, the box-arrows represent channels, and the text-boxes represents function blocks (combinatorial circuits).

The processor is a simple pipelined design with instructions retiring in program order. It consists of a fetch-decode-issue ring with a fixed number of to-

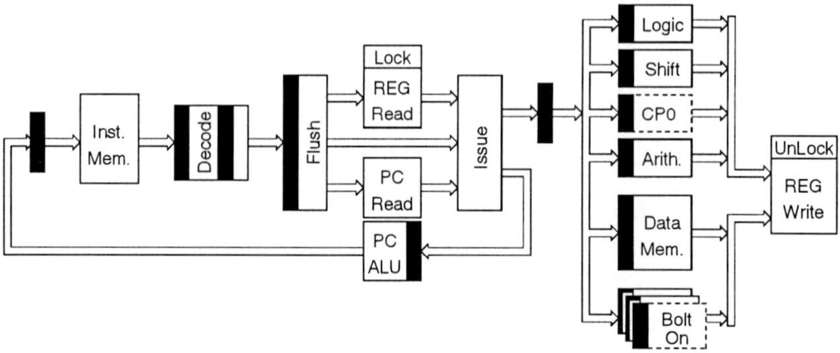

Figure 3.13. Architecture of the ARISC microprocessor.

kens. This ensures a fixed instruction prefetch depth. The issue stage forks decoded instructions into the execute pipeline and initiates the fetch of one more instruction. Register forwarding is avoided by a locking mechanism: when an instruction is issued for execution the destination register is locked until the write-back has taken place. If a subsequent instruction has a read-after-write data hazard this instruction is stalled until the register is unlocked. The tokens flowing in the design contain all operands and control signals related to the execution of an instruction, i.e. similar to what is stored in a pipeline stage in a synchronous processor. For further information the interested reader is referred to [23]. Other asynchronous microprocessors are based on similar principles.

3.8.3 A fine-grain pipelined vector multiplier

The GCD circuit and the ARISC presented in the preceding sections use bit-parallel communication channels. An example of a static data-flow structure that uses 1-bit channels and fine grain pipelining is the serial-parallel vector multiplier design reported in [124, 125]. Here all necessary word-level synchronization is performed implicitly by the function blocks. The large number of interacting rings and pipeline segments in the static data-flow representation of the design makes it rather complex. After reading the next chapter on performance analysis the interested reader may want to look at this design; it contains several interesting optimizations.

3.9. Summary

This chapter developed a high-level view of asynchronous design that is equivalent to RTL (register transfer level) in synchronous design – static data flow structures. The next chapter address performance analysis at this level of abstraction.

Chapter 4

PERFORMANCE

In this chapter we will address the performance analysis and optimization of asynchronous circuits. The material extends and builds upon the "static data-flow structures view" introduced in the previous chapter.

4.1. Introduction

In a synchronous circuit, performance analysis and optimization is a matter of finding the longest latency signal path between two registers; this determines the period of the clock signal. The global clock partitions the circuit into many combinatorial circuits that can be analyzed individually. This is known as static timing analysis and it is a rather simple task, even for a large circuit.

For an asynchronous circuit, performance analysis and optimization is a global and therefore much more complex problem. The use of handshaking makes the timing in one component dependent on the timing of its neighbours, which again depends on the timing of their neighbours, etc. Furthermore, the performance of a circuit does not depend only on its structure, but also on how it is initialized and used by its environment. The performance of an asynchronous circuit can even exhibit transients and oscillations.

We will first develop a qualitative understanding of the dynamics of the token flow in asynchronous circuits. A good understanding of this is essential for designing circuits with good performance. We will then introduce some quantitative performance parameters that characterize individual pipeline stages and pipelines and rings composed of identical pipeline stages. Using these parameters one can make first-level design decisions. Finally we will address how more complex and irregular structures can be analyzed.

The following text represents a major revision of material from [124] and it is based on original work by Ted Williams [153, 154, 155]. If consulting these references the reader should be aware of the exact definition of a token. Throughout this book a token is defined as a valid data value *or* an empty data value, whereas in the cited references (that deal exclusively with 4-phase handshaking) a token is a valid-empty data pair. The definition used here accentuates the similarity between a token in an asynchronous circuit and the token in

J. Sparsø and S. Furber (eds.), Principles of Asynchronous Circuit Design, 41–56.

a Petri net. Furthermore it provides some unification between 4-phase handshaking and 2-phase handshaking – 2-phase handshaking is the same game, but without empty-tokens.

In the following we will assume 4-phase handshaking, and the examples we provide all use bundled-data circuits. It is left as an exercise for the reader to make the simple adaptations that are necessary for dealing with 2-phase handshaking.

4.2. A qualitative view of performance

4.2.1 Example 1: A FIFO used as a shift register

The fundamental concepts can be illustrated by a simple example: a FIFO composed of a number of latches in which there are N valid tokens separated by N empty tokens, and whose environment alternates between reading a token from the FIFO and writing a token into the FIFO (see figure 4.1(a)). In this way the nomber of tokens in the FIFO is invariant. This example is relevant because many designs use FIFOs in this way, and because it models the behaviour of shift registers as well as rings – structures in which the number of tokens is also invariant.

A relevant performance figure is the throughput, which is the rate at which tokens are input to or output from the shift register. This figure is proportional to the time it takes to shift the contents of the chain of latches one position to the right.

Figure 4.1(b) illustrates the behaviour of an implementation in which there are $2N$ latches per valid token and figure 4.1(c) illustrates the behaviour of an implementation in which there are $3N$ latches per valid token. In both examples the number of valid tokens in the FIFO is $N = 3$, and the only difference between the two situations in figure 4.1(b) and 4.1(c) is the number of bubbles.

In figure 4.1(b) at time t_1 the environment reads the valid token, $D1$, as indicated by the solid channel symbol. This introduces a bubble that enables data transfers to take place one at a time ($t_2 - t_5$). At time t_6 the environment inputs a valid token, $D4$, and at this point all elements have been shifted one position to the right. Hence, the time used to move all elements one place to the right is proportional to the number of tokens, in this case $2N = 6$ time steps.

Adding more latches increases the number of bubbles, which again increases the number of data transfers that can take place simultaneously, thereby improving the performance. In figure 4.1(c) the shift register has $3N$ stages and therefore one bubble per valid-empty token-pair. The effect of this is that N data transfers can occur simultaneously and the time used to move all elements one place to the right is constant; 2 time steps.

If the number of latches was increased to $4N$ there would be one token per bubble, and the time to move all tokens one step to the right would be only

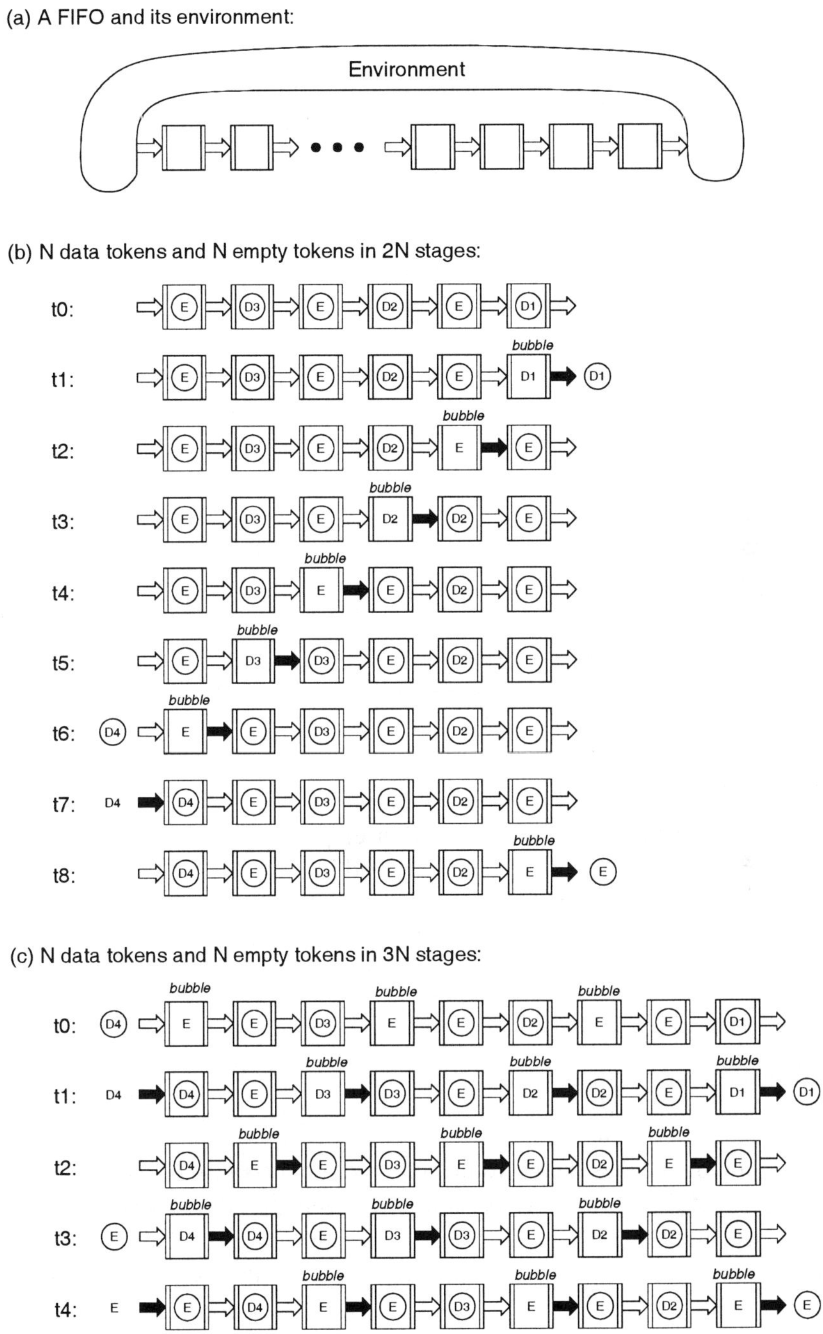

Figure 4.1. A FIFO and its environment. The environment alternates between reading a token from the FIFO and writing at token into the FIFO.

1 time step. In this situation the pipeline is half full and the latches holding bubbles act as slave latches (relative to the latches holding tokens). Increasing the number of bubbles further would not increase the performance further. Finally, it is interesting to notice that the addition of just one more latch holding a bubble to figure 4.1(b) would double the performance. The asynchronous designer has great freedom in trading more latches for performance.

As the number of bubbles in a design depends on the number of latches per token, the above analysis illustrates that performance optimization of a given circuit is primarily a task of structural modification – circuit level optimization like transistor sizing is of secondary importance.

4.2.2 Example 2: A shift register with parallel load

In order to illustrate another point – that the distribution of tokens and bubbles in a circuit can vary over time, depending on the dynamics of the circuit and its environment – we offer another example: a shift register with parallel load. Figure 4.2 shows an initial design of a 4-bit shift register. The circuit has a bit-parallel input channel, $din[3{:}0]$, connecting it to a data producing environment. It also has a 1-bit data channel, *do*, and a 1-bit control channel, ctl, connecting it to a data consuming environment. Operation is controlled by the data consuming environment which may request the circuit to: ($ctl = 0$) perform a parallel load *and* to provide the least significant bit from the bit-parallel channel on the *do* channel, or ($ctl = 1$) to perform a right shift and provide the next bit on the *do* channel. In this way the data consuming environment always inputs a control token (valid or empty) to which the circuit always responds by outputting a data token (valid or empty). During a parallel load, the previous content of the shift register is steered into the "dead end" sink-latches. During a right shift the constant 0 is shifted into the most significant position – corresponding to a logical right shift. The data consuming environment is not required to read all the input data bits, and it may continue reading zeros beyond the most significant input data bit.

The initial design shown in figure 4.2 suffers from two performance limiting inexpediencies: firstly, it has the same problem as the shift register in figure 4.1(b) – there are too few bubbles, and the peak data rate on the bit-serial output reduces linearly with the length of the shift register. Secondly, the control signal is forked to all of the MUXes and DEMUXes in the design. This implies a high fan-out of the request and data signals (which requires a couple of buffers) and synchronization of all the individual acknowledge signals (which requires a C-element with many inputs, possibly implemented as a tree of C-elements). The first problem can be avoided by adding a 3rd latch to the datapath in each stage of the circuit corresponding to the situation in

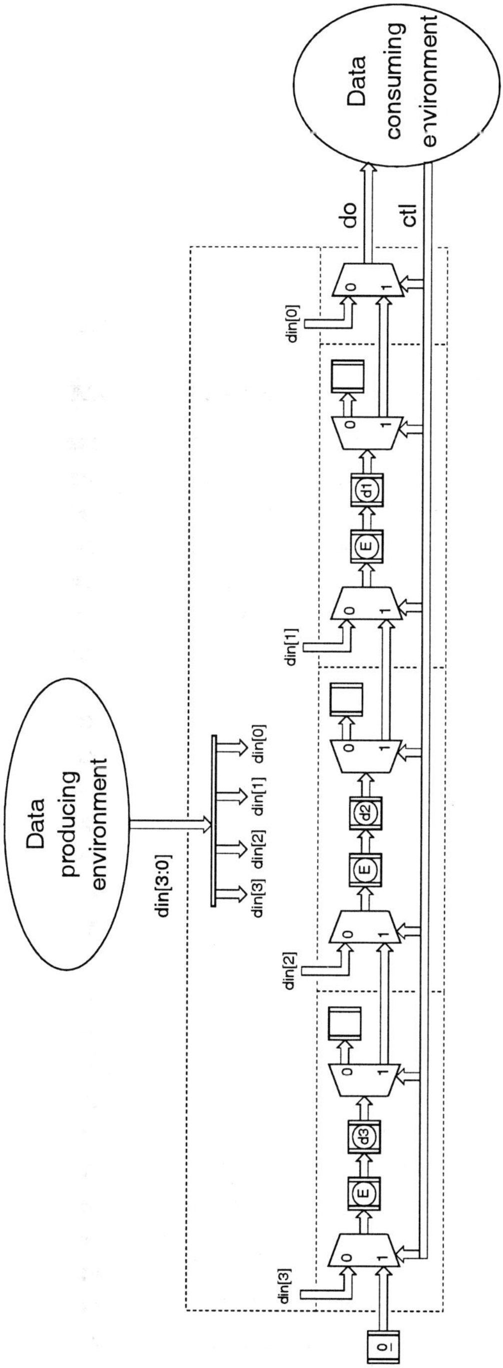

Figure 4.2. **Initial design of the shift register with parallel load.**

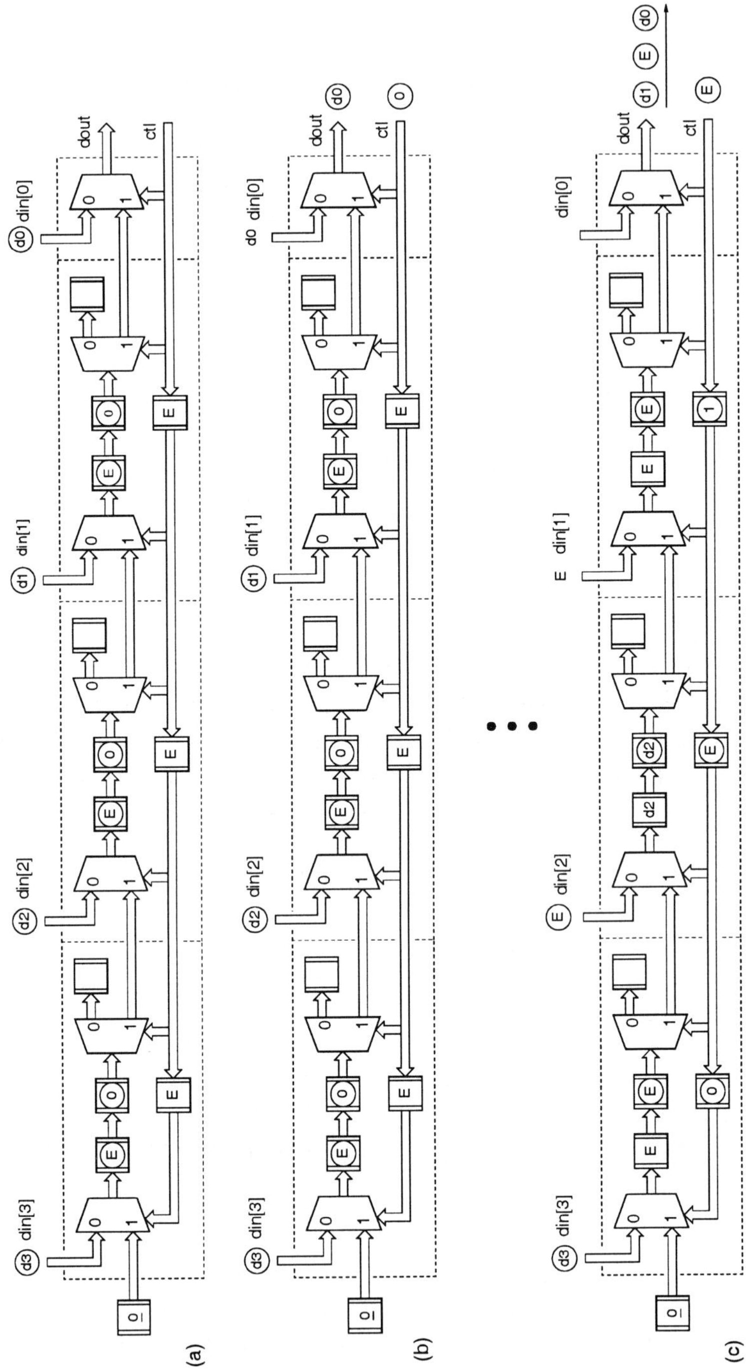

Figure 4.3. Improved design of the shift register with parallel load.

figure 4.1(c), but if the extra latches are added to the control path instead, as shown in figure 4.3(a) on page 46, they will solve both problems.

This improved design exhibits an interesting and illustrative dynamic behaviour: initially, the data latches are densely packed with tokens and all the control latches contain bubbles, figure 4.3(a). The first step of the parallel load cycle is shown in figure 4.3(b), and figure 4.3(c) shows a possible state after the data consuming environment has read a couple of bits. The most-significant stage is just about to perform its "parallel load" and the bubbles are now in the chain of data latches. If at this point the data consuming environment paused, the tokens in the control path would gradually disappear while tokens in the datapath would pack again. Note that at any time the total number of tokens in the circuit is constant!

4.3. Quantifying performance

4.3.1 Latency, throughput and wavelength

When the overall structure of a design is being decided, it is important to determine the optimal number of latches or pipeline stages in the rings and pipeline fragments from which the design is composed. In order to establish a basis for first order design decisions, this section will introduce some quantitative performance parameters. We will restrict the discussion to 4-phase handshaking and bundled-data circuit implementations and we will consider rings with only a single valid token. Subsection 4.3.4, which concludes this section on performance parameters, will comment on adapting to other protocols and implementation styles.

The performance of a pipeline is usually characterized by two parameters: *latency* and *throughput* (or its inverse called *period* or *cycle time*). For an asynchronous pipeline a third parameter, the *dynamic wavelength*, is important as well. With reference to figure 4.4 and following [153, 154, 155] these parameters are defined as follows:

Latency: The latency is the delay from the input of a data item until the corresponding output data item is produced. When data flows in the forward direction, acknowledge signals propagate in the reverse direction. Consequently two parameters are defined:

- The *forward latency*, L_f, is the delay from new data on the input of a stage ($Data[i-1]$ or $Req[i-1]$) to the production of the corresponding output ($Data[i]$ or $Req[i]$) provided that the acknowledge signals are in place when data arrives. $L_{f.V}$ and $L_{f.E}$ denote the latencies for propagating a valid token and an empty token respectively. It is assumed that these latencies are constants, i.e. that they are independent of the value of the data. [As forward propa-

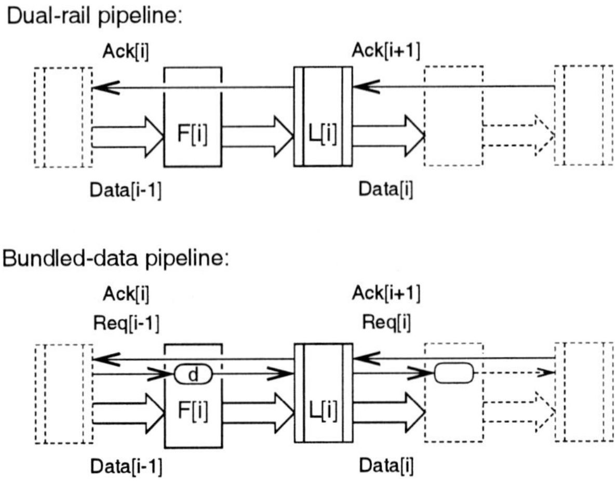

Figure 4.4. Generic pipelines for definition of performance parameters.

gation of an empty token does not "compute" it may be desirable to minimize $L_{f.E}$. In the 4-phase bundled-data approach this can be achieved through the use of an asymmetric delay element.]

- The *reverse latency*, L_r, is the delay from receiving an acknowledge from the succeeding stage (Ack[i+1]) until the corresponding acknowledge is produced to the preceding stage (Ack[i]) provided that the request is in place when the acknowledge arrives. $L_r\downarrow$ and $L_r\uparrow$ denote the latencies of propagating $Ack\downarrow$ and $Ack\uparrow$ respectively.

Period: The period, P, is the delay between the input of a valid token (followed by its succeeding empty token) and the input of the next valid token, i.e. a complete handshake cycle. For a 4-phase protocol this involves: (1) forward propagation of a valid data value, (2) reverse propagation of acknowledge, (3) forward propagation of the empty data value, and (4) reverse propagation of acknowledge. Therefore a lower bound on the period is:

$$P = L_{f.V} + L_r\uparrow + L_{f.E} + L_r\downarrow \tag{4.1}$$

Many of the circuits we consider in this book are symmetric, i.e. $L_{f.V} = L_{f.E}$ and $L_r\uparrow = L_r\downarrow$, and for these circuits the period is simply:

$$P = 2L_f + 2L_r \tag{4.2}$$

We will also consider circuits where $L_{f.V} > L_{f.E}$ and, as we will see in section 4.4.1 and again in section 7.3, the actual implementation of the latches may lead to a period that is larger than the minimum possible given by equation 4.1. In section 4.4.1 we analyze a pipeline whose period is:

$$P = 2L_r + 2L_{f.V} \tag{4.3}$$

Throughput: The *throughput, T*, is the number of valid tokens that flow through a pipeline stage per unit time: $T = 1/P$

Dynamic wavelength: The dynamic wavelength, W_d, of a pipeline is the number of pipeline stages that a forward-propagating token passes through during P:

$$W_d = \frac{P}{L_f} \tag{4.4}$$

Explained differently: W_d is the distance – measured in pipeline stages – between successive valid or empty tokens, when they flow unimpeded down a pipeline. Think of a valid token as the crest of a wave and its associated empty token as the trough of the wave. If $L_{f.V} \neq L_{f.E}$ the average forward latency $L_f = \frac{1}{2}(L_{f.V} + L_{f.E})$ should be used in the above equation.

Static spread: The static spread, S, is the distance – measured in pipeline stages – between successive valid (or empty) tokens in a pipeline that is full (i.e. contains no bubbles). Sometimes the term *occupancy* is used; this is the inverse of S.

4.3.2 Cycle time of a ring

The parameters defined above are local performance parameters that characterize the implementation of individual pipeline stages. When a number of pipeline stages are connected to form a ring, the following parameter is relevant:

Cycle time: The cycle time of a ring, T_{Cycle}, is the time it takes for a token (valid or empty) to make one round trip through all of the pipeline stages in the ring. To achieve maximum performance (i.e. minimum cycle time), the number of pipeline stages per valid token must match the dynamic wavelength, in which case $T_{Cycle} = P$. If the number of pipeline stages is smaller, the cycle time will be limited by the lack of bubbles, and if there are more pipeline stages the cycle time will be limited by the forward latency through the pipeline stages. In [153, 154, 155] these two modes of operation are called *bubble limited* and *data limited*, respectively.

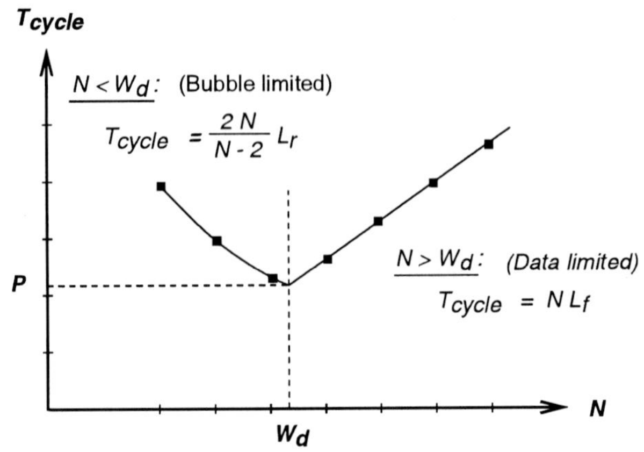

Figure 4.5. Cycle time of a ring as a function of the number of pipeline stages in it.

The cycle time of an N-stage ring in which there is one valid token, one empty token and $N-2$ bubbles can be computed from one of the following two equations (illustrated in figure 4.5):

- When $N \geq W_d$ the cycle time is limited by the forward latency through the N stages:

$$T_{Cycle}(DataLimited) = N \times L_f \tag{4.5}$$

If $L_{f.V} \neq L_{f.E}$ use $L_f = max\{L_{f.V}; L_{f.E}\}$.

- When $N \leq W_d$ the cycle time is limited by the reverse latency. With N pipeline stages, one valid token and one empty token, the ring contains $N-2$ bubbles, and as a cycle involves $2N$ data transfers (N valid and N empty), the cycle time becomes:

$$T_{Cycle}(BubbleLimited) = \frac{2N}{N-2} L_r \tag{4.6}$$

If $L_r\uparrow \neq L_r\downarrow$ use $L_r = \frac{1}{2}(L_r\uparrow + L_r\downarrow)$

For the sake of completeness it should be mentioned that a third possible mode of operation called *control limited* exists for some circuit configurations [153, 154, 155]. This is, however, not relevant to the circuit implementation configurations presented in this book.

The topic of performance analysis and optimization has been addressed in some more recent papers [31, 90, 91, 37] and in some of these the term "slack matching" is used (referring to the process of balancing the timing of forward flowing tokens and backward flowing bubbles).

4.3.3 Example 3: Performance of a 3-stage ring

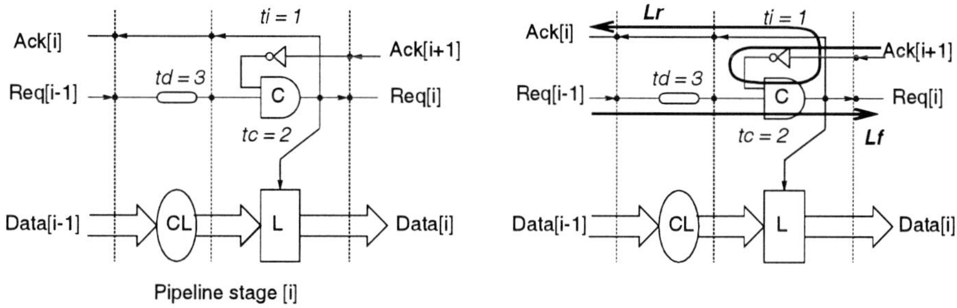

Figure 4.6. A simple 4-phase bundled-data pipeline stage, and an illustration of its forward and reverse latency signal paths.

Let us illustrate the above by a small example: a 3-stage ring composed of identical 4-phase bundled-data pipeline stages that are implemented as illustrated in figure 4.6(a). The data path is composed of a latch and a combinatorial circuit, *CL*. The control part is composed of a C-element and an inverter that controls the latch and a delay element that matches the delay in the combinatorial circuit. Without the combinatorial circuit and the delay element we have a simple FIFO stage. For illustrative purposes the components in the control part are assigned the following latencies: C-element: $t_c = 2$ ns, inverter: $t_i = 1$ ns, and delay element: $t_d = 3$ ns.

Figure 4.6(b) shows the signal paths corresponding to the forward and reverse latencies, and table 4.1 lists the expressions and the values of these parameters. From these figures the period and the dynamic wavelength for the two circuit configurations are calculated. For the FIFO, $W_d = 5.0$ stages, and for the pipeline, $W_d = 3.2$. A ring can only contain an integer number of stages and if W_d is not integer it is necessary to analyze rings with $\lfloor W_d \rfloor$ and $\lceil W_d \rceil$

Table 4.1. Performance of different simple ring configurations.

	FIFO		Pipeline	
Parameter	*Expression*	*Value*	*Expression*	*Value*
L_r	$t_c + t_i$	3 ns	$t_c + t_i$	3 ns
L_f	t_c	2 ns	$t_c + t_d$	5 ns
$P = 2L_f + 2L_r$	$4t_c + 2t_i$	10 ns	$4t_c + 2t_i + 2t_d$	16 ns
W_d		5 stages		3.2 stages
T_{Cycle} (3 stages)	$6\ L_r$	18 ns	$6\ L_r$	18 ns
T_{Cycle} (4 stages)	$4\ L_r$	12 ns	$4\ L_f$	20 ns
T_{Cycle} (5 stages)	$3.3\ L_r = 5\ L_f$	10 ns	$5\ L_f$	25 ns
T_{Cycle} (6 stages)	$6\ L_f$	12 ns	$6\ L_f$	30 ns

stages and determine which yields the smallest cycle time. Table 4.1 shows the results of the analysis including cycle times for rings with 3 to 6 stages.

4.3.4 Final remarks

The above presentation made a number of simplifying assumptions: (1) only rings and pipelines composed of identical pipeline stages were considered, (2) it assumed function blocks with symmetric delays (i.e. circuits where $L_{f.V} = L_{f.E}$), (3) it assumed function blocks with constant latencies (i.e. ignoring the important issue of data-dependent latencies and average-case performance), (4) it considered rings with only a single valid token, and (5) the analysis considered only 4-phase handshaking and bundled-data circuits.

For 4-phase dual-rail implementations (where request is embedded in the data encoding) the performance parameter equations defined in the previous section apply without modification. For designs using a 2-phase protocol, some straightforward modifications are necessary: there are no empty tokens and hence there is only one value for the forward latency L_f and one value for the reverse latency L_r. It is also a simple matter to state expressions for the cycle time of rings with more tokens.

It is more difficult to deal with data-dependent latencies in the function blocks and to deal with non-identical pipeline stages. Despite these deficiencies the performance parameters introduced in the previous sections are very useful as a basis for first-order design decisions.

4.4. Dependency graph analysis

When the pipeline stages incorporate different function blocks, or function blocks with asymmetric delays, it is a more complex task to determine the critical path. It is necessary to construct a graph that represents the dependencies between signal transitions in the circuit, and to analyze this graph and identify the critical path cycle [19, 153, 154, 155]. This can be done in a systematic or even mechanical way but the amount of detail makes it a complex task.

The nodes in such a *dependency graph* represent rising or falling signal transitions, and the edges represent dependencies between the signal transitions. Formally, a dependency is a marked graph [28]. Let us look at a couple of examples.

4.4.1 Example 4: Dependency graph for a pipeline

As a first example let us consider a (very long) pipeline composed of identical stages using a function block with asymmetric delays causing $L_{f.E} < L_{f.V}$. Figure 4.7(a) shows a 3-stage section of this pipeline. Each pipeline stage has the following latency parameters:

$$\begin{aligned} L_{f.V} &= t_{d(0\to 1)} + t_c = 5 \text{ ns} + 2 \text{ ns} = 7 \text{ ns} \\ L_{f.E} &= t_{d(1\to 0)} + t_c = 1 \text{ ns} + 2 \text{ ns} = 3 \text{ ns} \\ L_r\uparrow = L_r\downarrow &= t_i + t_c = 3 \text{ ns} \end{aligned}$$

There is a close relationship between the circuit diagram and the dependency graph. As signals alternate between rising transitions (↑) and falling transitions (↓) – or between valid and empty data values – the graph has two nodes per circuit element. Similarly the graph has two edges per wire in the circuit. Figure 4.7(b) shows the two graph fragments that correspond to a pipeline stage, and figure 4.7(c) shows the dependency graph that corresponds to the 3 pipeline stages in figure 4.7(a).

A label outside a node denotes the circuit delay associated with the signal transition. We use a particular style for the graphs that we find illustrative: the nodes corresponding to the forward flow of valid and empty data values are organized as two horizontal rows, and nodes representing the reverse flowing acknowledge signals appear as diagonal segments connecting the rows.

The cycle time or period of the pipeline is the time from a signal transition until the same signal transition occurs again. The cycle time can therefore be determined by finding the longest simple cycle in the graph, i.e. the cycle with the largest accumulated circuit delay which does not contain a sub-cycle. The dotted cycle in figure 4.7(c) is the longest simple cycle. Starting at point A the corresponding period is:

$$\begin{aligned} P &= \underbrace{t_{D(0\to 1)} + t_C}_{L_{f.V}} + \underbrace{t_I + t_C}_{L_r\downarrow} + \underbrace{t_{D(1\to 0)} + t_C}_{L_{f.V}} + \underbrace{t_I + t_C}_{L_r\uparrow} \\ &= 2L_r + 2L_{f.V} \;=\; 20 \text{ ns} \end{aligned}$$

Note that this is the period given by equation 4.3 on page 49. An alternative cycle time candidate is the following:

$$\underbrace{R_{[i]}\uparrow; Req_{[i]}\uparrow;}_{L_{f.V}} \underbrace{A_{[i-1]}\downarrow; Req_{[i-1]}\downarrow;}_{L_r\downarrow} \underbrace{R_{[i]}\downarrow; Req_{[i]}\downarrow;}_{L_{f.E}} \underbrace{A_{[i-1]}\uparrow; Req_{[i-1]}\uparrow;}_{L_r\uparrow}$$

and the corresponding period is:

$$P = 2L_r + L_{f.V} + L_{f.E} = 16 \text{ ns}$$

Note that this is the minimum possible period given by equation 4.1 on page 48. The period is determined by the longest cycle which is 20 ns. Thus, this example illustrates that for some (simple) latch implementations it may not be possible to reduce the cycle time by using function blocks with asymmetric delays ($L_{f.E} < L_{f.V}$).

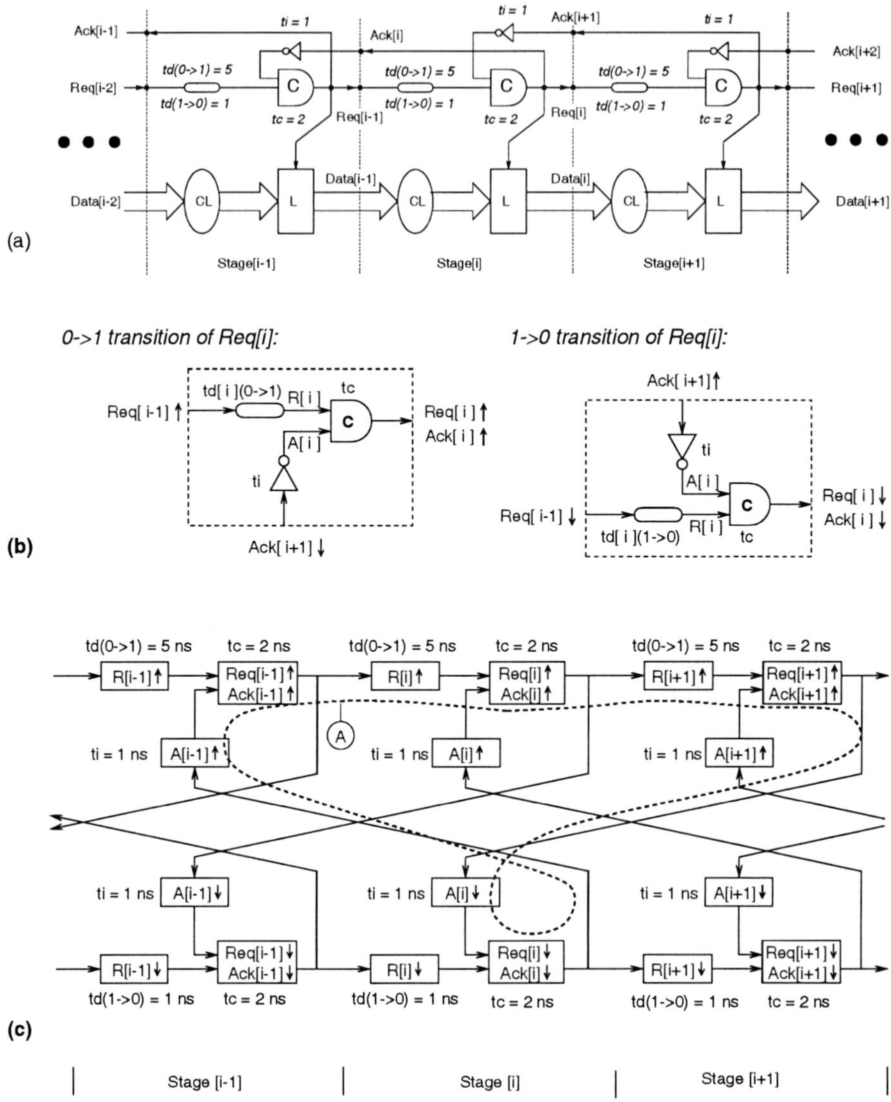

Figure 4.7. Data dependency graph for a 3-stage section of a pipeline: (a) the circuit diagram, (b) the two graph fragments corresponding to a pipeline stage, and (c) the resulting data-dependency graph.

4.4.2 Example 5: Dependency graph for a 3-stage ring

As another example of dependency graph analysis let us consider a three stage 4-phase bundled-data ring composed of different pipeline stages, figure 4.8(a): stage 1 with a combinatorial circuit that is matched by a symmetric

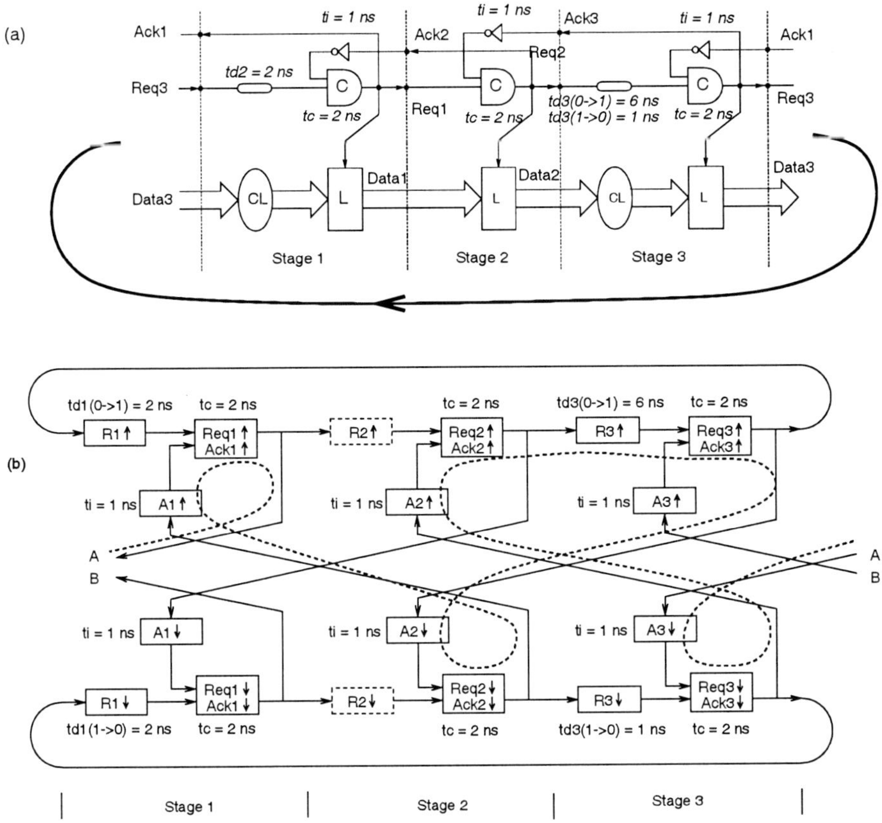

Figure 4.8. Data dependency graph for an example 3-stage ring: (a) the circuit diagram for the ring and (b) the resulting data-dependency graph.

delay element, stage 2 without combinatorial logic, and stage 3 with a combinatorial circuit that is matched by an asymmetric delay element.

The dependency graph is similar to the dependency graph for the 3-stage pipeline from the previous section. The only difference is that the output port of stage 3 is connected to the input port of stage 1, forming a closed graph. There are several "longest simple cycle" candidates:

1 A cycle corresponding to the forward flow of valid-tokens:

$$(\text{R1}\uparrow;\ \text{Req1}\uparrow;\ \text{R2}\uparrow;\ \text{Req2}\uparrow;\ \text{R3}\uparrow;\ \text{Req3}\uparrow)$$

For this cycle, the cycle time is $T_{Cycle} = 14$ ns.

2 A cycle corresponding to the forward flow of empty-tokens:

$$(\text{R1}\downarrow;\ \text{Req1}\downarrow;\ \text{R2}\downarrow;\ \text{Req2}\downarrow;\ \text{R3}\downarrow;\ \text{Req3}\downarrow)$$

For this cycle, the cycle time is $T_{Cycle} = 9$ ns.

3 A cycle corresponding to the backward flowing bubble:

(A1↑; Req1↑; A3↓; Req3↓; A2↑; Req2↑; A1↓; Req1↓; A3↑; Req3↑; A2↓; Req2↓)

For this cycle, the cycle time is $T_{Cycle} = 6L_r = 18$ ns.

The 3-stage ring contains one valid-token, one empty-token and one bubble, and it is interesting to note that the single bubble is involved in six data transfers, and therefore makes two reverse round trips for each forward round trip of the valid-token.

4 There is, however, another cycle with a slightly longer cycle time, as illustrated in figure 4.8(b). It is the cycle corresponding to the backward-flowing bubble where the sequence:

(A1↓; Req1↓; A3↑) is replaced by (R3↑)

For this cycle the cycle time is $T_{Cycle} = 6L_r = 20$ ns.

A dependency graph analysis of a 4-stage ring is very similar. The only difference is that there are two bubbles in the ring. In the dependency graph this corresponds to the existence of two "bubble cycles" that do not interfere with each other.

The dependency graph approach presented above assumes a closed circuit that results in a closed dependency graph. If a component such as a pipeline fragment is to be analyzed it is necessary to include a (dummy) model of its environment as well – typically in the form of independent and eager token producers and token consumers, i.e. dummy circuits that simply respond to handshakes. Figure 2.15 on page 24 illustrated this for a single pipeline stage control circuit.

Note that a dependency graph as introduced above is similar to a signal transition graph (STG) which we will introduce more carefully in chapter 6.

4.5. Summary

This chapter addressed the performance analysis of asynchronous circuits at several levels: firstly, by providing a qualitative understanding of performance based on the dynamics of tokens flowing in a circuit; secondly, by introducing quantitative performance parameters that characterize pipelines and rings composed of identical pipeline stages and, thirdly, by introducing dependency graphs that enable the analysis of pipelines and rings composed of non-identical stages.

At this point we have covered the design and performance analysis of asynchronous circuits at the "static data-flow structures" level, and it is time to address low-level circuit design principles and techniques. This will be the topic of the next two chapters.

Chapter 5

HANDSHAKE CIRCUIT IMPLEMENTATIONS

In this chapter we will address the implementation of handshake components. First, we will consider the basic set of components introduced in section 3.3 on page 32: (1) the latch, (2) the unconditional data-flow control elements join, fork and merge, (3) function blocks, and (4) the conditional flow control elements MUX and DEMUX. In addition to these basic components we will also consider the implementation of mutual exclusion elements and arbiters and touch upon the (unavoidable) problem of metastability. The major part of the chapter (sections 5.3–5.6) is devoted to the implementation of function blocks and the material includes a number of fundamental concepts and circuit implementation styles.

5.1. The latch

As mentioned previously, the role of latches is: (1) to provide storage for valid and empty tokens, and (2) to support the flow of tokens via handshaking with neighbouring latches. Possible implementations of the handshake latch were shown in chapter 2: Figure 2.9 on page 18 shows how a 4-phase bundled-data handshake latch can be implemented using a conventional latch and a control circuit (the figure shows several such examples assembled into pipelines). In a similar way figure 2.11 on page 20 shows the implementation of a 2-phase bundled-data latch, and figures 2.12-2.13 on page 21 show the implementation of a 4-phase dual-rail latch.

A handshake latch can be characterized in terms of the *throughput*, the *dynamic wavelength* and the *static spread* of a FIFO that is composed of identical latches. Common to the two 4-phase latch designs mentioned above is that a FIFO will fill with every other latch holding a valid token and every other latch holding an empty token (as illustrated in figure 4.1(b) on page 43). Thus, the static spread for these FIFOs is $S = 2$.

A 2-phase implementation does not involve empty tokens and consequently it may be possible to design a latch whose static spread is $S = 1$. Note, however, that the implementation of the 2-phase bundled-data handshake latch in

J. Sparsø and S. Furber (eds.), Principles of Asynchronous Circuit Design, 57–80.

figure 2.11 on page 20 involves several level-sensitive latches; the utilization of the level sensitive latches is no better.

Ideally, one would want to pack a valid token into every level-sensitive latch, and in chapter 7 we will address the design of 4-phase bundled-data handshake latches that have a smaller static spread.

5.2. Fork, join, and merge

Possible 4-phase bundled-data and 4-phase dual-rail implementations of the fork, join, and merge components are shown in figure 5.1. For simplicity the figure shows a fork with two output channels only, and join and merge components with two input channels only. Furthermore, all channels are assumed to be 1-bit channels. It is, of course, possible to generalize to three or more inputs and outputs respectively, and to extend to n-bit channels. Based on the explanation given below this should be straightforward, and it is left as an exercise for the reader.

4-phase fork and join A fork involves a C-element to combine the acknowledge signals on the output channels into a single acknowledge signal on the input channel. Similarly a 4-phase bundled-data join involves a C-element to combine the request signals on the input channels into a single request signal on the output channel. The 4-phase dual-rail join does not involve any active components as the request signal is encoded into the data.

The particular fork in figure 5.1 duplicates the input data, and the join concatenates the input data. This happens to be the way joins and forks are mostly used in our static data-flow structures, but there are many alternatives: for example, the fork could split the input data which would make it more symmetric to the join in figure 5.1. In any case the difference is only in how the input data is transferred to the output. From a control point of view the different alternatives are identical: a join synchronizes several input channels and a fork synchronizes several output channels.

4-phase merge The implementation of the merge is a little more elaborate. Handshakes on the input channels are mutually exclusive, and the merge simply relays the active input handshake to the output channel.

Let us consider the implementation of the 4-phase bundled-data merge first. It consists of an asynchronous control circuit and a multiplexer that is controlled by the input request. The control circuit is explained below.

The request signals on the input channels are mutually exclusive and may simply be ORed together to produce the request signal on the output channel.

For each input channel, a C-element produces an acknowledge signal in response to an acknowledge on the output channel provided that the input channel has valid data. For example, the C-element driving the x_{ack} signal is set high

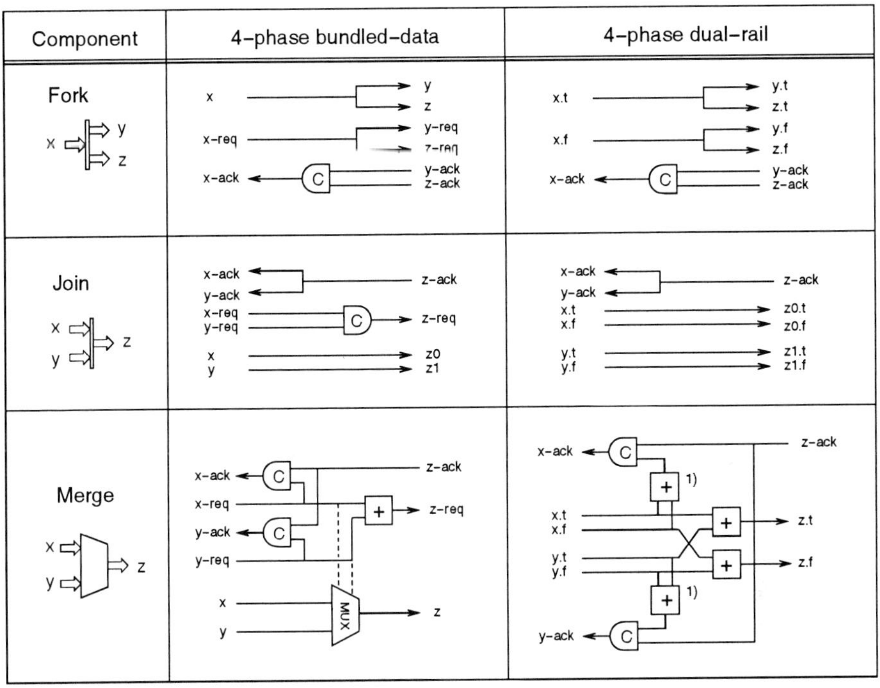

Figure 5.1. 4-phase bundled-data and 4-phase dual-rail implementations of the fork, join and merge components.

when x_{req} and z_{ack} have both gone high, and it is reset when both signals have gone low again. As z_{ack} goes low in response to x_{req} going low, it will suffice to reset the C-element in response to z_{ack} going low. This optimization is possible if asymmetric C-elements are available, figure 5.2. Similar arguments applies for the C-element that drives the y_{ack} signal. A more detailed introduction to generalized C-elements and related state-holding devices is given in chapter 6, sections 6.4.1 and 6.4.5.

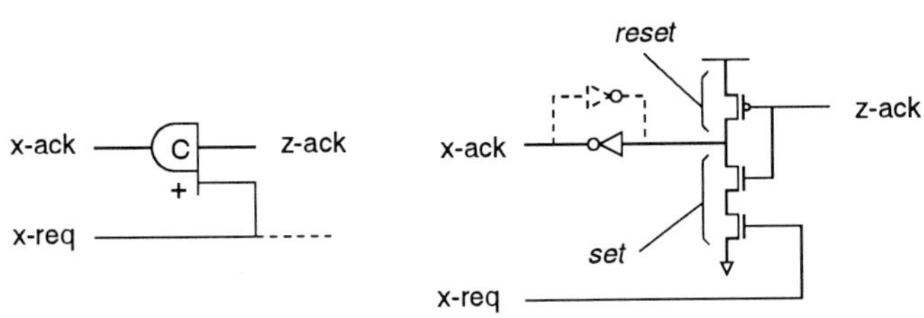

Figure 5.2. A possible implementation of the upper asymmetric C-element in the 4-phase bundled-data merge in figure 5.1.

The implementation of the 4-phase dual-rail merge is fairly similar. As request is encoded into the data signals an OR gate is used for each of the two output signals $z.t$ and $z.f$. Acknowledge on an input channel is produced in response to an acknowledge on the output channel provided that the input channel has valid data. Since the example assumes 1-bit wide channels, the latter is established using an OR gate (marked "1"), but for N-bit wide channels a completion detector (as shown in figure 2.13 on page 21) would be required.

2-phase fork, join and merge Finally a word about 2-phase bundled-data implementations of the fork, join and merge components: the implementation of 2-phase bundled-data fork and join components is identical to the implementation of the corresponding 4-phase bundled-data components (assuming that all signals are initially low).

The implementation of a 2-phase bundled-data merge, on the other hand, is complex and rather different, and it provides a good illustration of why the implementation of some 2-phase bundled-data components is complex. When observing an individual request or acknowledge signal the transitions will obviously alternate between rising and falling, but since nothing is known about the sequence of handshakes on the input channels there is no relationship between the polarity of a request signal transition on an input channel and the polarity of the corresponding request signal transition on the output channel. Similarly there is no relationship between the polarity of an acknowledge signal transition on the output channel and the polarity of the corresponding acknowledge signal transition on the input channel channel. This calls for some kind of storage element on each request and acknowledge signal produced by the circuit. This brings complexity, as does the associated control logic.

5.3. Function blocks – The basics

This section will introduce the fundamental principles of function block design, and subsequent sections will illustrate function block implementations for different handshake protocols. The running example will be an N-bit ripple carry adder.

5.3.1 Introduction

A function block is the asynchronous equivalent of a combinatorial circuit: it computes one or more output signals from a set of input signals. The term "function block" is used to stress the fact that we are dealing with circuits with a purely functional behaviour.

However, in addition to computing the desired function(s) of the input signals, a function block must also be transparent to the handshaking that is implemented by its neighbouring latches. This transparency to handshaking is

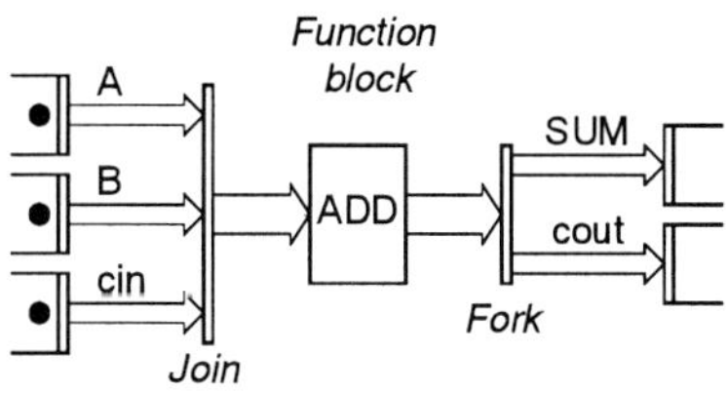

Figure 5.3. A function block whose operands and results are provided on separate channels requires a join of the inputs and a fork on the output.

what makes function blocks different from combinatorial circuits and, as we will see, there are greater depths to this than is indicated by the word "transparent" – in particular for function blocks that implicitly indicate completion (which is the case for circuits using dual-rail signals).

The most general scenario is where a function block receives its operands on separate channels and produces its results on separate channels, figure 5.3. The use of several independent input and output channels implies a join on the input side and a fork on the output side, as illustrated in the figure. These can be implemented separately, as explained in the previous section, or they can be integrated into the function block circuitry. In what follows we will restrict the discussion to a scenario where all operands are provided on a single channel and where all results are provided on a single channel.

We will first address the issue of handshake transparency and then review the fundamentals of ripple carry addition, in order to provide the necessary background for discussing the different implementation examples that follow. A good paper on the design of function blocks is [97].

5.3.2 Transparency to handshaking

The general concepts are best illustrated by considering a 4-phase dual-rail scenario – function blocks for bundled data protocols can be understood as a special case. Figure 5.4(a) shows two handshake latches connected directly and figure 5.4(b) shows the same situation with a function block added between the two latches. The function block must be transparent to the handshaking. Informally this means that if observing the signals on the ports of the latches, one should see the same sequence of handshake signal transitions; the only difference should be some slow-down caused by the latency of the function block.

A function block is obviously not allowed to produce a request on its output before receiving a request on its input; put the other way round, a request on the output of the function block should indicate that all of the inputs are valid and that all (relevant) internal signals and all output signals have been computed.

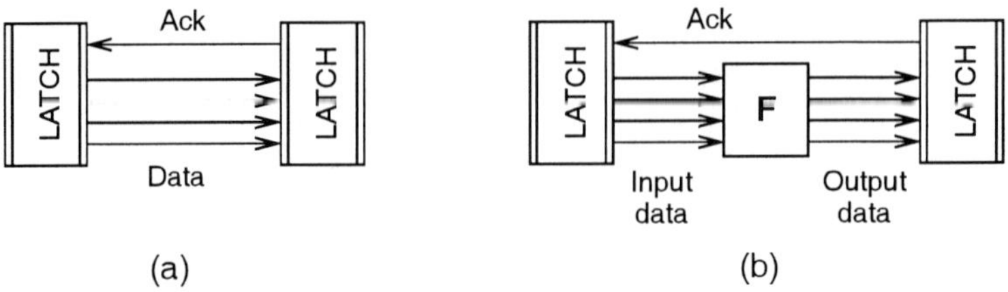

Figure 5.4. (a) Two latches connected directly by a handshake channel and (b) the same situation with a function block added between the latches. The handshaking as seen by the latches in the two situations should be the same, i.e. the function block must be designed such that it is transparent to the handshaking.

(Here we are touching upon the principle of indication once again.) In 4-phase protocols a symmetric set of requirements apply for the return-to-zero part of the handshaking.

Function blocks can be characterized as either *strongly indicating* or *weakly indicating* depending on how they behave with respect to this handshake transparency. The signalling that can be observed on the channel between the two

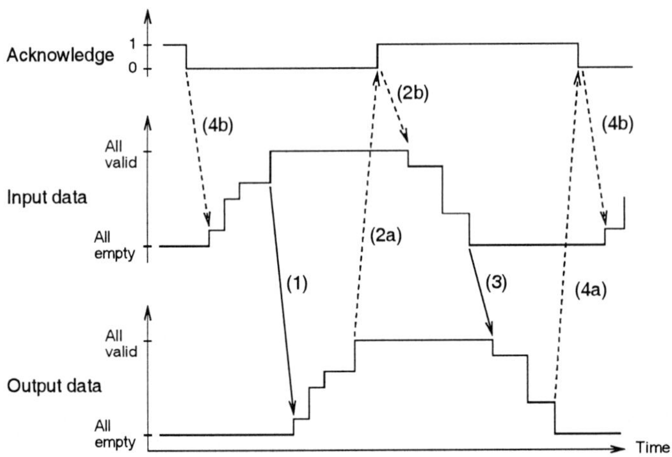

(1)	"All inputs become defined"	$\preceq$	"Some outputs become defined"
(2)	"All outputs become defined"	$\preceq$	"Some inputs become undefined"
(3)	"All inputs become undefined"	$\preceq$	"Some outputs become undefined"
(4)	"All outputs become undefined"	$\preceq$	"Some inputs become defined"

Figure 5.5. Signal traces and event orderings for a strongly indicating function block.

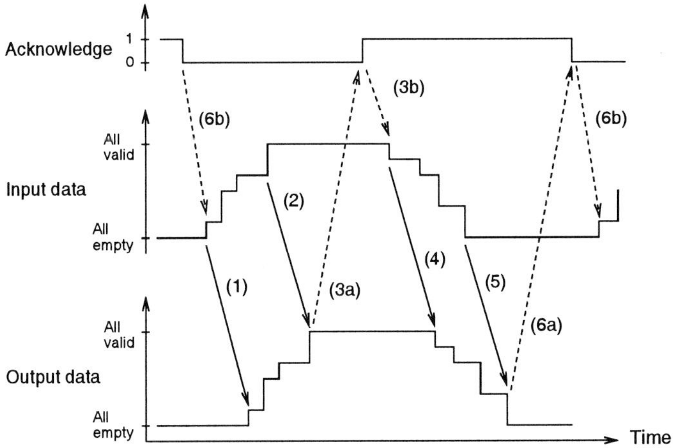

(1)	"Some inputs become defined"	$\preceq$	"Some outputs become defined"
(2)	"All inputs become defined"	$\preceq$	"All outputs become defined"
(3)	"All outputs become defined"	$\preceq$	"Some inputs become undefined"
(4)	"Some inputs become undefined"	$\preceq$	"Some outputs become undefined"
(5)	"All inputs become undefined"	$\preceq$	"All outputs become undefined"
(6)	"All outputs become undefined"	$\preceq$	"Some inputs become defined"

Figure 5.6. Signal traces and event orderings for a weakly indicating function block.

latches in figure 5.4(a) was illustrated in figure 2.3 on page 13. We can illustrate the handshaking for the situation in figure 5.4(b) in a similar way.

- A function block is *strongly indicating*, as illustrated in figure 5.5, if (1) it waits for all of its inputs to become valid before it starts to compute and produce valid outputs, and if (2) it waits for all of its inputs to become empty before it starts to produce empty outputs.

- A function block is *weakly indicating*, as illustrated in figure 5.6, if (1) it starts to compute and produce valid outputs as soon as possible, i.e. when some but not all input signals have become valid, and if (2) it starts to produce empty outputs as soon as possible, i.e. when some but not all input signals have become empty.

For a weakly indication function block to behave correctly, it is necessary to require that it never produces all valid outputs until after all inputs have become valid, and that it never produces all empty outputs until after all inputs have become empty. This behaviour is identical to Seitz's weak conditions in [121]. In [121] Seitz further explains that it can be proved that if the individual components satisfy the weak conditions then any "valid combinatorial circuit structure" of function blocks also satisfies the weak conditions, i.e. that function blocks may be combined to form larger function blocks. By "valid combinatorial circuit structure" we mean a structure where no components have inputs or outputs left unconnected and where there are no feed-back signal paths. Strongly indicating function blocks have the same property – a "valid combinatorial circuit structure" of strongly indicating function blocks is itself a strongly indicating function block.

Notice that both weakly and strongly indicating function blocks exhibit a hysteresis-like behaviour in the valid-to-empty and empty-to-valid transitions: (1) some/all outputs must remain valid until after some/all inputs have become empty, and (2) some/all outputs must remain empty until after some/all inputs have become valid. It is this hysteresis that ensures handshake transparency, and the implementation consequence is that one or more state holding circuits (normally in the form of C-elements) are needed.

Finally, a word about the 4-phase bundled-data protocol. Since $Req\uparrow$ is equivalent to "all data signals are valid" and since $Req\downarrow$ is equivalent to "all data signals are empty," a 4-phase bundled-data function block can be categorized as strongly indicating.

As we will see in the following, strongly indicating function blocks have worst-case latency. To obtain actual case latency weakly indicating function blocks must be used. Before addressing possible function block implementation styles for the different handshake protocols it is useful to review the basics of binary ripple-carry addition, the running example in the following sections.

5.3.3 Review of ripple-carry addition

Figure 5.7 illustrates the implementation principle of a ripple-carry adder. A 1-bit full adder stage implements:

$$s = a \oplus b \oplus c \tag{5.1}$$

$$d = ab + ac + bc \tag{5.2}$$

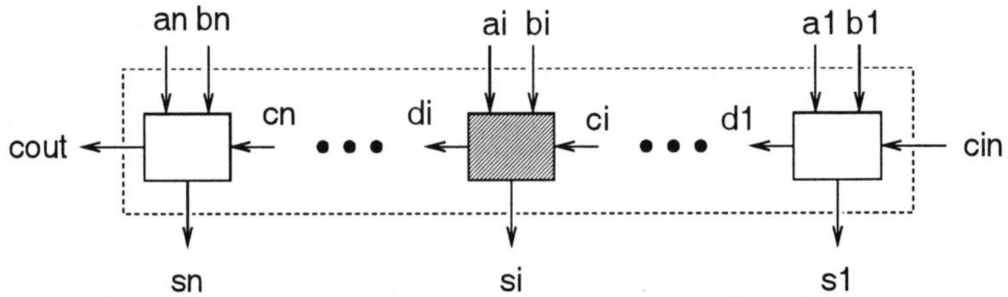

Figure 5.7. A ripple-carry adder. The carry output of one stage d_i is connected to the carry input of the next stage c_{i+1}.

In many implementations inputs a and b are recoded as:

$$p = a \oplus b \qquad \text{(``propagate'' carry)} \tag{5.3}$$
$$g = ab \qquad \text{(``generate'' carry)} \tag{5.4}$$
$$k = \overline{a}\overline{b} \qquad \text{(``kill'' carry)} \tag{5.5}$$

...and the output signals are computed as follows:

$$s = p \oplus c \tag{5.6}$$
$$d = g + pc \qquad \text{or alternatively} \tag{5.7a}$$
$$\overline{d} = k + p\overline{c} \tag{5.7b}$$

For a ripple-carry adder, the worst case critical path is a carry rippling across the entire adder. If the latency of a 1-bit full adder is t_{add} the worst case latency of an N-bit adder is $N \cdot t_{add}$. This is a very rare situation and in general the longest carry ripple during a computation is much shorter. Assuming random and uncorrelated operands the average latency is $log(N) \cdot t_{add}$ and, if numerically small operands occur more frequently, the average latency is even less. Using normal Boolean signals (as in the bundled-data protocols) there is no way to know when the computation has finished and the resulting performance is thus worst-case.

By using dual-rail carry signals $(d.t, d.f)$ it is possible to design circuits that indicate completion as part of the computation and thus achieve actual case latency. The crux is that a dual-rail carry signal, d, conveys one of the following 3 messages:

$(d.t, d.f) = (0,0)$	=	Empty	"The carry has not been computed yet" (possibly because it depends on c)
$(d.t, d.f) = (1,0)$	=	True	"The carry is 1"
$(d.t, d.f) = (0,1)$	=	False	"The carry is 0"

Consequently it is possible for a 1-bit adder to output a valid carry without waiting for the incoming carry if its inputs make this possible ($a = b = 0$ or $a = b = 1$). This idea was first put forward in 1955 in a paper by Gilchrist [52]. The same idea is explained in [62, pp. 75-78] and in [121].

5.4. Bundled-data function blocks

5.4.1 Using matched delays

A bundled-data implementation of the adder in figure 5.7 is shown in figure 5.8. It is composed of a traditional combinatorial circuit adder and a matching delay element. The delay element provides a constant delay that matches the *worst case* latency of the combinatorial adder. This includes the worst case

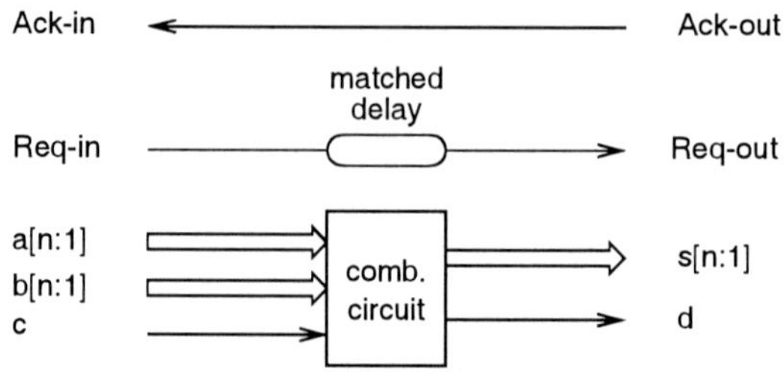

Figure 5.8. **A 4-phase bundled data implementation of the N–bit handshake adder from figure 5.7.**

critical path in the circuit – a carry rippling across the entire adder – as well as the worst case operating conditions. For reliable operation some safety margin is needed.

In addition to the combinatorial circuit itself, the delay element represents a design challenge for the following reasons: to a first order the delay element will track delay variations that are due to the fabrication process spread as well as variations in temperature and supply voltage. On the other hand, wire delays can be significant and they are often beyond the designer's control. Some design policy for matched delays is obviously needed. In a full custom design environment one may use a dummy circuit with identical layout but with weaker transistors. In a standard cell automatic place and route environment one will have to accept a fairly large safety margin or do post-layout timing analysis and trimming of the delays. The latter sounds tedious but it is similar to the procedure used in synchronous design where setup and hold times are checked and delays trimmed after layout.

In a 4-phase bundled-data design an asymmetric delay element may be preferable from a performance point of view, in order to perform the return-to-zero part of the handshaking as quickly as possible. Another issue is the power consumption of the delay element. In the ARISC processor design reported in [23] the delay elements consumed 10 % of the total power.

5.4.2 Delay selection

In [105] Nowick proposed a scheme called "speculative completion". The basic principle is illustrated in figure 5.9. In addition to the desired function some additional circuitry is added that selects among several matched delays. The estimate must be conservative, i.e. on the safe side. The estimation can be based on the input signals and/or on some internal signals in the circuit that implements the desired function.

For an N-bit ripple-carry adder the propagate signals (c.f. equation 5.3) that form the individual 1-bit full adders (c.f. figure 5.7) may be used for the estimation. As an example of the idea consider a 16-bit adder. If $p_8 = 0$ the

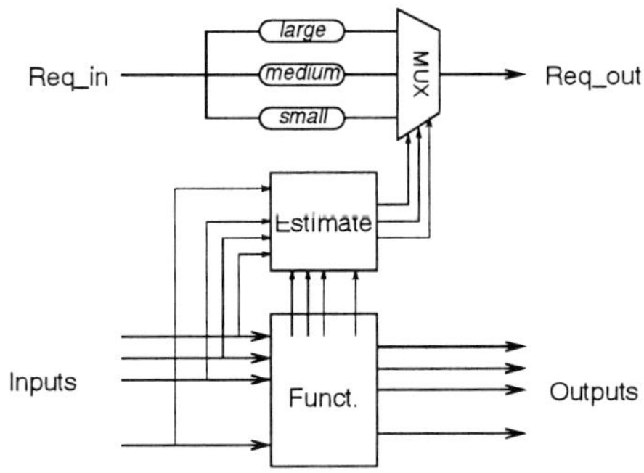

Figure 5.9. The basic idea of "speculative completion".

longest carry ripple can be no longer than 8 stages, and if $p_{12} \wedge p_8 \wedge p_4 = 0$ the longest carry ripple can be no longer than 4 stages. Based on such simple estimates a sufficiently large matched delay is selected. Again, if a 4-phase protocol is used, asymmetric delay elements are preferable from a performance point of view.

To the designer the trade-off is between an aggressive estimate with a large circuit overhead (area and power) or a less aggressive estimate with less overhead. For more details on the implementation and the attainable performance gains the reader is is referred to [105, 107].

5.5. Dual-rail function blocks

5.5.1 Delay insensitive minterm synthesis (DIMS)

In chapter 2 (page 22 and figure 2.14) we explained the implementation of an AND gate for dual-rail signals. Using the same basic topology it is possible to implement other simple gates such as OR, EXOR, etc. An inverter involves no active circuitry as it is just a swap of the two wires.

Arbitrary functions can be implemented by combining gates in exactly the same way as when one designs combinatorial circuits for a synchronous circuit. The handshaking is implicitly taken care of and can be ignored when composing gates and implementing Boolean functions. This has the important implication that existing logic synthesis techniques and tools may be used, the only difference is that the basic gates are implemented differently.

The dual-rail AND gate in figure 2.14 is obviously rather inefficient: 4 C-elements and 1 OR gate totaling approximately 30 transistors – a factor five greater than a normal AND gate whose implementation requires only 6 transistors. By implementing larger functions the overhead can be reduced. To illustrate this figure 5.10(b)-(c) shows the implementation of a 1-bit full adder. We will discuss the circuit in figure 5.10(d) shortly.

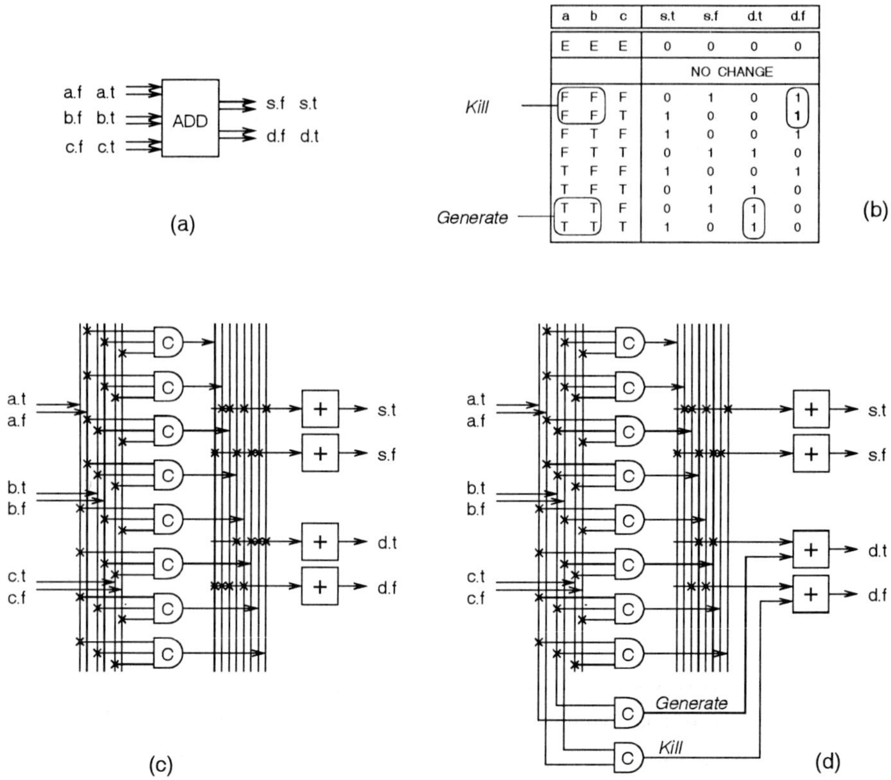

	a	b	c	s.t	s.f	d.t	d.f
	E	E	E	0	0	0	0
				NO CHANGE			
Kill	F	F	F	0	1	0	1
Kill	F	F	T	1	0	0	1
	F	T	F	1	0	0	1
	F	T	T	0	1	1	0
	T	F	F	1	0	0	1
	T	F	T	0	1	1	0
Generate	T	T	F	0	1	1	0
Generate	T	T	T	1	0	1	0

Figure 5.10. A 4-phase dual-rail full-adder: (a) Symbol, (b) truth table, (c) DIMS implementation and (d) an optimization that makes the full adder weakly indicating.

The PLA-like structure of the circuit in figure 5.10(c) illustrates a general principle for implementing arbitrary Boolean functions. In [124] we called this approach DIMS – Delay-Insensitive Minterm Synthesis – because the circuits are delay-insensitive and because the C-elements in the circuits generate all minterms of the input variables. The truth tables have 3 groups of rows specifying the output when the input is: (1) the empty codeword to which the circuit responds by setting the output empty, (2) an intermediate codeword which does not affect the output, or (3) a valid codeword to which the circuit responds by setting the output to the proper valid value.

The fundamental ideas explained above all go back to David Muller's work in the late 1950s and early 1960s [93, 92]. While [93] develops the fundamental theorem for the design of speed-independent circuits, [92] is a more practical introduction including a design example: a bit-serial multiplier using latches and gates as explained above.

Referring to section 5.3.2, the DIMS circuits as explained here can be categorized as strongly indicating, and hence they exhibit worst case latency. In an N-bit ripple-carry adder the empty-to-valid and valid-to-empty transitions will ripple in strict sequence from the least significant full adder to the most significant one.

If we change the full-adder design slightly as illustrated in figure 5.10(d) a valid d may be produced before the c input is valid ("kill" or "generate"), and an N-bit ripple-carry adder built from such full adders will exhibit actual-case latency – the circuits are weakly indicating function blocks.

The designs in figure 5.10(c) and 5.10(d), and ripple-carry adders built from these full adders, are all symmetric in the sense that the latency of propagating an empty value is the same as the latency of propagating the preceding valid value. This may be undesirable. Later in section 5.5.4 we will introduce an elegant design that propagates empty values in constant time (with the latency of 2 full adder cells).

5.5.2 Null Convention Logic

The C-elements and OR gates from the previous sections can be seen as n-of-n and 1-of-n threshold gates with hysteresis, figure 5.11. By using arbitrary m-of-n threshold gates with hysteresis – an idea proposed by Theseus Logic, Inc., [39] – it is possible to reduce the implementation complexity. An m-of-n threshold gate with hysteresis will set its output high when any m inputs have gone high and it will set its output low when all its inputs are low. This elegant circuit implementation idea is the key element in Theseus Logic's Null Convention Logic. At the higher levels of design NCL is no different from the data-flow view presented in chapter 3 and NCL has great similarities to the circuit design styles presented in [92, 122, 124, 97]. Figure 5.11 shows that

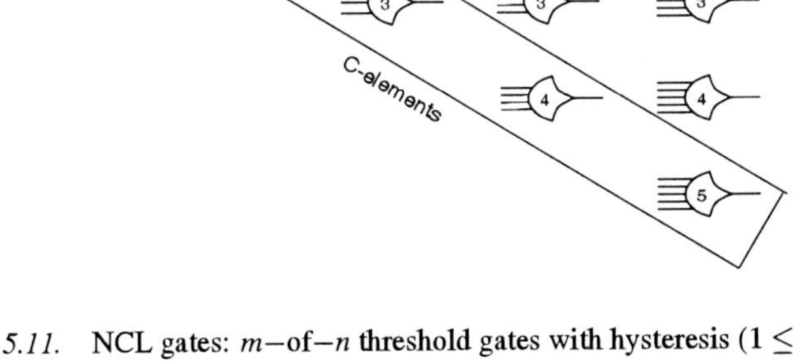

Figure 5.11. NCL gates: m–of–n threshold gates with hysteresis ($1 \leq m \leq n$).

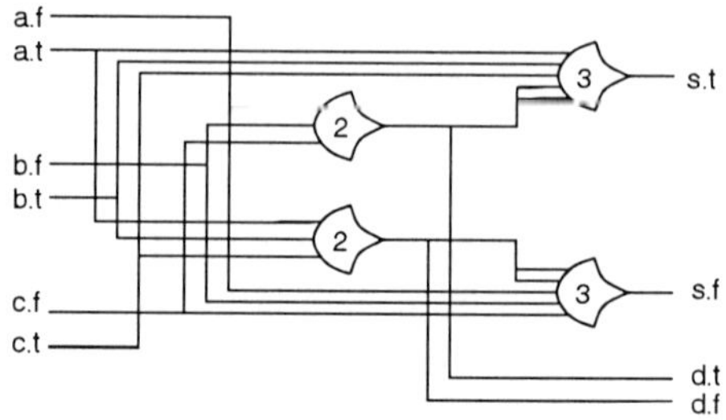

Figure 5.12. A full adder using NCL gates.

OR gates and C-elements can be seen as special cases in the world of threshold gates. The digit inside a gate symbol is the threshold of the gate. Figure 5.12 shows the implementation of a dual-rail full adder using NCL threshold gates. The circuit is weakly indicating.

5.5.3 Transistor-level CMOS implementations

The last two adder designs we will introduce are based on CMOS transistor-level implementations using dual-rail signals. Dual-rail signals are essentially what are produced by precharged differential logic circuits that are used in memory structures and in logic families like DCVSL, figure 5.13 [151, 55].

In a bundled-data design the precharge signal can be the request signal on the input channel to the function block. In a dual-rail design the precharge p-type transistors may be replaced by transistor networks that detect when all

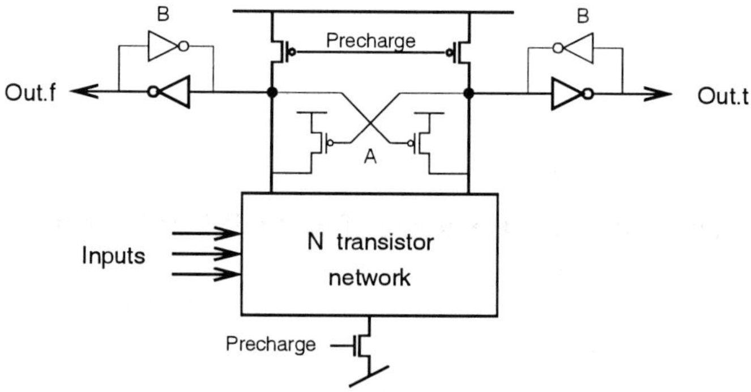

Figure 5.13. A precharged differential CMOS combinatorial circuit. By adding the cross-coupled p-type transistors labeled "A" or the (weak) feedback-inverters labeled "B" the circuit becomes (pseudo)static.

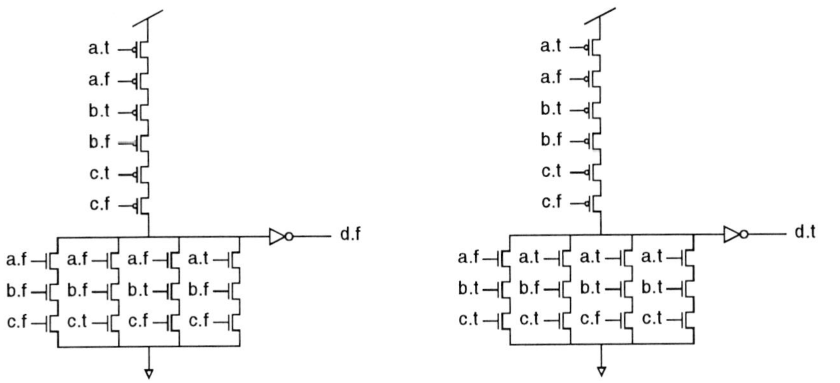

Figure 5.14. Transistor-level implementation of the carry signal for the strongly indicating full adder from figure 5.10(c).

inputs are empty. Similarly the pull down n-type transistor signal paths should only conduct when the required input signals are valid.

Transistor implementations of the DIMS and NCL gates introduced above are thus straightforward. Figure 5.14 shows a transistor-level implementation of a carry circuit for a strongly-indicating full adder. In the pull-down circuit each column of transistors corresponds to a minterm. In general when implementing DCVSL gates it is possible to share transistors in the two pull-down networks, but in this particular case it has not been done in order to illustrate better the relationship between the transistor implementation and the gate implementation in figure 5.10(c).

The high stacks of p-type transistors are obviously undesirable. They may be replaced by a single transistor controlled by an "all empty" signal generated elsewhere. Finally, we mention that the weakly-indicating full adder design presented in the next section includes optimizations that minimize the p-type transistor stacks.

5.5.4 Martin's adder

In [85] Martin addresses the design of dual-rail function blocks in general and he illustrates the ideas using a very elegant dual-rail ripple-carry adder. The adder has a small transistor count, it exhibits actual case latency when adding valid data, and it propagates empty values in constant time – the adder represents the ultimate in the design of weakly indicating function blocks.

Looking at the weakly-indicating transistor-level carry circuit in figure 5.14 we see that d remains valid until a, b, and c are all empty. If we designed a similar sum circuit its output s would also remain valid until a, b, and c are all empty. The weak conditions in figure 5.6 only require that one output remains

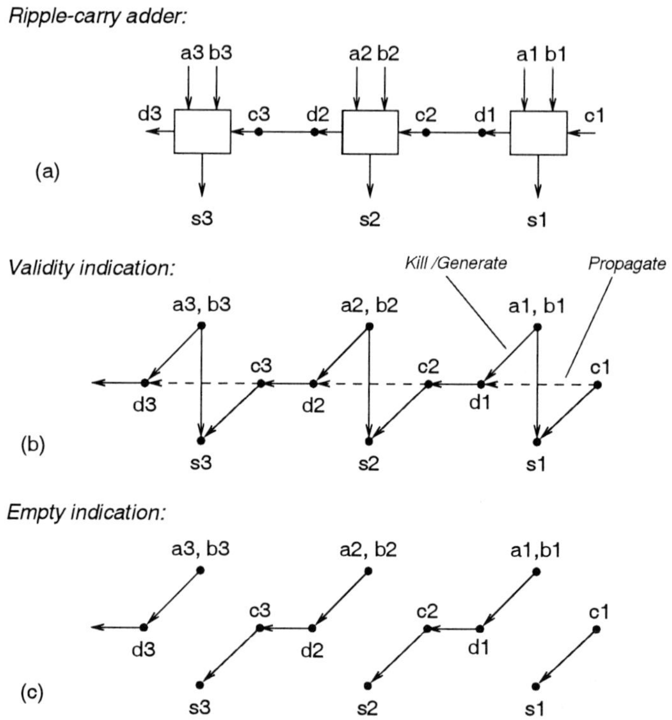

Figure 5.15. (a) A 3-stage ripple-carry adder and graphs illustrating how valid data (b) and empty data (c) propagate through the circuit (Martin [85]).

valid until all inputs have become invalid. Hence it is allowed to split the indication of a, b and c being empty among the carry and the sum circuits.

In [85] Martin uses some very illustrative directed graphs to express how the output signals indicate when input signals and internal signals are valid or empty. The nodes in the graphs are the signals in the circuit and the directed edges represent indication dependencies. Solid edges represent *guaranteed* dependencies and dashed edges represent *possible* dependencies. Figure 5.15(a) shows three full adder stages of a ripple-carry adder, and figures 5.15(b) and 5.15(c) show how valid and empty inputs respectively propagate through the circuit.

The propagation and indication of valid values is similar to what we discussed above in the other adder designs, but the propagation and indication of empty values is different and exhibits constant latency. When the outputs $d3$, $s3$, $s2$, and $s1$ are all valid this indicates that all input signals and all internal carry signals are valid. Similarly when the outputs $d3$, $s3$, $s2$, and $s1$ are all empty this indicates that all input signals and all internal carry signals are empty – the ripple-carry adder satisfies the weak conditions.

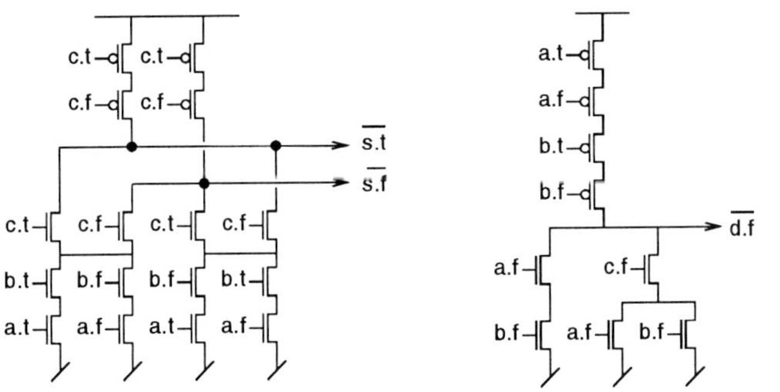

Figure 5.16. The CMOS transistor implementation of Martin's adder [85, Fig. 3].

The corresponding transistor implementation of a full adder is shown in figure 5.16. It uses 34 transistors, which is comparable to a traditional combinatorial circuit implementation.

The principles explained above apply to the design of function blocks in general. "Valid/empty indication (or acknowledgement), dependency graphs" as shown in figure 5.15 are a very useful technique for understanding and designing circuits with low latency and the weakest possible indication.

5.6. Hybrid function blocks

The final adder we will present has 4-phase bundled-data input and output channels and a dual-rail carry chain. The design exhibits characteristics similar to Martin's dual-rail adder presented in the previous section: actual case latency when propagating valid data, constant latency when propagating empty data, and a moderate transistor count. The basic structure of this hybrid adder is shown in figure 5.17. Each full adder is composed of a carry circuit and a sum circuit. Figure 5.18(a)-(b) shows precharged CMOS implementations of the two circuits. The idea is that the circuits precharge when $Req_{in} = 0$, evaluate when $Req_{in} = 1$, detect when all carry signals are valid and use this information to indicate completion, i.e. $Req_{out}\uparrow$. If the latency of the completion detector does not exceed the latency in the sum circuit in a full adder then a matched delay element is needed as indicated in figure 5.17.

The size and latency of the completion detector in figure 5.17 grows with the size of the adder, and in wide adders the latency of the completion detector may significantly exceed the latency of the sum circuit. An interesting optimization that reduces the completion detector overhead – possibly at the expense of a small increase in overall latency ($Req_{in}\uparrow$ to $Req_{out}\uparrow$) – is to use a mix of strongly and weakly indicating function blocks [101]. Following the naming convention established in figure 5.7 on page 64 we could make, for example,

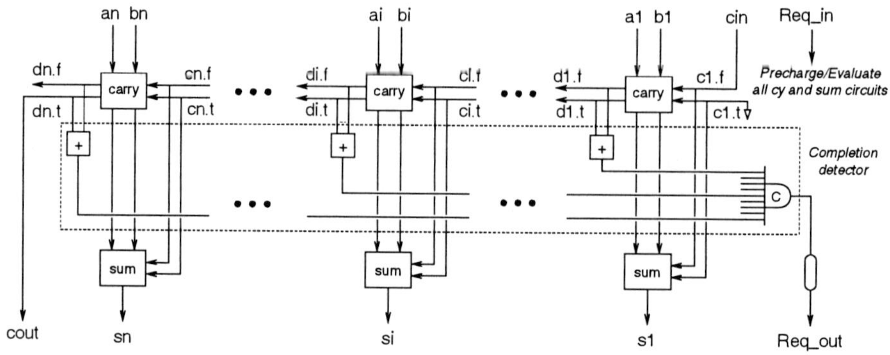

Figure 5.17. Block diagram of a hybrid adder with 4-phase bundled-data input and output channels and with an internal dual-rail carry chain.

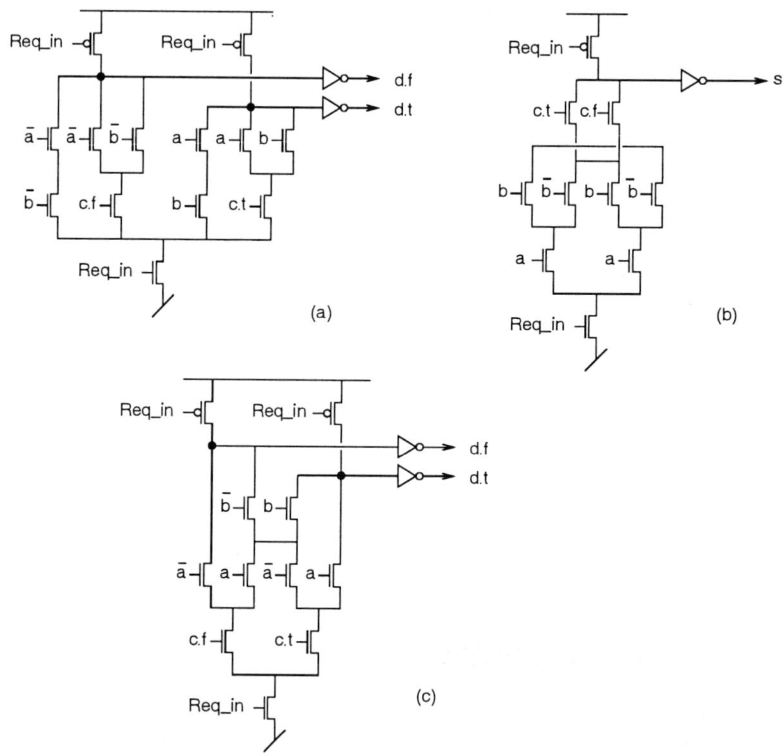

Figure 5.18. The CMOS transistor implementation of a full adder for the hybrid adder in figure 5.17: (a) a weakly indicating carry circuit, (b) the sum circuit and (c) a strongly indicating carry circuit.

adders 1, 4, 7, ... weakly indicating and all other adders strongly indicating. In this case only the carry signals out of stages 3, 6, 9, ... need to be checked to detect completion. For $i = 3, 6, 9, \ldots$ d_i indicates the completion of d_{i-1} and d_{i-2} as well. Many other schemes for mixing strongly and weakly indicating full adders are possible. The particular scheme presented in [101] exploited the fact that typical-case operands (sampled audio) are numerically small values, and the design detects completion from a single carry signal.

Summary – function block design

The previous sections have explained the basics of how to implement function blocks and have illustrated this using a variety of ripple-carry adders. The main points were "transparency to handshaking" and "actual case latency" through the use of weakly-indicating components.

Finally, a word of warning to put things into the right perspective: to some extent the ripple-carry adders explained above over-sell the advantages of average-case performance. It is easy to get carried away with elegant circuit designs but it may not be particularly relevant at the system level:

- In many systems the worst-case latency of a ripple-carry adder may simply not be acceptable.
- In a system with many concurrently active components that synchronize and exchange data at high rates, the slowest component at any given time tends to dominate system performance; the average-case performance of a system may not be nearly as good as the average-case latency of its individual components.
- In many cases addition is only one part of a more complex compound arithmetic operation. For example, the final design of the asynchronous filter bank presented in [103] did not use the ideas presented above. Instead we used entirely strongly-indicating full adders because this allowed an efficient two-dimensional precharged compound add-multiply-accumulate unit to be implemented.

5.7. MUX and DEMUX

Now that the principles of function block design have been covered we are ready to address the implementation of the MUX and DEMUX components, c.f. figure 3.3 on page 32. Let's recapitulate their function: a MUX will synchronize the control channel and relay the data and the handshaking of the selected input channel to the output data channel. The other input channel is ignored (and may have a request pending). Similarly a DEMUX will synchronize the control and the data input channels and steer the input to the selected output channel. The other output channel is passive and in the idle state.

If we consider only the "active" channels then the MUX and the DEMUX can be understood and designed as function blocks – they must be transparent to the handshaking in the same way as function blocks. The control channel and the (selected) input data channel are first joined and then an output is produced. Since no data transfer can take place without the control channel and the (selected) input data channel both being active, the implementations become strongly indicating function blocks.

Let's consider implementations using 4-phase protocols. The simplest and most intuitive designs use a dual-rail control channel. Figure 5.19 shows the implementation of the MUX and the DEMUX using the 4-phase bundled-data

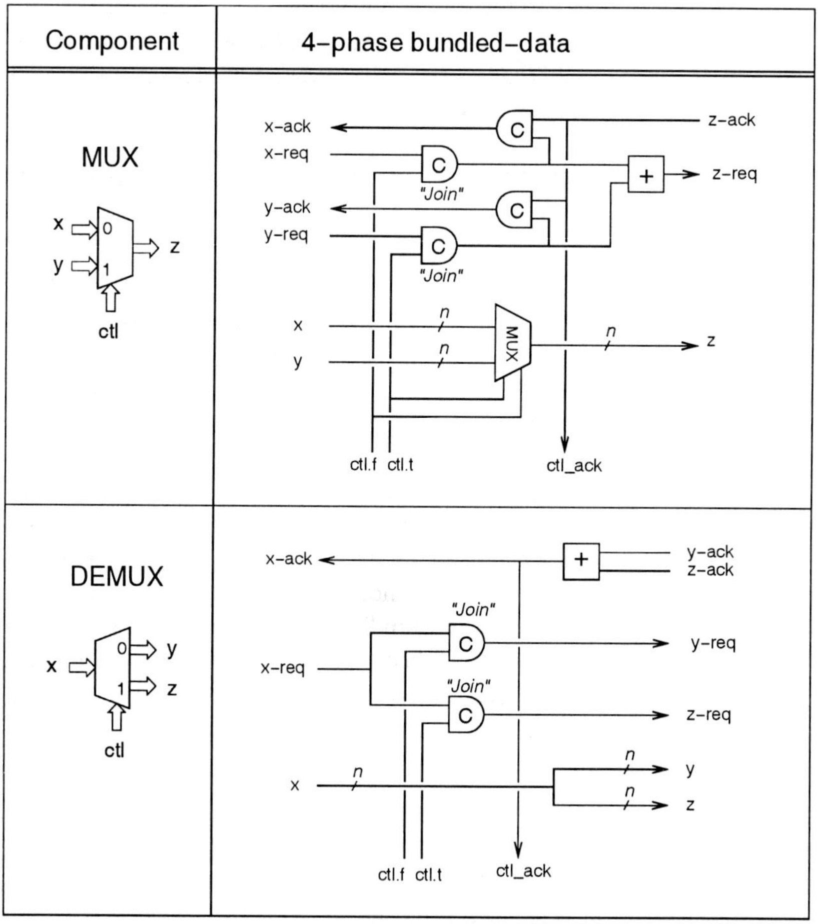

Figure 5.19. Implementation of MUX and DEMUX. The input and output data channels *x*, *y*, and *z* use the 4-phase bundled-data protocol and the control channel *ctl* uses the 4-phase dual-rail protocol (in order to simplify the design).

protocol on the input and output data channels and the 4-phase dual-rail protocol on the control channel. In both circuits *ctl.t* and *ctl.f* can be understood as two mutually exclusive requests that select between the two alternative input-to-output data transfers, and in both cases *ctl.t* and *ctl.f* are joined with the relevant input requests (at the C-elements marked "Join"). The rest of the MUX implementation is then similar to the 4-phase bundled-data MERGE in figure 5.1 on page 59. The rest of the DEMUX should be self explanatory; the handshaking on the two output ports are mutually exclusive and the acknowledge signals y_{ack} and z_{ack} are ORed to form $x_{ack} = ctl_{ack}$.

All 4-phase dual-rail implementations of the MUX and DEMUX components are rather similar, and all 4-phase bundled-data implementations may be obtained by adding 4-phase bundled-data to 4-phase dual-rail protocol conversion circuits on the control input. At the end of chapter 6, an all 4-phase bundled-data MUX will be one of the examples we use to illustrate the design of speed-independent control circuits.

5.8. Mutual exclusion, arbitration and metastability

5.8.1 Mutual exclusion

Some handshake components (including MERGE) require that the communication along several (input) channels is mutually exclusive. For the simple static data-flow circuit structures we have considered so far this has been the case, but in general one may encounter situations where a resource is shared between several independent parties/processes.

The basic circuit needed to deal with such situations is a mutual exclusion element (MUTEX), figure 5.20 (we will explain the implementation shortly). The input signals $R1$ and $R2$ are two requests that originate from two independent sources, and the task of the MUTEX is to pass these inputs to the corresponding outputs $G1$ and $G2$ in such a way that at most one output is active at any given time. If only one input request arrives the operation is trivial. If one input request arrives well before the other, the latter request is blocked until the first request is de-asserted. The problem arises when both input sig-

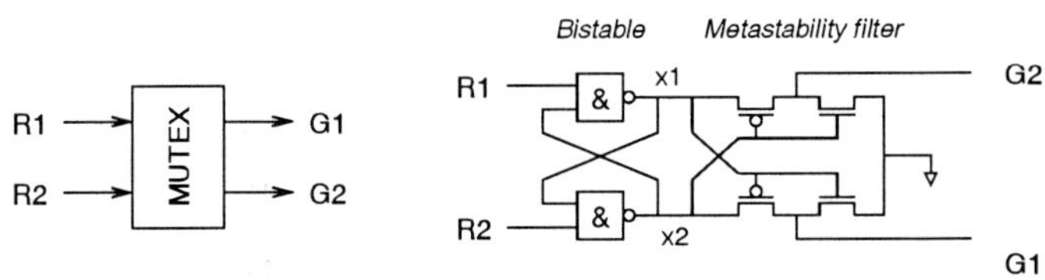

Figure 5.20. The mutual exclusion element: symbol and possible implementation.

nals are asserted at the same time. Then the MUTEX is required to make an arbitrary decision, and this is where metastability enters the picture.

The problem is exactly the same as when a synchronous circuit is exposed to an asynchronous input signal (one that does not satisfy set-up and hold time requirements). For a clocked flip-flop that is used to *synchronize* an asynchronous input signal, the question is whether the data signal made its transition before or after the active edge of the clock. As with the MUTEX the question is again which signal transition occured first, and as with the MUTEX a random decision is needed if the transition of the data signal coincides with the active edge of the clock signal.

The fundamental problem in a MUTEX and in a synchronizer flip-flop is that we are dealing with a bi-stable circuit that receives requests to enter each of its two stable states at the same time. This will cause the circuit to enter a metastable state in which it may stay for an unbounded length of time before randomly settling in one of its stable states. The problem of synchronization is covered in most textbooks on digital design and VLSI, and the analysis of metastability that is presented in these textbooks applies to our MUTEX component as well. A selection of references is: [95, sect. 9.4] [53, sect. 5.4 and 6.5] [151, sect. 5.5.7] [115, sect. 6.2.2 and 9.4-5] [150, sect. 8.9].

For the synchronous designer the problem is that metastability may persist beyond the time interval that has been allocated to recover from potential metastability. It is simply not possible to obtain a decision within a bounded length of time. The asynchronous designer, on the other hand, will eventually obtain a decision, but there is no upper limit on the time he will have to wait for the answer. In [22] the terms "time safe" and "value safe" are introduced to denote and classify these two situations.

A possible implementation of the MUTEX, as shown in figure 5.20, involves a pair of cross coupled NAND gates and a metastability filter. The cross coupled NAND gates enable one input to block the other. If both inputs are asserted at the same time, the circuit becomes metastable with both signals $x1$ and $x2$ halfway between supply and ground. The metastability filter prevents these undefined values from propagating to the outputs; $G1$ and $G2$ are both kept low until signals $x1$ and $x2$ differ by more than a transistor threshold voltage.

The metastability filter in figure 5.20 is a CMOS transistor-level implementation from [83]. An NMOS predecessor of this circuit appeared in [121]. Gate-level implementations are also possible: the metastability filter can be implemented using two buffers whose logic thresholds have been made particularly high (or low) by "trimming" the strengths of the pull-up and pull-down transistor paths ([151, section 2.3]). For example, a 4-input NAND gate with all its inputs tied together implements a buffer with a particularly high logic threshold. The use of this idea in the implementation of mutual exclusion elements is described in [6, 139].

5.8.2 Arbitration

The MUTEX can be used to build a handshake arbiter that can be used to control access to a resource that is shared between several autonomous independent parties. One possible implementation is shown in figure 5.21.

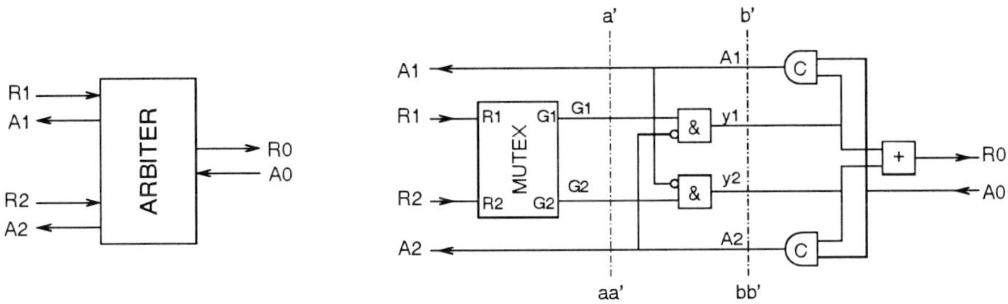

Figure 5.21. The handshake arbiter: symbol and possible implementation.

The MUTEX ensures that signals $G1$ and $G2$ at the a'–aa' interface are mutually exclusive. Following the MUTEX are two AND gates whose purpose it is to ensure that handshakes on the $(y1,A1)$ and $(y2,A2)$ channels at the b'–bb' interface are mutually exclusive: $y2$ can only go high if $A1$ is low and $y1$ can only go high if signal $A2$ is low. In this way, if handshaking is in progress along one channel, it blocks handshaking on the other channel. As handshaking along channels $(y1,A1)$ and $(y2,A2)$ are mutually exclusive the rest of the arbiter is simply a MERGE, c.f., figure 5.1 on page 59. If data needs to be passed to the shared resource a multiplexer is needed in exactly the same way as in the MERGE. The multiplexer may be controlled by signals $y1$ and/or $y2$.

5.8.3 Probability of metastability

Let us finally take a quantitative look at metastability: if $P(met_t)$ denotes the probability of the MUTEX being metastable for a period of time of t or longer (within an observation interval of one second), and if this situation is considered a failure, then we may calculate the mean time between failure as:

$$MTBF = \frac{1}{P(met_t)} \tag{5.8}$$

The probability $P(met_t)$ may be calculated as:

$$P(met_t) = P(met_t | met_{t=0}) \cdot P(met_{t=0}) \tag{5.9}$$

where:

- $P(met_t | met_{t=0})$ is the probability that the MUTEX is still metastable at time t given that it was metastable at time $t = 0$.
- $P(met_{t=0})$ is the probability that the MUTEX will enter metastability within a given observation interval.

The probability $P(met_{t=0})$ can be calculated as follows: the MUTEX will go metastable if its inputs $R1$ and $R2$ are exposed to transitions that occur almost simultaneously, i.e. within some small time window Δ. If we assume that the two input signals are uncorrelated and that they have average switching frequencies f_{R1} and f_{R2} respectively, then:

$$P(met_{t=0}) = \frac{1}{\Delta \cdot f_{R1} \cdot f_{R2}} \tag{5.10}$$

which can be understood as follows: within an observation interval of one second the input signal $R2$ makes $1/f_{R2}$ attempts at hitting one of the $1/f_{R1}$ time intervals of duration Δ where the MUTEX is vulnerable to metastability.

The probability $P(met_t | met_{t=0})$ is determined as:

$$P(met_t | met_{t=0}) = e^{-t/\tau} \tag{5.11}$$

where τ expresses the ability of the MUTEX to exit the metastable state spontaneously. This equation can be explained in two different ways and experimental results have confirmed its correctness. One explanation is that the cross coupled NAND gates have no memory of how long they have been metastable, and that the only probability distribution that is "memoryless" is an exponential distribution. Another explanation is that a small-signal model of the cross-coupled NAND gates at the metastable point has a single dominating pole.

Combining equations 5.8–5.11 we obtain:

$$MTBF = \frac{e^{t/\tau}}{\Delta \cdot f_{R1} \cdot f_{R2}} \tag{5.12}$$

Experiments and simulations have shown that this equation is reasonably accurate provided that t is not very small, and experiments or simulations may be used to determine the two parameters Δ and τ. Representative values for good circuit designs implemented in a 0.25 μm CMOS process are $\Delta = 30ps$ and $\tau = 25ps$.

5.9. Summary

This chapter addressed the implementation of the various handshake components: latch, fork, join, merge, function blocks, mux, demux, mutex and arbiter). A significant part of the material addressed principles and techniques for implementing function blocks.

Chapter 6

SPEED-INDEPENDENT CONTROL CIRCUITS

This chapter provides an introduction to the design of asynchronous sequential circuits and explains in detail one well-developed specification and synthesis method: the synthesis of speed-independent control circuits from signal transition graph specifications.

6.1. Introduction

Over time many different formalisms and theories have been proposed for the design of asynchronous control circuits (e.g. sequential circuits or state machines). The multitude of approaches arises from the combination of: (a) different specification formalisms, (b) different assumptions about delay models for gates and wires, and (c) different assumptions about the interaction between the circuit and its environment. Full coverage of the topic is far beyond the scope of this book. Instead we will first present some of the basic assumptions and characteristics of the various design methods and give pointers to relevant literature and then we will explain in detail one method: the design of speed-independent circuits from signal transition graphs – a method that is supported by a well-developed public domain tool, Petrify.

A good starting point for further reading is a book by Myers [95]. It provides in-depth coverage of the various formalisms, methods, and theories for the design of asynchronous sequential circuits and it provides a comprehensive list of references.

6.1.1 Asynchronous sequential circuits

To start the discussion figure 6.1 shows a generic synchronous sequential circuit and two alternative asynchronous control circuits: a Huffman style fundamental mode circuit with buffers (delay elements) in the feedback signals, and a Muller style input-output mode circuit with wires in the feedback path.

The synchronous circuit is composed of a set of registers holding the current state and a combinational logic circuit that computes the output signals and the next state signals. When the clock ticks the next state signals are copied into the registers thus becoming the current state. Reliable operation only requires that

J. Sparsø and S. Furber (eds.), Principles of Asynchronous Circuit Design, 81–113.

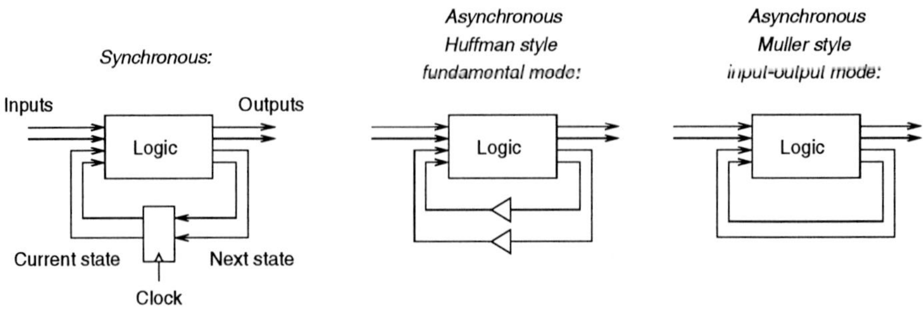

Figure 6.1. (a) A synchronous sequential circuit. (b) A Huffman style asynchronous sequential circuit with buffers in the feedback path, and (c) a Muller style asynchronous sequential circuit with wires in the feedback path.

the next state output signals from the combinational logic circuit are stable in a time window around the rising edge of the clock, an interval that is defined by the setup and hold time parameters of the register. Between two clock ticks the combinational logic circuit is allowed to produce signals that exhibit hazards. The only thing that matters is that the signals are ready and stable when the clock ticks.

In an asynchronous circuit there is no clock and all signals have to be valid at all times. This implies that at least the output signals that are seen by the environment must be free from all hazards. To achieve this, it is sometimes necessary to avoid hazards on internal signals as well. This is why the synthesis of asynchronous sequential circuits is difficult. Because it is difficult researchers have proposed different methods that are based on different (simplifying) assumptions.

6.1.2 Hazards

For the circuit designer a hazard is an unwanted glitch on a signal. Figure 6.2 shows four possible hazards that may be observed. A circuit that is in a stable state does not spontaneously produce a hazard – hazards are related to the dynamic operation of a circuit. This again relates to the dynamics of the input signals as well as the delays in the gates and wires in the circuit. A discussion of hazards is therefore not possible without stating precisely which delay model is being used and what assumptions are made about the interaction between the circuit and its environment. There are greater theoretical depths in this area than one might think at a first glance.

Gates are normally assumed to have delays. In section 2.5.3 we also discussed wire delays, and in particular the implications of having different delays in different branches of a forking wire. In addition to gate and wire delays it is also necessary to specify which delay model is being used.

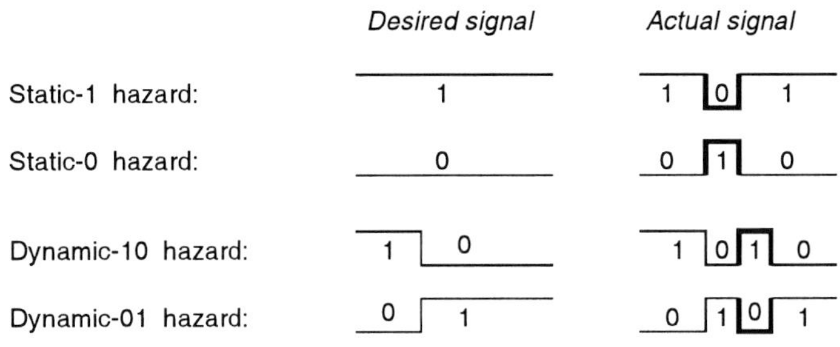

Figure 6.2. Possible hazards that may be observed on a signal.

6.1.3 Delay models

A *pure delay* that simply shifts any signal waveform later in time is perhaps what first comes to mind. In the hardware description language VHDL this is called a *transport delay*. It is, however, not a very realistic model as it implies that the gates and wires have infinitely high bandwidth. A more realistic delay model is the *inertial delay* model. In addition to the time shifting of a signal waveform, an inertial delay suppresses short pulses. In the inertial delay model used in VHDL two parameters are specified, the *delay time* and the *reject time*, and pulses shorter than the reject time are filtered out. The inertial delay model is the default delay model used in VHDL.

These two fundamental delay models come in several flavours depending on how the delay time parameter is specified. The simplest is a *fixed delay* where the delay is a constant. An alternative is a *min-max delay* where the delay is unknown but within a lower and upper bound: $t_{min} \leq t_{delay} \leq t_{max}$. A more pessimistic model is the *unbounded delay* where delays are positive (i.e. not zero), unknown and unbounded from above: $0 < t_{delay} < \infty$. This is the delay model that is used for gates in speed-independent circuits.

It is intuitive that the inertial delay model and the min-max delay model both have properties that help filter out some potential hazards.

6.1.4 Fundamental mode and input-output mode

In addition to the delays in the gates and wires, it is also necessary to formalize the interaction between the circuit being designed and its environment. Again, strong assumptions may simplify the design of the circuit. The design methods that have been proposed over time all have their roots in one of the following assumptions:

Fundamental mode: The circuit is assumed to be in a state where all input signals, internal signals, and output signals are stable. In such a stable state the environment is allowed to change one input signal. After that, the environment is not allowed to change the input signals again until the entire circuit has stabilized. Since internal signals such as state variables are unknown to the environment, this implies that the longest delay in the circuit must be calculated and the environment is required to keep the input signals stable for at least this amount of time. For this to make sense, the delays in gates and wires in the circuit have to be bounded from above. The limitation on the environment is formulated as an absolute time requirement.

The design of asynchronous sequential circuits based on fundamental mode operation was pioneered by David Huffman in the 1950s [59, 60].

Input-output mode: Again the circuit is assumed to be in a stable state. Here the environment is allowed to change the inputs. When the circuit has produced the corresponding output (and it is allowable that there are no output changes), the environment is allowed to change the inputs again. There are no assumptions about the internal signals and it is therefore possible that the next input change occurs before the circuit has stabilized in response to the previous input signal change.

The restrictions on the environment are formulated as causal relations between input signal transitions and output signal transitions. For this reason the circuits are often specified using trace based methods where the designer specifies all possible sequences of input and output signal transitions that can be observed on the interface of the circuit. Signal transition graphs, introduced later, are such a trace-based specification technique.

The design of asynchronous sequential circuits based on the input-output mode of operation was pioneered by David Muller in the 1950s [93, 92]. As mentioned in section 2.5.1, these circuits are speed-independent.

6.1.5 Synthesis of fundamental mode circuits

In the classic work by Huffman the environment was only allowed to change one input signal at a time. In response to such an input signal change, the combinational logic will produce new outputs, of which some are fed back, figure 6.1(b). In the original work it was further required that only one feedback signal changes (at a time) and that the delay in the feedback buffer is large enough to ensure that the entire combinational circuit has stabilized before it sees the change of the feedback signal. This change may, in turn, cause the combinational logic to produce new outputs, etc. Eventually through a sequence of single signal transitions the circuit will reach a stable state where the environment is again allowed to make a single input change. Another way of expressing this behaviour is to say that the circuit starts out in a stable state

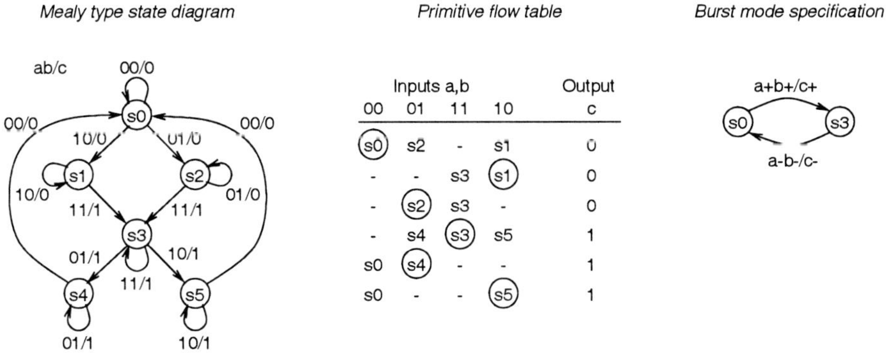

Figure 6.3. **Some alternative specifications of a Muller C-element: a Mealy state diagram, a primitive flow table, and a burst-mode state diagram.**

(which is defined to be a state that will persist until an input signal changes). In response to an input signal change the circuit will step through a sequence of transient, unstable states, until it eventually settles in a new stable state. This sequence of states is such that from one state to the next only one variable changes.

The interested reader is encouraged to consult [75], [133] or [95] and to specify and synthesize a C-element. The following gives a flavour of the design process and the steps involved:

- The design *may* start with a state graph specification that is very similar to the specification of a synchronous sequential circuit. This is optional. Figure 6.3 shows a Mealy type state graph specification of the C-element.

The classic design process involves the following steps:

- The intended sequential circuit is specified in the form of a primitive flow table (a state table with one row per stable state). Figure 6.3 shows the primitive flow table specification of a C-element.
- A minimum-row reduced flow table is obtained by merging compatible states in the primitive flow table.
- The states are encoded.
- Boolean equations for output variables and state variables are derived.

Later work has generalized the fundamental mode approach by allowing a restricted form of multiple-input and multiple-output changes. This approach is called *burst mode* [32, 27]. When in a stable state, a burst-mode circuit will wait for a set of input signals to change (in arbitrary order). After such an input burst has completed the machine computes a burst of output signals and new values of the internal variables. The environment is not allowed to produce a new input burst until the circuit has completely reacted to the previous burst – fundamental mode is still assumed, but only between bursts of input changes. For comparison, figure 6.3 also shows a burst-mode specification of a C-element. Burst-mode circuits are specified using state graphs that are very similar to those used in the design of synchronous circuits. Several mature tools for synthesizing burst-mode controllers have been developed in academia [40, 160]. These tools are available in the public domain.

6.2. Signal transition graphs

The rest of this chapter will be devoted to the specification and synthesis of speed-independent control circuits. These circuits operate in input-output mode and they are naturally specified using signal transition graphs, (STGs). An STG is a petri net and it can be seen as a formalization of a timing diagram. The synthesis procedure that we will explain in the following consists of: (1) Capturing the behaviour of the intended circuit and its environment in an STG. (2) Generating the corresponding state graph, and adding state variables if needed. (3) Deriving Boolean equations for the state variables and outputs.

6.2.1 Petri nets and STGs

Briefly, a Petri net [3, 113, 94] is a graph composed of directed arcs and two types of nodes: transitions and places. Depending on the interpretation that is assigned to places, transitions and arcs, Petri nets can be used to model and analyze many different (concurrent) systems. Some places can be marked with tokens and the Petri net model can be "executed" by firing transitions. A transition is enabled to fire if there are tokens on all of its input places, and an enabled transition must eventually fire. When a transition fires, a token is removed from each input place and a token is added to each output place. We will show an example shortly. Petri nets offer a convenient way of expressing choice and concurrency.

It is important to stress that there are many variations of and extensions to Petri nets – Petri nets are a family of related models and not a single, unique and well defined model. Often certain restrictions are imposed in order to make the analysis for certain properties practical. The STGs we will consider in the following belong to such a restricted subclass: an STG is a 1-bounded Petri net in which only simple forms of input choice are allowed. The exact meaning of

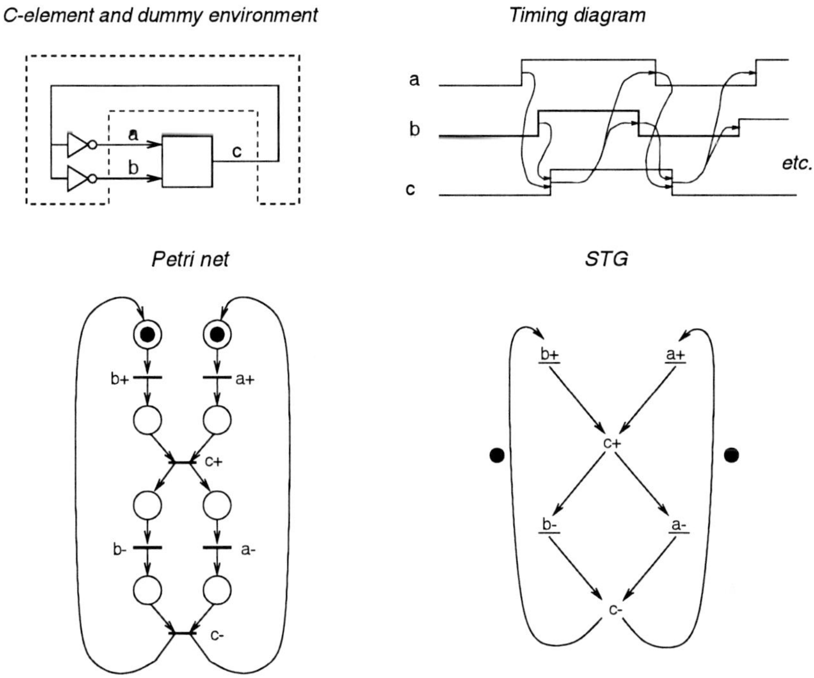

Figure 6.4. A C-element and its 'well behaved' dummy environment, its specification in the form of a timing diagram, a Petri net, and an STG formalization of the timing diagram.

"1-bounded" and "simple forms of input choice" will be defined at the end of this section.

In an STG the transitions are interpreted as signal transitions and the places and arcs capture the causal relations between the signal transitions. Figure 6.4 shows a C-element and a 'well behaved' dummy environment that maintains the input signals until the C-element has changed its outputs. The intended behaviour of the C-element could be expressed in the form of a timing diagram as shown in the figure. Figure 6.4 also shows the corresponding Petri net specification. The Petri net is marked with tokens on the input places to the $a+$ and $b+$ transitions, corresponding to state $(a,b,c) = (0,0,0)$. The $a+$ and $b+$ transitions may fire in any order, and when they have both fired the $c+$ transition becomes enabled to fire, etc. Often STGs are drawn in a simpler form where most places have been omitted. Every arc that connects two transitions is then thought of as containing a place. Figure 6.4 shows the STG specification of the C-element.

A given marking of a Petri net corresponds to a possible state of the system being modeled, and by executing the Petri net and identifying all possible markings it is possible to derive the corresponding state graph of the system. The state graph is generally much more complex than the corresponding Petri net.

An STG describing a meaningful circuit enjoys certain properties, and for the synthesis algorithms used in tools like Petrify to work, additional properties and restrictions may be required. An STG is a Petri net with the following characteristics:

1 **Input free choice:** The selection among alternatives must only be controlled by mutually exclusive inputs.

2 **1-bounded:** There must never be more than one token in a place.

3 **Liveness:** The STG must be free from deadlocks.

An STG describing a meaningful speed-independent circuit has the following characteristics:

4 **Consistent state assignment:** The transitions of a signal must strictly alternate between $+$ and $-$ in any execution of the STG.

5 **Persistency:** If a signal transition is enabled it must take place, i.e. it must not be disabled by another signal transition. The STG specification of the circuit must guarantee persistency of internal signals (state variables) and output signals, whereas it is up to the environment to guarantee persistency of the input signals.

In order to be able to synthesize a circuit implementation the following characteristic is required:

6 **Complete state coding (CSC):** Two or more different markings of the STG must not have the same signal values (i.e. correspond to the same state). If this is not the case, it is necessary to introduce extra state variables such that different markings correspond to different states. The synthesis tool Petrify will do this automatically.

6.2.2 Some frequently used STG fragments

For the newcomer it may take a little practice to become familiar with specifying and designing circuits using STGs. This section explains some of the most frequently used templates from which one can construct complete specifications.

The basic constructs are: *fork, join, choice* and *merge,* see figure 6.5. The choice is restricted to what is called *input free choice*: the transitions following the choice place must represent mutually exclusive input signal transitions. This requirement is quite natural; we will only specify and design deterministic circuits. Figure 6.6 shows an example Petri net that illustrates the use

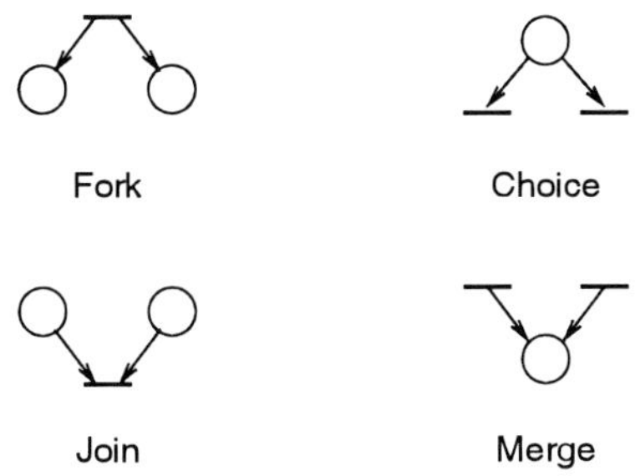

Figure 6.5. Petri net fragments for fork, join, free choice and merge constructs.

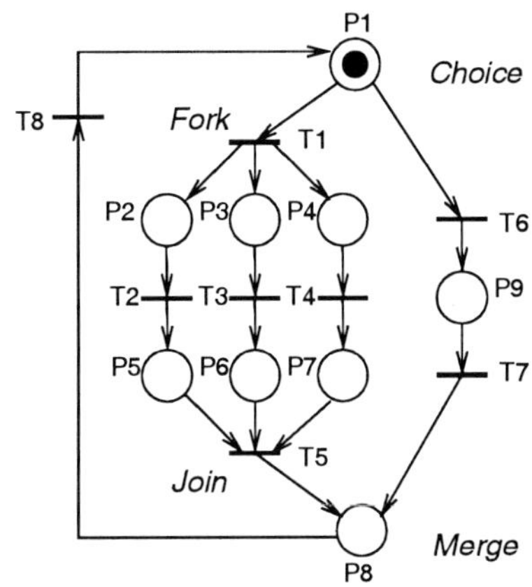

Figure 6.6. An example Petri net that illustrates the use of fork, join, free choice and merge.

of fork, join, free choice and merge. The example system will either perform transitions $T6$ and $T7$ in sequence, or it will perform $T1$ followed by the concurrent execution of transitions $T2$, $T3$ and $T4$ (which may occur in any order), followed by $T5$.

Towards the end of this chapter we will design a 4-phase bundled-data version of the MUX component from figure 3.3 on page 32. For this we will need some additional constructs: a *controlled choice* and a Petri net fragment for the input end of a bundled-data channel.

Figure 6.7 shows a Petri net fragment where place $P1$ and transitions $T3$ and $T4$ represent a controlled choice: a token in place $P1$ will engage in either transition $T3$ or transition $T4$. The choice is controlled by the presence of a token in either $P2$ or $P3$. It is crucial that there can never be a token in both these places at the same time, and in the example this is ensured by the mutually exclusive input signal transitions $T1$ and $T2$.

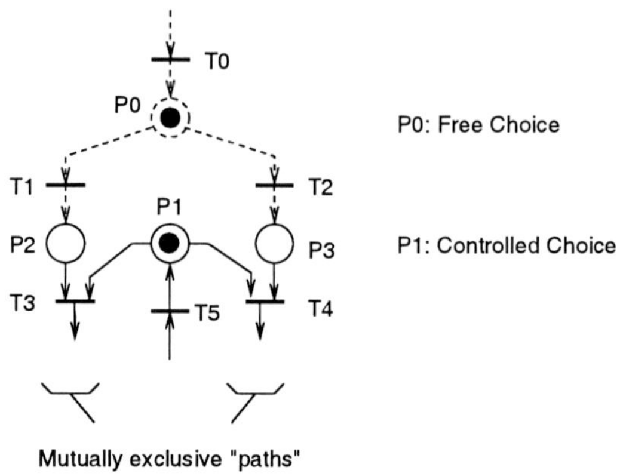

Figure 6.7. A Petri net fragment including a controlled choice.

Figure 6.8 shows a Petri net fragment for a component with a one-bit input channel using a 4-phase bundled-data protocol. It could be the control channel used in the MUX and DEMUX components introduced in figure 3.3 on page 32. The two transitions *dummy*1 and *dummy*2 do not represent transitions on the three signals in the channel, they are dummy transitions that facilitate expressing the specification. These dummy transitions represent an extension to the basic class of STGs.

Note also that the four arcs connecting:

place *P*5 and transition *Ctl*+
place *P*5 and transition *Ctl*−
place *P*6 and transition *dummy*2
place *P*7 and transition *dummy*1

have arrows at both ends. This is a shorthand notation for an arc in each direction. Note also that there are several instances where a place is both an input place and a output place for a transition. Place *P*5 and transition *Ctl*+ is an example of this.

The overall structure of the Petri net fragment can be understood as follows: at the top is a sequence of transitions and places that capture the handshaking on the *Req* and *Ack* signals. At the bottom is a loop composed of places *P*6 and *P*7 and transitions *Ctl*+ and *Ctl*− that captures the control signal changing between high and low. The absence of a token in place *P*5 when *Req* is high expresses the fact that *Ctl* is stable in this period. When *Req* is low and a token is present in place *P*5, *Ctl* is allowed to make as many transitions as it desires. When *Req*+ fires, a token is put in place *P*4 (which is a controlled choice place). The *Ctl* signal is now stable, and depending on its value one of the two transitions *dummy*1 or *dummy*2 will become enabled and eventually

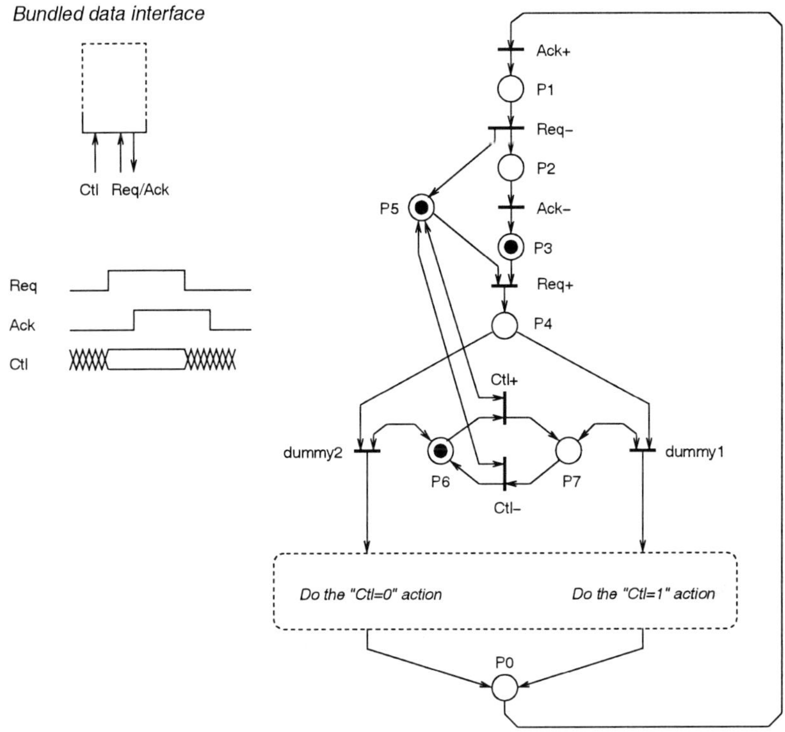

Figure 6.8. A Petri net fragment for a component with a one-bit input channel using a 4-phase bundled-data protocol.

fire. At this point the intended input-to-output operation that is not included in this example may take place, and finally the handshaking on the control port finishes ($Ack+$; $Req-$; $Ack-$).

6.3. The basic synthesis procedure

The starting point for the synthesis process is an STG that satisfies the requirements listed on page 88. From this STG the corresponding state graph is derived by identifying all of the possible markings of the STG that are reachable given its initial marking. The last step of the synthesis process is to derive Boolean equations for the state variables and output variables.

We will go through a number of examples by hand in order to illustrate the techniques used. Since the state of a circuit includes the values of all of the signals in the circuit, the computational complexity of the synthesis process can be large, even for small circuits. In practice one would always use one of the CAD tools that has been developed – for example Petrify that we will introduce later.

6.3.1 Example 1: a C-element

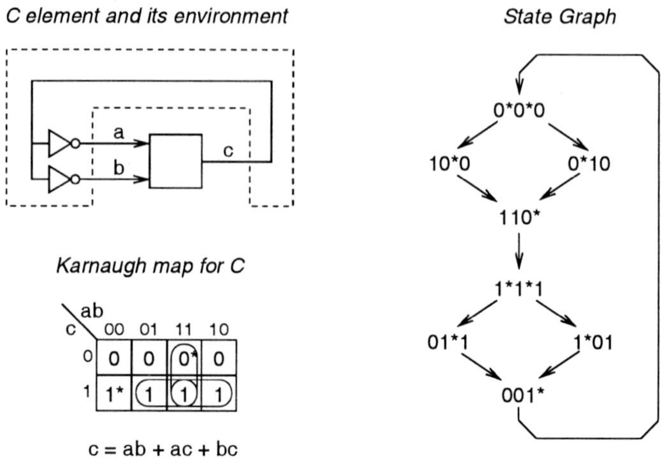

Figure 6.9. State graph and Boolean equation for the C-element STG from figure 6.4.

Figure 6.9 shows the state graph corresponding to the STG specification in figure 6.4 on page 87. Variables that are excited in a given state are marked with an asterisk. Also shown in figure 6.9 is the Karnaugh map for output signal c. The Boolean expression for c must cover states in which $c = 1$ and states where it is excited, $c = 0^*$ (changing to 1). In order to better distinguish excited variables from stable ones in the Karnaugh maps, we will use R (rising) instead of 0^* and F (falling) instead of 1^* throughout the rest of this book.

It is comforting to see that we can successfully derive the implementation of a known circuit, but the C-element is really too simple to illustrate all aspects of the design process.

6.3.2 Example 2: a circuit with choice

The following example provides a better illustration of the synthesis procedure, and in a subsequent section we will come back to this example and explain more efficient implementations. The example is simple – the circuit has only 2 inputs and 2 outputs – and yet it brings forward all relevant issues. The example is due to Chris Myers of the University of Utah who presented it in his 1996 course EE 587 "Asynchronous VLSI System Design." The example has roots in the papers [12, 13].

Figure 6.10 shows a specification of the circuit. The circuit has two inputs a and b and two outputs c and d, and the circuit has two alternative behaviours as illustrated in the timing diagram. The corresponding STG specification is shown in figure 6.11 along with the state graph for the circuit. The STG in-

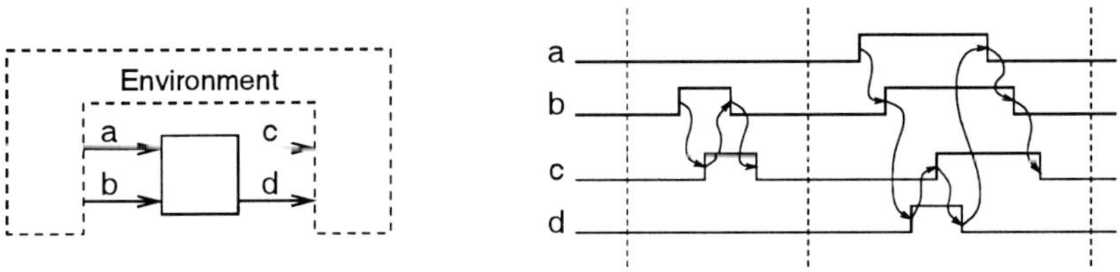

Figure 6.10. The example circuit from [12, 13].

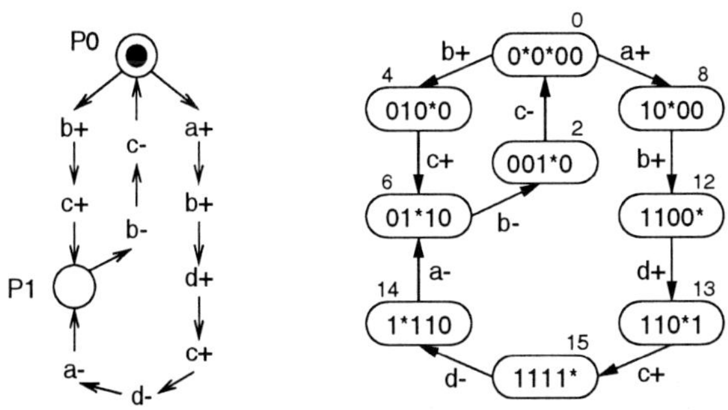

Figure 6.11. The STG specification and the corresponding state graph.

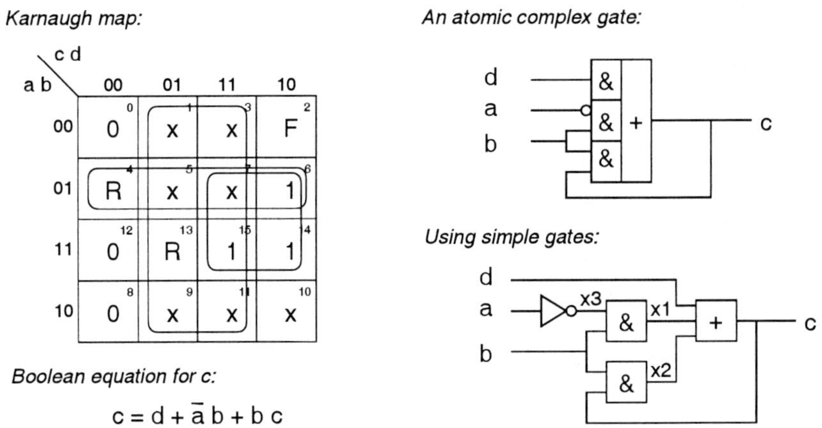

Figure 6.12. The Karnaugh map, the Boolean equation, and two alternative gate-level implementations of output signal *c*.

cludes only the free choice place $P0$ and the merge place $P1$. All arcs that directly connect two transitions are assumed to include a place. The states in the state diagram have been labeled with decimal numbers to ease filling out the Karnaugh maps.

The STG satisfies all of the properties 1-6 listed on page 88 and it is thus possible to proceed and derive Boolean equations for output signals c and d. [Note: In state 0 both inputs are marked to be excited, $(a,b) = (0^*,0^*)$, and in states 4 and 8 one of the signals is still 0 but no longer excited. This is a problem of notation only. In reality only one of the two variables is excited in state 0, but we don't know which one. Furthermore, the STG is only required to be persistent with respect to the internal signals and the output signals. Persistency of the input signals must be guaranteed by the environment].

For output c, figure 6.12 shows the Karnaugh map, the Boolean equation and two alternative gate implementations: one using a single atomic And-Or-Invert gate, and one using simple AND and OR gates. Note that there are states that are not reachable by the circuit. In the Karnaugh map these states correspond to don't cares. The implementation of output signal d is left as an exercise for the reader ($d = ab\overline{c}$).

6.3.3 Example 2: Hazards in the simple gate implementation

The STG in figure 6.10 satisfies all of the implementation conditions 1-6 (including persistency), and consequently an implementation where each output signal is implemented by a single atomic complex gate is hazard free. In the case of c we need a complex And-Or gate with inversion of input signal a. In general such an atomic implementation is not feasible and it is necessary to decompose the implementation into a structure of simpler gates. Unfortunately this will introduce extra variables, and these extra variables may not satisfy the persistency requirement that an excited signal transition must eventually fire. Speed-independence preserving logic decomposition is therefore a very interesting and relevant topic [20, 76].

The implementation of c using simple gates that is shown in figure 6.12 is not speed-independent; it may exhibit both static and dynamic hazards, and it provides a good illustration of the mechanisms behind hazards. The problem is that the signals $x1$, $x2$ and $x3$ are not included in the original STG and state graph. A detailed analysis that includes these signals would *not* satisfy the persistency requirement. Below we explain possible failure sequences that may cause a static-1 hazard and a dynamic-10 hazard on output signal c. Figure 6.13 illustrates the discussion.

A static-1 hazard may occur when the circuit goes through the following sequence of states: 12, 13, 15, 14. The transition from state 12 to state 13

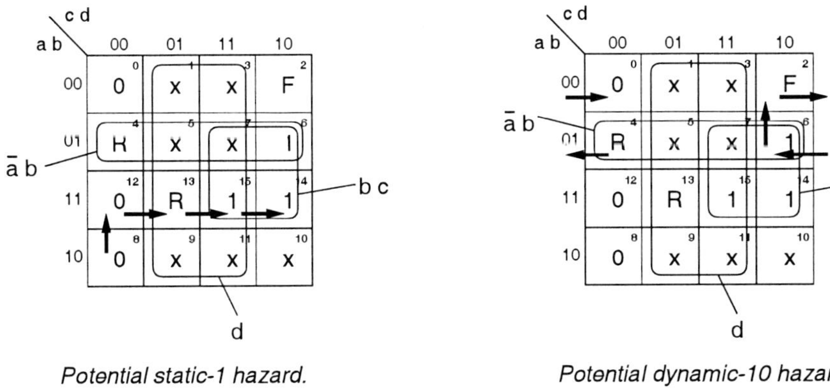

Figure 6.13. The Karnaugh maps for output signal c showing state sequences that may lead to hazards.

corresponds to d going high and the transition from state 15 to state 14 corresponds to d going low again. In state 13 c is excited (R) and it is supposed to remain high throughout states 13, 15, 14, and 6. States 13 and 15 are covered by the cube d, and state 14 is covered by cube bc that is supposed to "take over" and maintain $c = 1$ after d has gone low. If the AND gate with output signal $x2$ that corresponds to cube bc is slow we have the problem - the static-1 hazard.

A dynamic-10 hazard may occur when the circuit goes through the following sequence of states: 4, 6, 2, 0. This situation corresponds to the upper AND gate (with output signal $x1$) and the OR gate relaying $b+$ into $c+$ and $b-$ into $c-$. However, after the $c+$ transition the lower AND gate, $x2$, becomes excited (R) as well, but the firing of this gate is not indicated by any other signal transition – the OR gate already has one input high. If the lower AND gate ($x2$) fires, it will later become excited (F) in response to $c-$. The net effect of this is that the lower AND gate ($x2$) may superimpose a 0-1-0 pulse onto the c output after the intended $c-$ transition has occured.

In the above we did not consider the inverter with input signal a and output signal $x3$. Since a is not an input to any other gate, this decomposition is SI.

In summary both types of hazard are related to the circuit going through a sequence of states that are covered by several cubes that are supposed to maintain the signal at the same (stable) level. The cube that "takes over" represents a signal that may not be indicated by any other signal. In essence it is the same problem that we touched upon in section 2.2 on page 14 and in section 2.4.3 on page 20 – an OR gate can only indicate when the first input goes high.

6.4. Implementations using state-holding gates

6.4.1 Introduction

During operation each variable in the circuit will go through a sequence of states where it is (stable) 0, followed by one or more states where it is excited (R), followed by a sequence of states where it is (stable) 1, followed by one or more states where it is excited (F), etc. In the above implementation we were covering all states where a variable, z, was high or excited to go high ($z = 1$ and $z = R = 0*$).

An alternative is to use a state-holding device such as a set-reset latch. The Boolean equations for the set and reset signals need only cover the $z = R = 0*$ states and the $z = F = 1*$ states respectively. This will lead to simpler equations and potentially simpler decompositions. Figure 6.14 shows the implementation template using a standard set-reset latch and an alternative solution based on a standard C-element. In the latter case the reset signal must be inverted. Later, in section 6.4.5, we will discuss alternative and more elaborate implementations, but for the following discussion the basic topologies in figure 6.14 will suffice.

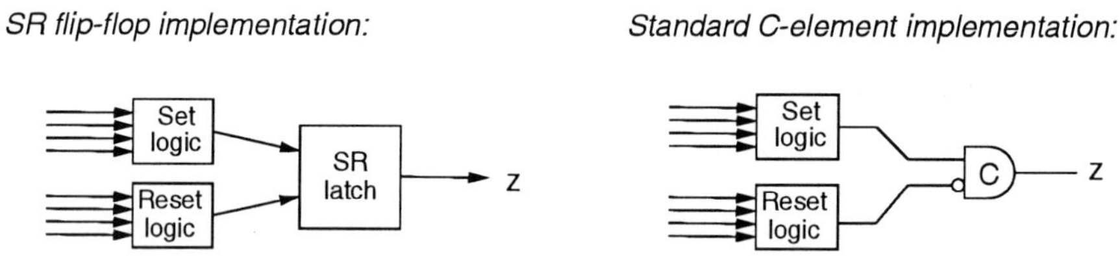

Figure 6.14. Possible implementation templates using (simple) state holding elements.

At this point it is relevant to mention that the equations for when to set and reset the state-holding element for signal z can be found by rewriting the original equation (that covers states in which $z = R$ and $z = 1$) into the following form:

$$z = \text{“Set”} + z \cdot \overline{\text{“Reset”}} \tag{6.1}$$

For signal c in the above example (figure 6.12 on page 93) we would get the following *set* and *reset functions:* $c_{set} = d + \overline{a}b$ and $c_{reset} = \overline{b}$ (which is identical to the result in figure 6.15 in section 6.4.3). Furthermore it is obvious that for all reachable states (only) the set and reset functions for a signal z must never be active at the same time:

$$\text{“Set”} \wedge \text{“Reset”} \equiv 0$$

The following sections will develop the idea of using state-holding elements and we will illustrate the techniques by re-implementing example 2 from the previous section.

6.4.2 Excitation regions and quiescent regions

The above idea of using a state-holding device for each variable can be formalized as follows:

An **excitation region**, ER, for a variable z is a maximally-connected set of states in which the variable is excited:

- ER(z+) denotes a region of states where $z = R = 0*$
- ER(z−) denotes a region of states where $z = F = 1*$

A **quiescent region**, QR, for a variable z is a maximally-connected set of states in which the variable is not excited:

- QR(z+) denotes a region of states where $z = 1$
- QR(z−) denotes a region of states where $z = 0$

For a given circuit the state space can be disjointly divided into one or more regions of each type.

The **set function** (cover) for a variable z:

- must contain all states in the ER(z+) regions
- may contain states from the QR(z+) regions
- may contain states not reachable by the circuit

The **reset function** (cover) for a variable z:

- must contain all states in the ER(z−) regions
- may contain states from the QR(z−) regions
- may contain states not reachable by the circuit

In section 6.4.4 below we will add what is known as the **monotonic cover constraint** or the **unique entry constraint** in order to avoid hazards:

- A cube (product term) in the set or reset function of a variable must only be entered through a state where the variable is excited.

Having mentioned this last constraint, we have above a complete recipe for the design of speed-independent circuits where each non-input signal is implemented by a state holding device. Let us continue with example 2.

6.4.3 Example 2: Using state-holding elements

Figure 6.15 illustrates the above procedure for example 2 from sections 6.3.2 and 6.3.3. As before, the Boolean equations (for the set and reset functions) may need to be implemented using atomic complex gates in order to ensure that the resulting circuit is speed-independent.

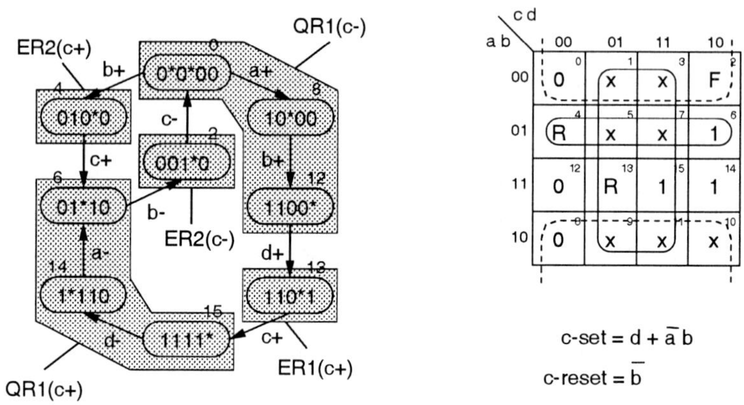

Figure 6.15. Excitation and quiescent regions in the state diagram for signal *c* in the example circuit from figure 6.10, and the corresponding derivation of equations for the set and reset functions.

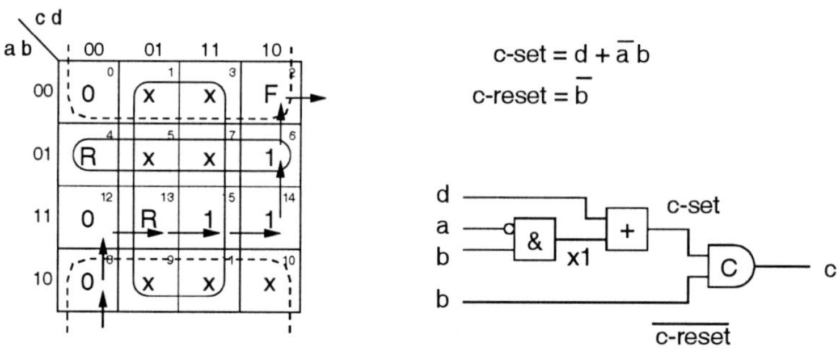

Figure 6.16. Implementation of *c* using a standard C-element and simple gates, along with the Karnaugh map from which the set and reset functions were derived.

6.4.4 The monotonic cover constraint

A standard C-element based implementation of signal *c* from above, with the set and reset functions implemented using simple gates, is shown in figure 6.16 along with the Karnaugh map from which the set and reset functions are derived. The set function involves two cubes *d* and $\overline{a}b$ that are input signals to an OR gate. This implementation may exhibit a dynamic-10 hazard on the

c_{set}-signal in a similar way to that discussed previously. The Karnaugh map in figure 6.16 shows the sequence of states that may lead to a malfunction: (8, 12, 13, 15, 14, 6, 0). Signal d is low in state 12, high in states 13 and 15, and low again in state 14. This sequence of states corresponds to a pulse on d. Through the OR gate this will create a pulse on the c_{set} signal that will cause c to go high. Later in state 2, c will go low again. This is the desired behaviour. The problem is that the internal signal $x1$ that corresponds to the other cube in the expression for c_{set} becomes excited ($x1 = R$) in state 6. If this AND gate is slow this may produce an unintended pulse on the c_{set} signal after c has been reset again.

If the cube $\overline{a}b$ (that covers states 4, 5, 7, and 6) is reduced to include only states 4 and 5 corresponding to $c_{set} = d + \overline{a}b\overline{c}$ we would avoid the problem. The effect of this modification is that the OR gate is never exposed to more than one input signal being high, and when this is the case we do not have problems with the principle of indication (c.f. the discussion of indication and dual-rail circuits in chapter 2). Another way of expressing this is that a cover cube must only be entered through states belonging to an excitation region. This requirement is known as:

- the **monotonic cover constraint:** only one product term in a sum-of-products implementation is allowed to be high at any given time. Obviously the requirement need only be satisfied in the states that are reachable by the circuit, or alternatively

- the **unique entry constraint:** cover cubes may only be entered through excitation region states.

6.4.5 Circuit topologies using state-holding elements

In addition to the set-reset flip-flop and the standard C-element based templates presented above, there are a number of alternative solutions for implementing variables using a state-holding device.

A popular approach is the *generalized C-element* that is available to the CMOS transistor-level designer. Here the state-holding mechanism and the set and reset functions are implemented in one (atomic) compound structure of n- and p-type transistors. Figure 6.17 shows a gate-level symbol for a circuit where $z_{set} = ab$ and $z_{reset} = \overline{b}\overline{c}$ along with dynamic and static CMOS implementations.

An alternative implementation that may be attractive to a designer using a standard cell library that includes (complex) And-Or-Invert gates is shown in figure 6.18. The circuit has the interesting property that it produces both the desired signal z and its complement $\overline{z}$ and during transitions it *never* produces $(z, \overline{z}) = (1, 1)$. Again, the example is a circuit where $z_{set} = ab$ and $z_{reset} = \overline{b}\overline{c}$.

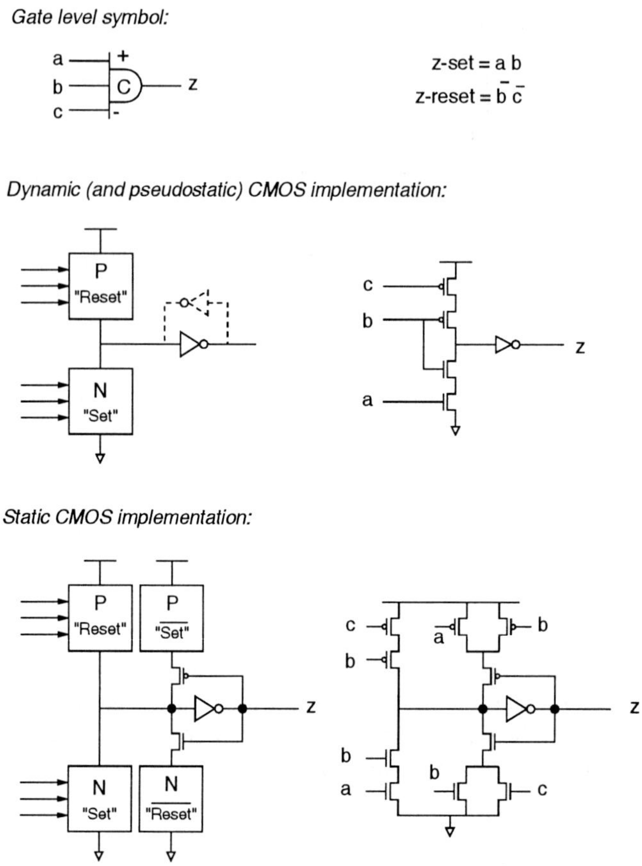

Figure 6.17. A generalized C-element: gate-level symbol, and some CMOS transistor implementations.

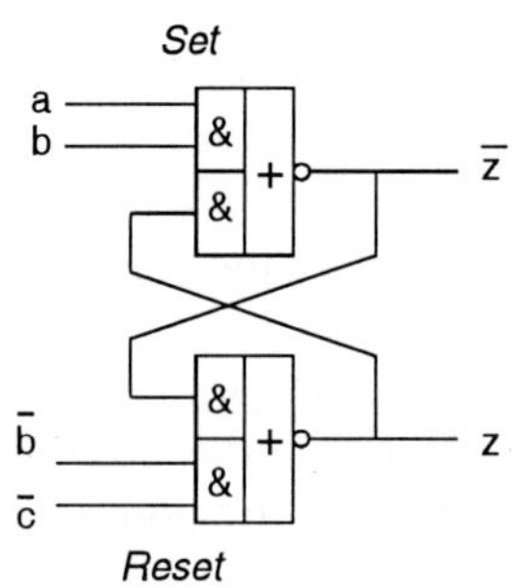

Figure 6.18. An SR implementation based on two complex And-Or-Invert gates.

6.5. Initialization

Initialization is an important aspect of practical circuit design, and unfortunately it has not been addressed in the above. The synthesis process *assumes* an initial state that corresponds to the initial marking of the STG, and the resulting synthesized circuit is a correct speed-independent implementation of the specification *provided that the circuit starts out in the same initial state.* Since the synthesized circuits generally use state-holding elements or circuitry with feedback loops it is necessary to actively force the circuit into the intended initial state.

Consequently, the designer has to do a manual post-synthesis hack and extend the circuit with an extra signal which, when active, sets all state-holding constructs into the desired state. Normally the circuits will not be speed-independent with respect to this initialization signal; it is assumed to be asserted for long enough to cause the desired actions before it is de-asserted.

For circuit implementations using state-holding elements such as set-reset latches and standard C-elements, initialization is trivial provided that these components have special clear/preset signals in addition to their normal inputs. In all other cases the designer has to add an initialization signal to the relevant Boolean equations explicitly. If the synthesis process is targeting a given cell library, the modified logic equations may need further logic decomposition, and as we have seen this may compromise speed-independence.

The fact that initialization is not included in the synthesis process is obviously a drawback, but normally one would implement a library of control circuits and use these as building blocks when designing circuits at the more abstract "static data-flow structures" level as introduced in chapter 3.

Initializing all control circuits as outlined above is a simple and robust approach. However, initialization of asynchronous circuits based on handshake components may also be achieved by an implicit approach that that exploits the function of the circuit to "propagate" initial signal values into the circuit. In Tangram (section 8.3, and chapter 13 in part III) this this is called self-initialization, [135].

6.6. Summary of the synthesis process

The previous sections have covered the basic theory for synthesizing SI control circuits from STG specifications. The style of presentation has deliberately been chosen to be an informal one with emphasis on examples and the intuition behind the theory and the synthesis procedure.

The theory has roots in work done by the following Universities and groups: University of Illinois [93], MIT [26, 24], Stanford [13], IMEC [145, 159], St. Petersburg Electrical Engineering Institute [146], and the multinational group of researchers who have developed the Petrify tool [29] that we will introduce in the next section. This author has attended several discussions from which it is clear that in some cases the concepts and theories have been developed independently by several groups, and I will refrain from attempting a precise history of the evolution. The reader who is interested in digging deeper into the subject is encouraged to consult the literature; in particular the book by Myers [95].

In summary the synthesis process outlined in the previous sections involves the following steps:

1 Specify the desired behaviour of the circuit and its (dummy) environment using an STG.

2 Check that the STG satisfies properties 1-5 on page 88: 1-bounded, consistent state assignment, liveness, only input free choice and controlled choice and persistency. An STG satisfying these conditions is a valid specification of an SI circuit.

3 Check that the specification satisfies property 6 on page 88: complete state coding (CSC). If the specification does not satisfy CSC it is necessary to add one or more state variables or to change the specification (which is often possible in 4-phase control circuits where the downgoing signal transitions can be shuffled around). Some tools (including Petrify) can insert state variables automatically, whereas re-shuffling of signals – which represents a modification of the specification – is a task for the designer.

4 Select an implementation template and derive the Boolean equations for the variables themselves, or for the set and reset functions when state holding devices are used. Also decide if these equations can be implemented in atomic gates (typically complex AOI-gates) or if they are to be implemented by structures of simpler gates. These decisions may be set by switches in the synthesis tools.

5 Derive the Boolean equations for the desired implementation template.

6 Manually modify the implementation such that the circuit can be forced into the desired initial state by an explicit reset or initialization signal.

7 Enter the design into a CAD tool and perform simulation and layout of the circuit (or the system in which the circuit is used as a component).

6.7. Petrify: A tool for synthesizing SI circuits from STGs

Petrify is a public domain tool for manipulating Petri nets and for synthesizing SI control circuits from STG specifications. It is available from `http://www.lsi.upc.es/~jordic/petrify/petrify.html`.

Petrify is a typical UNIX program with many options and switches. As a circuit designer one would probably prefer a push-button synthesis tool that accepts a specification and produces a circuit. Petrify can be used this way but it is more than this. If you know how to play the game it is an interactive tool for specifying, checking, and manipulating Petri nets, STGs and state graphs. In the following section we will show some examples of how to design speed-independent control circuits.

Input to Petrify is an STG described in a simple textual format. Using the program draw_astg that is part of the Petrify distribution (and that is based on the graph visualization package 'dot' developed at AT&T) it is possible to produce a drawing of the STGs and state graphs. The graphs are "nice" but the topological organization may be very different from how the designer thinks about the problem. Even the simple task of checking that an STG entered in textual form is indeed the intended STG may be difficult.

To help ease this situation a graphical STG entry and simulation tool called VSTGL (Visual STG Lab) has been developed at the Technical University of Denmark. To help the designer obtain a correct specification VSTGL includes an interactive simulator that allows the designer to add tokens and to fire transitions. It also carries out certain simple checks on the STG.

VSTGL is available from `http://vstgl.sourceforge.net/` and it is the result of a small student project done by two 4th year students. VSTGL is stable and reliable, though naming of signal transitions may seem a bit awkward.

Petrify can solve CSC violations by inserting state variables, and it can be controlled to target the implementation templates introduced in section 6.4:

- The **-cg** option will produce a complex-gate circuit (one where each non-input signal is implemented in a single complex gate).
- The **-gc** option will produce a generalized C-element circuit. The outputs from Petrify are the Boolean equations for the set and reset functions for each non-input signal.
- The **-gcm** option will produce a generalized C-element solution where the set and reset functions satisfy the monotonic cover requirement. Consequently the solution can also be mapped onto a standard C-element implementation where the *set* and *reset* functions are implemented using simple AND and OR gates.
- The **-tm** option will cause Petrify to perform technology mapping onto a gate library that can be specified by the user. Technology mapping can obviously not be combined with the `-cg` and `-gc` options.

Petrify comes with a manual and some examples. In the following section we will go through some examples drawn from the previous chapters of the book.

6.8. Design examples using Petrify

In the following we will illustrate the use of Petrify by specifying and synthesizing: (a) example 2 – the circuit with choice, (b) a control circuit for the 4-phase bundled-data implementation of the latch from figure 3.3 on page 32 and (c) a control circuit for the 4-phase bundled-data implementation of the MUX from figure 3.3 on page 32. For all of the examples we will assume push channels only.

6.8.1 Example 2 revisited

As a first example, we will synthesize the different versions of example 2 that we have already designed manually. Figure 6.19 shows the STG as it is entered into VSTGL. The corresponding textual input to Petrify (the ex2.g file) and the STG as it may be visualized by Petrify are shown in figure 6.20. Note in figure 6.20 that an index is added when a signal transition appears more than once in order to facilitate the textual input.

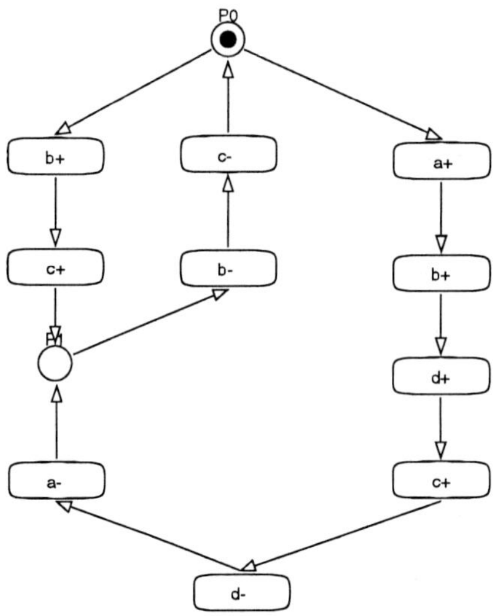

Figure 6.19. The STG of example 2 as it is entered into VSTGL.

Using complex gates

```
> petrify ex2.g -cg -eqn ex2-cg.eqn

The STG has CSC.
# File generated by petrify 4.0 (compiled 22-Dec-98 at 6:58 PM)
# from <ex2.g> on 6-Mar-01 at 8:30 AM
    ....
```

```
.model ex2
.inputs a b
.outputs c d
.graph
P0 a+ b+
c+ P1
b+ c+
P1 b-
c- P0
b- c-
a- P1
d- a-
c+/1 d-
a+ b+/1
b+/1 d+
d+ c+/1
.marking { P0 }
.end
```

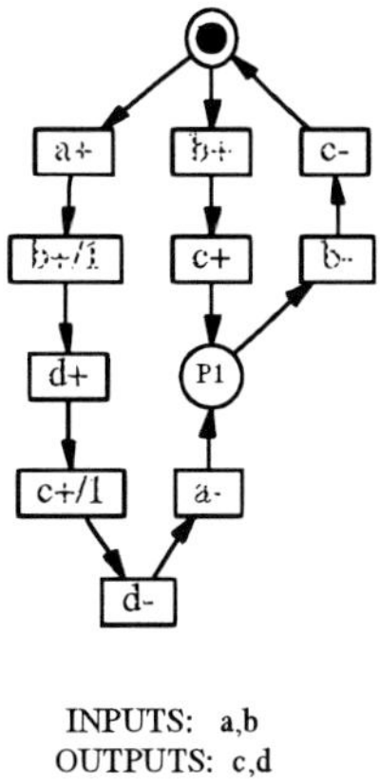

Figure 6.20. The textual description of the STG for example 2 and the drawing of the STG that is produced by Petrify.

```
# The original TS had (before/after minimization) 9/9 states
# Original STG:   2 places,  10 transitions,  13 arcs  ...
# Current STG:    4 places,   9 transitions,  18 arcs  ...
# It is a Petri net with 1 self-loop places
      ...

> more ex2-cg.eqn

# EQN file for model ex2
# Generated by petrify 4.0 (compiled 22-Dec-98 at 6:58 PM)
# Outputs between brackets "[out]" indicate a feedback to input "out"
# Estimated area = 7.00

INORDER = a b c d;
OUTORDER = [c] [d];
[c] = b (c + a') + d;
[d] = a b c';
```

Using generalized C-elements:

```
> petrify ex2.g -gc -eqn ex2-gc.eqn
   ...
> more ex2-gc.eqn

# EQN file for model ex2
# Generated by petrify 4.0 (compiled 22-Dec-98 at 6:58 PM)
# Outputs between brackets "[out]" indicate a feedback to input "out"
# Estimated area = 12.00
```

```
INORDER = a b c d;
OUTORDER = [c] [d];
[0] = a' b + d;
[1] = a b c';
[d] = d c' + [1];          # mappable onto gC
[c] = c b + [0];           # mappable onto gC
```

The equations for the generalized C-elements should be "interpreted" according to equation 6.1 on page 96

Using standard C-elements and set/reset functions that satisfy the **monotonic cover constraint**:

```
> petrify ex2.g -gcm -eqn ex2-gcm.eqn
  ...
> more ex2-gcm.eqn
```

```
# EQN file for model ex2
# Generated by petrify 4.0 (compiled 22-Dec-98 at 6:58 PM)
# Outputs between brackets "[out]" indicate a feedback to input "out"
# Estimated area = 10.00

INORDER = a b c d;
OUTORDER = [c] [d];
[0] = a' b c' + d;
[d] = a b c';
[c] = c b + [0];           # mappable onto gC
```

Again, the equations for the generalized C-element should be "interpreted" according to equation 6.1 on page 96.

6.8.2 Control circuit for a 4-phase bundled-data latch

Figure 6.21 shows an asynchronous handshake latch with a dummy environment on its left and right side. The latch can be implemented using a normal N-bit wide transparent latch and the control circuit we are about to design. A driver may be needed for the latch control signal *Lt*. In order to make the latch controller robust and independent of the delay in this driver, we may feed the buffered signal (*Lt*) back such that the controller knows when the signal has been presented to the latch. Figure 6.21 also shows fragments of the STG specification – the handshaking of the left and right hand environments and ideas about the behaviour of the latch controller. Initially *Lt* is low and the latch is transparent, and when new input data arrives they will flow through the latch. In response to *Rin*+, and provided that the right hand environment is ready for another handshake ($Aout = 0$), the controller may generate *Rout*+

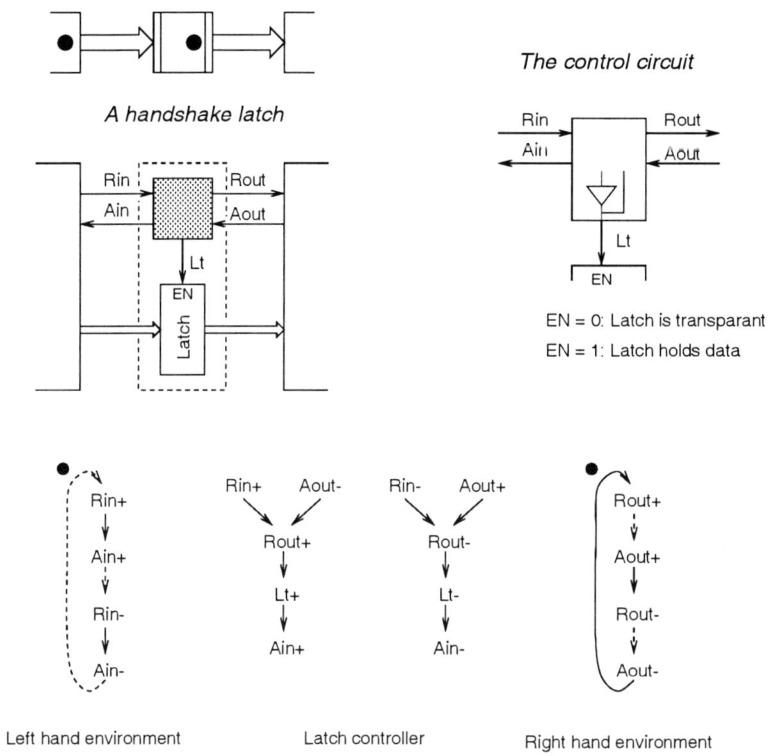

Figure 6.21. A 4-phase bundled-data handshake latch and some STG fragments that capture ideas about its behaviour.

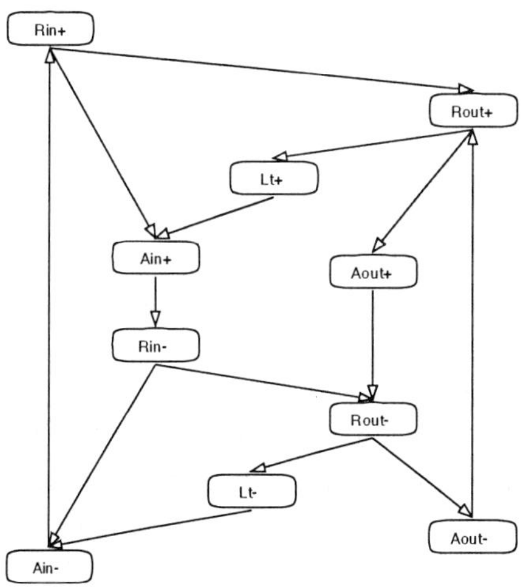

Figure 6.22. The resulting STG for the latch controller (as input to VSTGL).

right away. Furthermore the data should be latched, *Lt*+, and an acknowledge sent to the left hand environment, *Ain*+. A symmetric scenario is possible in response to *Rin*− when the latch is switched back into the transparent mode. Combining these STG fragments results in the STG shown in figure 6.22. Running Petrify yields the following:

```
>petrify lctl.g -cg -eqn lctl-cg.eqn

The STG has CSC.
# File generated by petrify 4.0 (compiled 22-Dec-98 at 6:58 PM)
# from <lctl.g> on 6-Mar-01 at 11:18 AM
...
# The original TS had (before/after minimization) 16/16 states
# Original STG:   0 places,  10 transitions,  14 arcs (  0 pt + ...
# Current STG:    0 places,  10 transitions,  12 arcs (  0 pt + ...
# It is a Marked Graph
.model lctl
.inputs  Aout Rin
.outputs  Lt Rout Ain
.graph
Rout+ Aout+ Lt+
Lt+ Ain+
Aout+ Rout-
Rin+ Rout+
Ain+ Rin-
Rin- Rout-
Ain- Rin+
Rout- Lt- Aout-
Aout- Rout+
Lt- Ain-
.marking { <Aout-,Rout+> <Ain-,Rin+> }
.end

>more lctl-cg.eqn

# EQN file for model lctl
# Generated by petrify 4.0 (compiled 22-Dec-98 at 6:58 PM)
# Outputs between brackets "[out]" indicate a feedback to input "out"
# Estimated area = 7.00

INORDER = Aout Rin Lt Rout Ain;
OUTORDER = [Lt] [Rout] [Ain];
[Lt] = Rout;
[Rout] = Rin (Rout + Aout') + Aout' Rout;
[Ain] = Lt;
```

The equation for `[Rout]` may be rewritten as:

```
[Rout] = Rin Aout' + Rout (Rin + Aout')
```

which can be recognized to be a C-element with inputs `Rin` and `Aout'`.

6.8.3 Control circuit for a 4-phase bundled-data MUX

After the above two examples, where we have worked out already well-known circuit implementations, let us now consider a more complex example that cannot (easily) be done by hand. Figure 6.23 shows the handshake multiplexer from figure 3.3 on page 32. It also shows how the handshake MUX can be implemented by a "regular" combinational circuit multiplexer and a control circuit. Below we will design a speed-independent control circuit for a 4-phase bundled-data MUX.

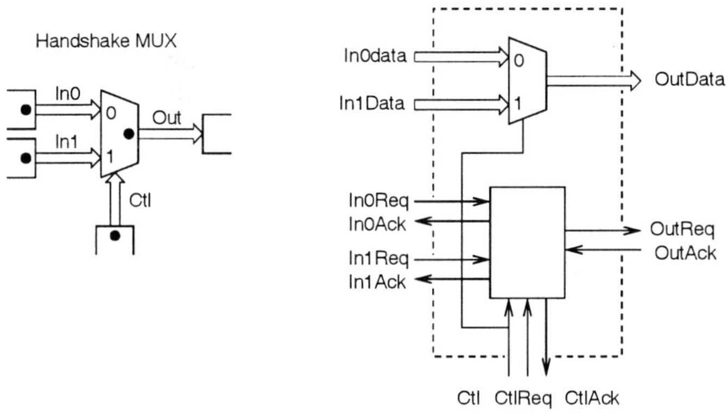

Figure 6.23. **The handshake MUX and the structure of a 4-phase bundled-data implementation.**

The MUX has three input channels and we *must* assume they are connected to three *independent* dummy environments. The dots remind us that the channels are push channels. When specifying the behaviour of the MUX control circuit and its (dummy) environment it is important to keep this in mind. A typical error when drawing STGs is to specify an environment with a more limited behaviour than the real environment. For each of the three input channels the STG has cycles involving ($Req+$;$Ack+$;$Req-$;$Ack-$; etc.), and each of these cycles is initialized to contain a token.

As mentioned previously, it is sometimes easier to deal with control channels using dual-rail (or in general $1-\text{of}-N$) data encodings since this implies dealing with one-hot (decoded) control signals. As a first step towards the STG for a MUX using entirely 4-phase bundled-data channels, figure 6.24 shows an STG for a MUX where the control channel uses dual-rail signals (*Ctl.t*, *Ctl.f* and *CtlAck*). This STG can then be combined with the STG-fragment for a 4-phase bundled-data channel from figure 6.8 on page 91, resulting in the STG in figure 6.25. The "intermediate" STG in figure 6.24 emphasizes the fact that the MUX can be seen as a controlled join – the two mutually exclusive and structurally identical halves are basically the STGs of a join.

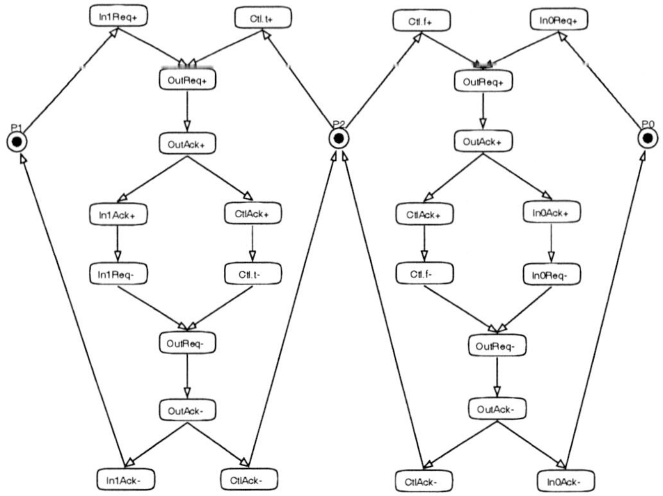

Figure 6.24. The STG specification of the control circuit for a 4-phase bundled-data MUX using a 4-phase dual-rail control channel. Combined with the STG fragment for a bundled-data (control) channel the resulting STG for an all 4-phase dual-rail MUX is obtained (figure 6.25).

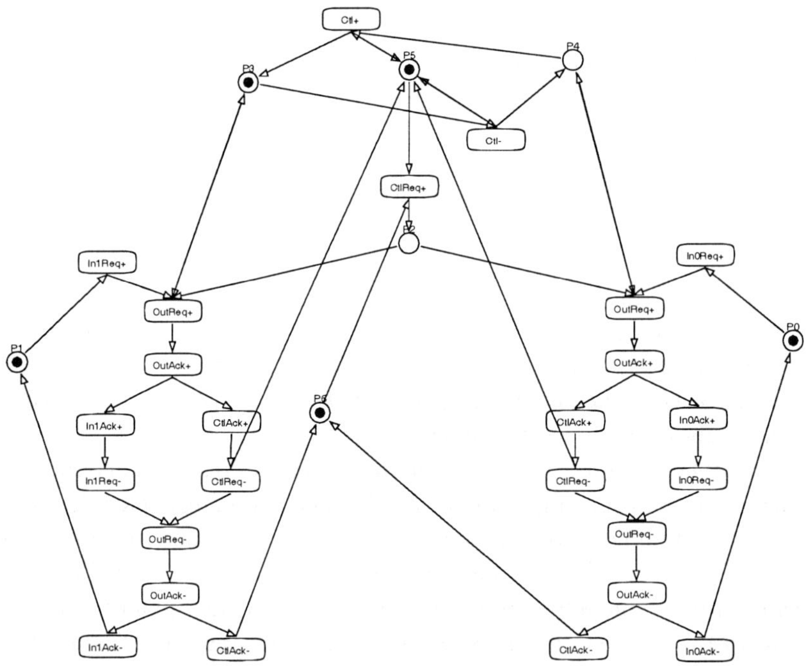

Figure 6.25. The final STG specification of the control circuit for the 4-phase bundled-data MUX. All channels, including the control channel, are 4-phase bundled-data.

Below is the result of running Petrify, this time with the `-o` option that writes the resulting STG (possibly with state signals added) in a file rather than to stdout.

```
> petrify MUX4p.g  o MUX4p-csc.g -gcm -eqn MUX4p-gcm.eqn

  State coding conflicts for signal In1Ack
  State coding conflicts for signal In0Ack
  State coding conflicts for signal OutReq
The STG has no CSC.
Adding state signal: csc0
The STG has CSC.
```

```
> more MUX4p-gcm.eqn

# EQN file for model MUX4p
# Generated by petrify 4.0 (compiled 22-Dec-98 at 6:58 PM)
# Outputs between brackets "[out]" indicate a feedback to input "out"
# Estimated area = 29.00

INORDER = In0Req OutAck In1Req Ctl CtlReq In1Ack In0Ack OutReq
          CtlAck csc0;
OUTORDER = [In1Ack] [In0Ack] [OutReq] [CtlAck] [csc0];
[In1Ack] = OutAck csc0';
[In0Ack] = OutAck csc0;
[2] = CtlReq (In1Req csc0' + In0Req Ctl');
[3] = CtlReq' (In1Req' csc0' + In0Req' csc0);
[OutReq] = OutReq [3]' + [2];        # mappable onto gC
[5] = OutAck' csc0;
[CtlAck] = CtlAck [5]' + OutAck;     # mappable onto gC
[7] = OutAck' CtlReq';
[8] = CtlReq Ctl;
[csc0] = csc0 [8]' + [7];            # mappable onto gC
```

As can be seen, the STG does not satisfy CSC (complete state coding) as several markings correspond to the same state vector, so Petrify adds an internal state-signal *csc*0. The intuition is that after *CtlReq*− the Boolean signal *Ctl* is no longer valid but the MUX control circuit has not yet finished its job. If the circuit can't see what to continue doing from its input signals it needs an internal state variable in which to keep this information. The signal *csc*0 is an active-low signal: it is set low if $Ctl = 0$ when *CtlReq*+ and it is set back to high when *OutAck* and *CtlReq* are both low. The fact that the signal *csc*0 is high when all channels are idle (all handshake signals are low) should be kept in mind when dealing with reset, c.f. section 6.5.

The exact details of how the state variable is added can be seen from the STG that includes *csc*0 which is produced by Petrify before it synthesizes the logic expressions for the circuit.

It is sometimes possible to avoid adding a state variable by re-shuffling signal transitions. It is not always obvious what yields the best solution. In principle more concurrency should improve performance, but it also results in a larger state-space for the circuit and this often tends to result in larger and slower circuits. A discussion of performance also involves the interaction with the environment. There is plenty of room for exploring alternative solutions.

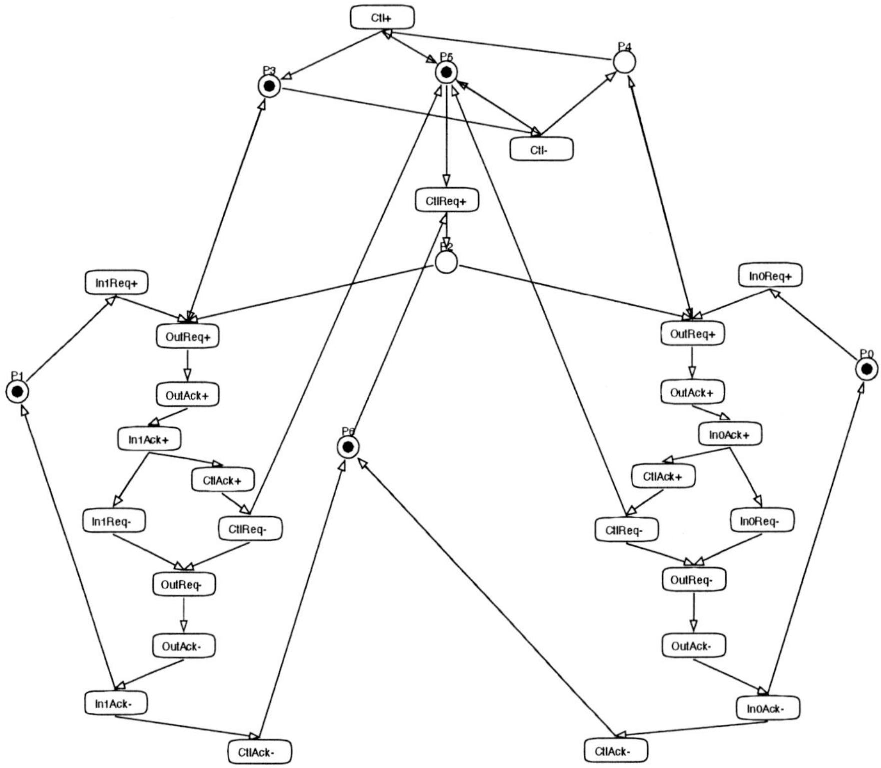

Figure 6.26. **The modified STG specification of the 4-phase bundled-data MUX control circuit.**

In figure 6.26 we have removed some concurrency from the MUX STG by ordering the transitions on $In0Ack/In1Ack$ and $CtlAck$ ($In0Ack+ \prec CtlAck+$, $In1Ack+ \prec CtlAck+$ etc.). This STG satisfies CSC and the resulting circuit is marginally smaller:

```
> more MUX4p-gcm.eqn

# EQN file for model MUX4pB
# Generated by petrify 4.0 (compiled 22-Dec-98 at 6:58 PM)
# Outputs between brackets "[out]" indicate a feedback to input "out"
```

```
# Estimated area = 27.00

INORDER = In0Req OutAck In1Req Ctl CtlReq In1Ack In0Ack OutReq CtlAck;
OUTORDER = [In1Ack] [In0Ack] [OutReq] [CtlAck];
[0] = Ctl CtlReq OutAck;
[1] = Ctl' CtlReq OutAck;
[2] = CtlReq (Ctl' In0Req + Ctl In1Req);
[3] = CtlReq' (In0Ack' In1Req' + In0Req' In0Ack);
[OutReq] = OutReq [3]' + [2];               # mappable onto gC
[CtlAck] = In1Ack + In0Ack;
[In1Ack] = In1Ack OutAck + [0];             # mappable onto gC
[In0Ack] = In0Ack OutAck + [1];             # mappable onto gC
```

6.9. Summary

This chapter has provided an introduction to the design of asynchronous sequential (control) circuits with the main focus on speed-independent circuits and specifications using STGs. The material was presented from a practical view in order to enable the reader to go ahead and design his or her own speed-independent control circuits. This, rather than comprehensiveness, has been our goal, and as mentioned in the introduction we have largely ignored important alternative approaches including burst-mode and fundamental-mode circuits.

Chapter 7

ADVANCED 4-PHASE BUNDLED-DATA PROTOCOLS AND CIRCUITS

The previous chapters have explained the basics of asynchronous circuit design. In this chapter we will address 4-phase bundled-data protocols and circuits in more detail. This will include: (1) a variety of channel types, (2) protocols with different data-validity schemes, and (3) a number of more sophisticated latch control circuits. These latch controllers are interesting for two reasons: they are very useful in optimizing the circuits for area, power and speed, and they are nice examples of the types of control circuits that can be specified and synthesized using the STG-based techniques from the previous chapter.

7.1. Channels and protocols

7.1.1 Channel types

So far we have considered only *push channels* where the sender is the active party that initiates the communication of data, and where the receiver is the passive party. The opposite situation, where the receiver is the active party that initiates the communication of data, is also possible, and such a channel is called a *pull channel.* A channel that carries no data is called a *nonput channel* and is used for synchronization purposes. Finally, it is also possible to communicate data from a receiver to a sender along with the acknowledge signal. Such a channel is called a *biput channel.* In a 4-phase bundled-data implementation data from the receiver is bundled with the acknowledge, and in a 4-phase dual-rail protocol the passive party will acknowledge the reception of a codeword by returning a codeword rather than just an an acknowledge signal. Figure 7.1 illustrates these four channel types (nonput, push, pull, and biput) assuming a bundled-data protocol. Each channel type may, of course, use any of the handshake protocols (2-phase or 4-phase) and data encodings (bundled-data, dual-rail, m–of–n, etc.) introduced previously.

J. Sparsø and S. Furber (eds.), Principles of Asynchronous Circuit Design, 115–122.

7.1.2 Data-validity schemes

For the bundled-data protocols it is also relevant to define the time interval in which data is valid, and figure 7.2 illustrates the different possibilities.

For a push channel the request signal carries the message "here is new data for you" and the acknowledge signal carries the information "thank you, I have absorbed the data, and you may release the data wires." Similarly, for a pull channel the request signal carries the message "please send new data" and the acknowledge signal carries the message "here is the data that you requested." It is the signal transitions on the request and acknowledge wires that are interpreted in this way. A 4-phase handshake involves two transitions on each wire and, depending on whether it is the rising or the falling transitions on the request and acknowledge signals that are interpreted in this way, several data-validity schemes emerge: early, broad, late and extended early.

Since 2-phase handshaking does not involve any redundant signal transitions there is only one data-validity scheme for each channel type (push or pull), as illustrated in figure 7.2.

It is common to all of the data-validity schemes that the data is valid some time before the event that indicates the start of the interval, and that it remains stable until some time after the event that indicates the end of the interval. Furthermore, all of the data-validity schemes express the requirements of the party that receives the data. The fact that a receiver signals "thank you, I have absorbed the data, and you may go ahead and release the data wires," does not mean that this actually happens – the sender may prolong the data-validity interval, and the receiver may even rely on this.

A typical example of this is the extended-early data-validity schemes in figure 7.2. On a push channel the data-validity interval begins some time before $Req \uparrow$ and ends some time after $Req \downarrow$.

7.1.3 Discussion

The above classification of channel types and handshake protocols stems mostly from Peeters' Ph.D. thesis [112]. The choice of channel type, handshake protocol and data-validity scheme obviously affects the implementation of the handshake components in terms of area, speed, and power. Just as a design may use a mix of different bundled-data and dual-rail protocols, it may also use a mix of channel types and data-validity schemes.

For example, a 4-phase bundled-data push channel using a broad or an extended-early data-validity scheme is a very convenient input to a function block that is implemented using precharged CMOS circuitry: the request signal may directly control the precharge and evaluate transistors because the broad and the extended-early data-validity schemes guarantee that the input data is stable during the evaluate phase.

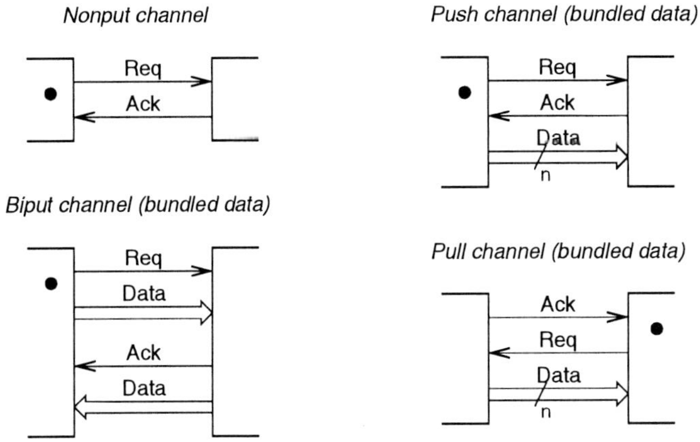

Figure 7.1. The four fundamental channel types: nonput, push, biput, and pull.

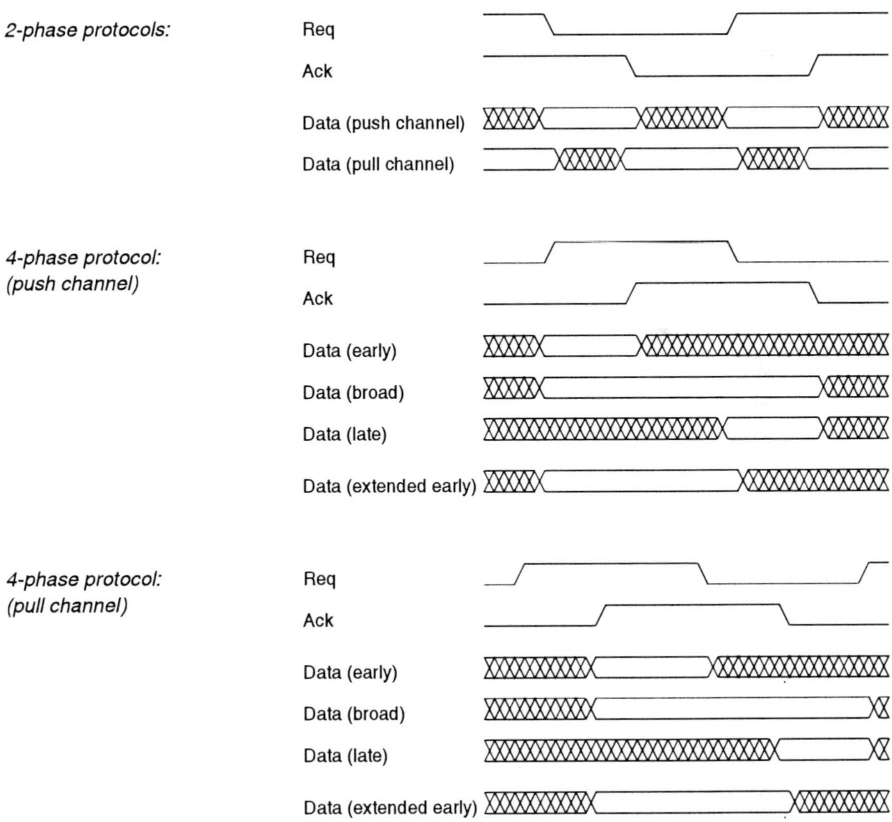

Figure 7.2. Data-validity schemes for 2-phase and 4-phase bundled data.

Another interesting option in a 4-phase bundled-data design is to use function blocks that assume a broad data validity scheme on the input channel and that produce a late data validity scheme on the output channel. Under these assumptions it is possible to use a *symmetric* delay element that matches only half of the latency of the combinatorial circuit. The idea is that the *sum* of the delay of $Req\uparrow$ and $Req\downarrow$ matches the latency of the combinatorial circuit, and that $Req\downarrow$ indicates valid output data. In [112, p.46] this is referred to as *true single phase* because the return-to-zero part of the handshaking is no longer redundant. This approach also has implications for the implementation of the components that connect to the function block.

It is beyond the scope of this text to enter into a discussion of where and when to use the different options. The interested reader is referred to [112, 77] for more details.

7.2. Static type checking

When designing circuits it is useful to think of the combination of channel type and data-validity scheme as being similar to a data type in a programming language, and to do some static type checking of the circuit being designed by asking questions like: "what types are allowed on the input ports of this handshake component?" and "what types are produced on the output ports of this handshake component?". The latter may depend on the type that was provided on the input port. A similar form of type checking for synchronous circuits using two-phase non-overlapping clocks has been proposed in [104] and used in the Genesil silicon compiler [67].

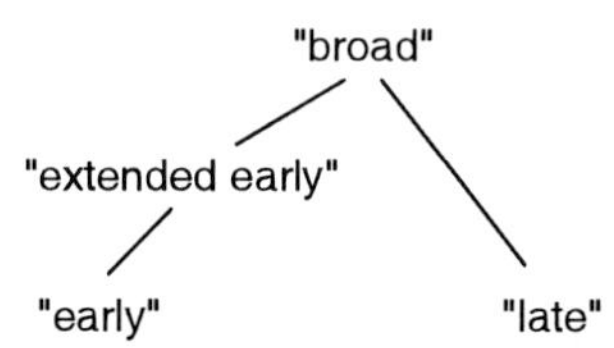

Figure 7.3. Hierarchical ordering of the four data-validity schemes for the 4-phase bundled-data protocol.

Figure 7.3 shows a hierarchical ordering of the four possible types (data validity schemes) for a 4-phase bundled-data push channel: "broad" is the strongest type and it can be used as input to circuits that require any of the weaker types. Similarly "extended early" may be used where only "early" is required. Circuits that are transparent to handshaking (function blocks, join, fork, merge, mux, demux) produce outputs whose type is at most as strong as their (weakest) input type. In general the input and output types are the same but there are examples where this is not the case. The only circuit that can produce outputs whose type is stronger than the input type is a latch. Let us look at some examples:

- A join that concatenates two inputs of type "extended early" produces outputs that are only "early.'

- From the STG fragments in figure 6.21 on page 107 it may be seen that the simple 4-phase bundled-data latch controller from the previous chapters (figure 2.9 on page 18) assumes "early" inputs and that it produces "extended-early" outputs.

- The 4-phase bundled-data MUX design in section 6.8.3 assumes "extended early" on its control input (the STG in figure 6.25 on page 110 specifies stable input from $CtlReq+$ to $CtlReq-$).

The reader is encouraged to continue this exercise and perhaps draw the associated timing diagrams from which the types of the outputs may be deduced. The type checking explained here is a very useful technique for debugging circuits that exhibit erronous behaviour.

7.3. More advanced latch control circuits

In previous chapters we have only considered 4-phase bundled-data handshake latches using a latch controller consisting of a C-element and an inverter (figure 2.9 on page 18). In [41] this circuit is called a *simple* latch controller, and in [77] it is called an *un-decoupled* latch controller.

When a pipeline or FIFO that uses the simple latch controller fills, every second handshake latch will be holding a valid token and every other hand-

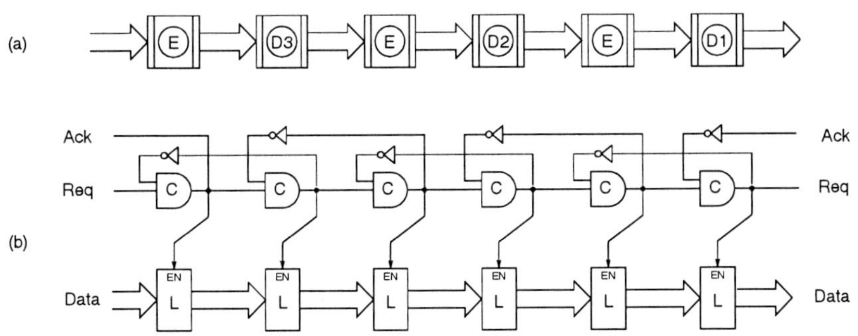

Figure 7.4. (a) A FIFO based on handshake latches, and (b) its implementation using simple latch controllers and level-sensitive latches. The FIFO fills with valid data in every other latch. A latch is transparent when $EN = 0$ and it is opaque (holding data) when $EN = 1$.

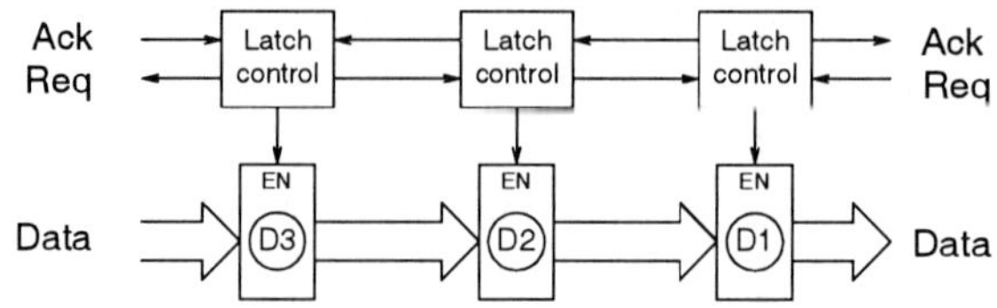

Figure 7.5. A FIFO where every level-sensitive latch holds valid data when the FIFO is full. The semi-decoupled and fully-decoupled latch controllers from [41] allow this behaviour.

shake latch will be holding an empty token as illustrated in figure 7.4(a) – the static spread of the pipeline is $S = 2$.

This token picture is a bit misleading. The empty tokens correspond to the return-to-zero part of the handshaking and in reality the latches are not "holding empty tokens" – they are transparent, and this represents a waste of hardware resource.

Ideally one would want to store a valid token in every level-sensitive latch as illustrated in figure 7.5 and just "add" the empty tokens to the data stream on the interfaces as part of the handshaking. In [41] Furber and Day explain the design of two such improved 4-phase bundled-data latch control circuits: a *semi-decoupled* and a *fully-decoupled* latch controller. In addition to these specific circuits the paper also provides a nice illustration of the use of STGs for designing control circuits following the approach explained in chapter 6. The three latch controllers have the following characteristics:

- The *simple or un-decoupled* latch controller has the problem that new input data can only be latched when the previous handshake on the output channel has completed, i.e., after $A_{out}\downarrow$. Furthermore, the handshakes on the input and output channels interact tightly: $R_{out}\uparrow \preceq A_{in}\uparrow$ and $R_{out}\downarrow \preceq A_{in}\downarrow$.

- The *semi-decoupled* latch controller relaxes these requirements somewhat: new inputs may be latched after $R_{out}\downarrow$, and the controller may produce $A_{in}\uparrow$ independently of the handshaking on the output channel – the interaction between the input and output channels has been relaxed to: $A_{out}\uparrow \preceq A_{in}\uparrow$.

- The *fully-decoupled* latch controller further relaxes these requirements: new inputs may be latched after $A_{out}\uparrow$ (i.e. as soon as the downstream latch has indicated that it has latched the current data), and the handshaking on the input channel may complete without any interaction with the output channel.

Another potential drawback of the simple latch controller is that it is unable to take advantage of function blocks with asymmetric delays as explained in

Latch controller	Static spread, S	Period, P
"Simple"	2	$2L_r + 2L_{f.V}$
"Semi-decoupled"	1	$2L_r + 2L_{f.V}$
"Fully-decoupled"	1	$2L_r + L_{f.V} + L_{f.E}$

Table 7.1. Summary of the characteristics of the latch controllers in [41].

section 4.4.1 on page 52. The fully-decoupled latch controller presented in [41] does not have this problem. Due to the decoupling of the input and output channels the dependency graph critical cycle that determines the period, P, only visits nodes related to two neighbouring pipeline stages and the period becomes minimum (c.f. section 4.4.1). Table 7.1 summarizes the characteristics of the simple, semi-decoupled and fully-decoupled latch controllers.

All of the above-mentioned latch controllers are "normally transparent" and this may lead to excessive power consumption because inputs that make multiple transitions before settling will propagate through several consecutive pipeline stages. By using "normally opaque" latch controllers every latch will act as a barrier. If a handshake latch that is holding a bubble is exposed to a token on its input, the latch controller switches the latch into the transparent mode, and when the input data have propagated safely into the latch, it will switch the latch back into the opaque mode in which it will hold the data. In the design of the asynchronous MIPS processor reported in [23] we experienced approximately a 50 % power reduction when using normally opaque latch controllers instead of normally transparent latch controllers.

Figure 7.6 shows the STG specification and the circuit implementation of the normally opaque latch controller used in [23]. As seen from the STG there is quite a strong interaction between the input and output channels, but the dependency graph critical cycle that determines the period only visits nodes related to two neighbouring pipeline stages and the period is minimum. It may be necessary to add some delay into the $Lt+$ to $Rout+$ path in order to ensure that input signals have propagated through the latch before $Rout+$. Furthermore the duration of the $Lt = 0$ pulse that causes the latch to be transparent is determined by gate delays in the latch controller itself, and the pulse must be long enough to ensure safe latching of the data. The latch controller assumes a broad data-validity scheme on its input channel and it provides a broad data-validity scheme on its output channel.

7.4. Summary

This chapter introduced a selection of channel types, data-validity schemes and a selection of latch controllers. The presentation was rather brief; the aim was just to present the basics and to introduce some of the many options and

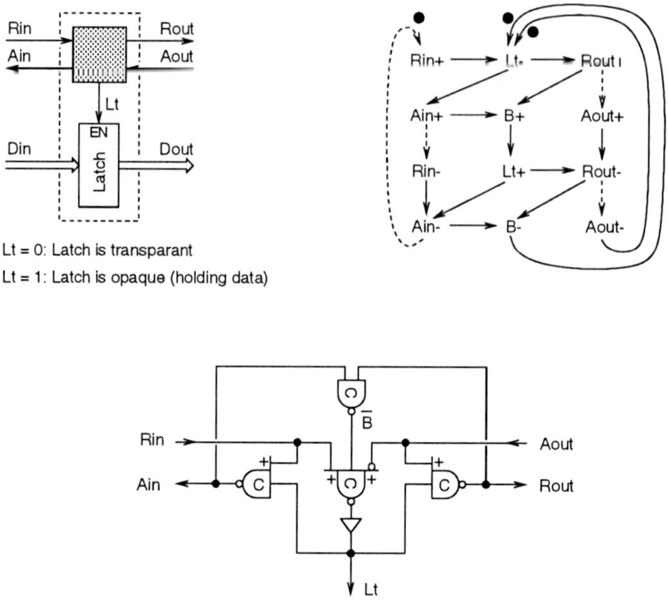

Figure 7.6. **The STG specification and the circuit implementation of the normally opaque fully-decoupled latch controller from [23].**

possibilities for optimizing the circuits. The interested reader is referred to the literature for more details.

Finally a warning: the static data-flow view of asynchronous circuits presented in chapter 3 (i.e. that valid and empty tokens are copied forward controlled by the handshaking between latch controllers) and the performance analysis presented in chapter 4 assume that all handshake latches use the simple normally transparent latch controller. When using semi-decoupled or fully-decoupled latch controllers, it is necessary to modify the token flow view, and to rework the performance analysis. To a first order one might substitute each semi-decoupled or fully-decoupled latch controller with a pair of simple latch controllers. Furthermore a ring need only include two handshake latches if semi-decoupled or fully-decoupled latch controllers are used.

Chapter 8

HIGH-LEVEL LANGUAGES AND TOOLS

This chapter addresses languages and CAD tools for the high-level modeling and synthesis of asynchronous circuits. The aim is briefly to introduce some basic concepts and a few representative and influential design methods. The interested reader will find more details elsewhere in this book (in Part II and chapter 13) as well as in the original papers that are cited in the text. In the last section we address the use of VHDL for the design of asynchronous circuits.

8.1. Introduction

Almost all work on the high-level modeling and synthesis of asynchronous circuits is based on the use of a language that belongs to the CSP family of languages, rather than one of the two industry-standard hardware description languages, VHDL and Verilog. Asynchronous circuits are highly *concurrent* and communication between modules is based on *handshake channels*. Consequently a hardware description language for asynchronous circuit design should provide efficient primitives supporting these two characteristics. The CSP language proposed by Hoare [57, 58] meets these requirements. CSP stands for "Communicating Sequential Processes" and its key characteristics are:

- Concurrent processes.
- Sequential and concurrent composition of statements within a process.
- Synchronous message passing over point-to-point channels (supported by the primitives send, receive and – possibly – probe).

CSP is a member of a large family of languages for programming concurrent systems in general: OCCAM [68], LOTOS [108, 16], and CCS [89], as well as languages defined specifically for designing asynchronous circuits: Tangram [142, 135], CHP [81], and Balsa [9, 10]. Further details are presented elsewhere in this book on Tangram (in Part III, chapter 13) and Balsa (in Part II).

In this chapter we first take a closer look at the CSP language constructs supporting communication and concurrency. This will include a few sample

J. Sparsø and S. Furber (eds.), Principles of Asynchronous Circuit Design, 123–152.

programs to give a flavour of this type of language. Following this we briefly explain two rather different design methods that both take a CSP-like program as the starting point for the design:

- At Philips Research Laboratories, van Berkel, Peeters, Kessels et al. have developed a proprietary language, Tangram, and an associated silicon compiler [142, 141, 135, 112]. Using a process called syntax-directed compilation, the synthesis tool maps a Tangram program into a structure of handshake components. Using these tools several significant asynchronous chips have been designed within Philips [137, 138, 144, 73, 74]. The last of these is a smart-card circuit that is described in chapter 13 on page 221.

- At Caltech Martin has developed a language CHP – Communicating Hardware Processes – and a set of tools that supports a partly manual, partly automated design flow that targets highly optimized transistor-level implementations of QDI 4-phase dual-rail circuits [80, 83].

CHP has a syntax that is similar to CSP (using various special symbols) whereas Tangram has a syntax that is more like a traditional programming language (using keywords); but in essence they are both very similar to CSP.

In the last section of this chapter we will introduce a VHDL-package that provides CSP-like message passing and explain an associated VHDL-based design flow that supports a manual step-wise refinement design process.

8.2. Concurrency and message passing in CSP

The "sequential processes" part of the CSP acronym denotes that each process is described by a program whose statements are executed in sequence one by one. A semicolon is used to separate statements (as in many other programming languages). The semicolon can be seen as an operator that combines statements into programs. In this respect a process in CSP is very similar to a process in VHDL. However, CSP also allows the parallel composition of statements within a process. The symbol "||" denotes parallel composition. This feature is not found in VHDL, whereas the fork-join construct in Verilog does allow statement-level concurrency within a process.

The "communicating" part of the CSP acronym refers to synchronous message passing using point-to-point channels as illustrated in figure 8.1, where two processes $P1$ and $P2$ are connected by a channel named C. Using a send statement, $C!x$, process $P1$ sends (denoted by the '!' symbol) the value of its variable x on channel C, and using a receive statement, $C?y$, process $P2$ receives (denoted by the '?' symbol) from channel C a value that is assigned to its variable y. The channel is memoryless and the transfer of the value of variable x in $P1$ into variable y in $P2$ is an atomic action. This has the effect

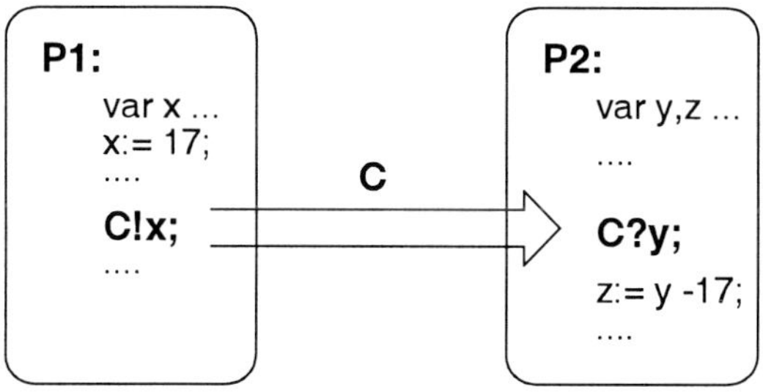

Figure 8.1. Two processes $P1$ and $P2$ connected by a channel C. Process $P1$ sends the value of its variable x to the channel C, and process $P2$ receives the value and assigns it to its variable y.

of synchronizing processes $P1$ and $P2$. Whichever comes first will wait for the other party, and the send and receive statements complete at the same time. The term *rendezvous* is sometimes used for this type of synchronization.

When a process executes a send (or receive) statement, it commits to the communication and suspends until the process at the other end of the channel performs its receive (or send) statement. This may not always be desirable, and Martin has extended CSP with a probe construct [79] which allows the process at the passive end of a channel to probe whether or not a communication is pending on the channel, without committing to any communication. The probe is a function which takes a channel name as its argument and returns a Boolean. The syntax for probing channel C is $\overline{C}$.

As an aside we mention that some languages for programming concurrent systems assume channels with (possibly unbounded) buffering capability. The implication of this is that the channel acts as a FIFO, and the communicating processes do not synchronize when they communicate. Consequently this form of communication is called asynchronous message passing.

Going back to our synchronous message passing, it is obvious that the physical implementation of a memoryless channel is simply a set of wires together with a protocol for synchronizing the communicating processes. It is also obvious that any of the protocols that we have considered in the previous chapters may be used. Synchronous message passing is thus a very useful language construct that supports the high-level modeling of asynchronous circuits by abstracting away the exact details of the data encoding and handshake protocol used on the channel.

Unfortunately both VHDL and Verilog lack such primitives. It is possible to write low-level code that implements the handshaking, but it is highly undesirable to mix such low-level details into code whose purpose is to capture the high-level behaviour of the circuit.

In the following section we will provide some small program examples to give a flavour of this type of language. The examples will be written in Tangram as they also serve the purpose of illustrating syntax-directed compilation in a subsequent section. The source code, handshake circuit figures, and fragments of the text have been kindly provided by Ad Peeters from Philips.

Manchester University has recently developed a similar language and synthesis tool that is available in the public domain [10], and is introduced in Part II of this book. Other examples of related work are presented in [17] and [21].

8.3. Tangram: program examples

This section provides a few simple Tangram program examples: a 2-place shift register, a 2-place ripple FIFO, and a greatest common divisor function.

8.3.1 A 2-place shift register

Figure 8.2 shows the code for a 2-place shift register named `ShiftReg`. It is a process with an input channel `In` and an output channel `Out`, both carrying variables of type `[0..255]`. There are two local variables `x` and `y` that are initialized to `0`. The process performs an unbounded repetition of a sequence of three statements: `out!y; y:=x; in?x`.

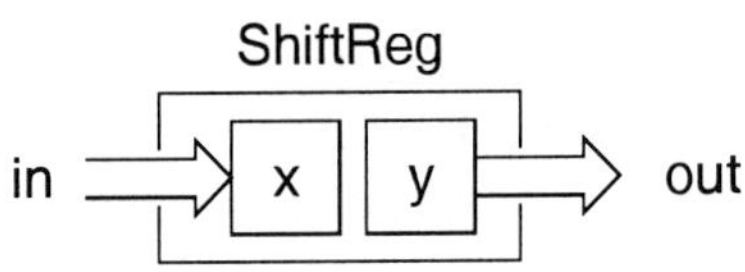


```
  T = type [0..255]
& ShiftReg : main proc(in? chan T & out! chan T).
  begin
  & var x,y: var T := 0
    |
    forever do
      out!y ; y:=x ; in?x
    od
  end
```

Figure 8.2. A Tangram program for a 2-place shift register.

8.3.2 A 2-place (ripple) FIFO

Figure 8.3 shows the Tangram program for a 2-place first-in first-out buffer named `Fifo`. It can be understood as two 1-place buffers that are operating in parallel and that are connected by a channel `c`. At first sight it appears very similar to the 2-place shift register presented above, but a closer examination will show that it is more flexible and exhibits greater concurrency.

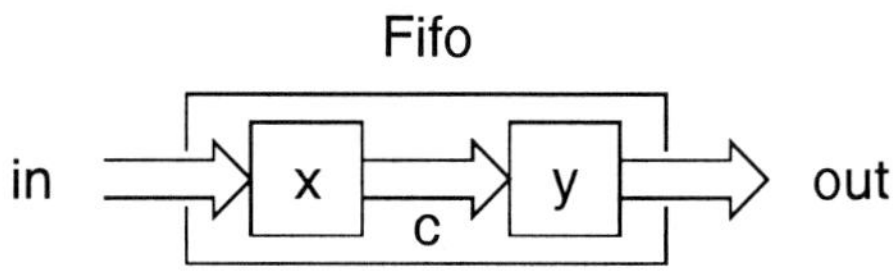


```
  T = type [0..255]
& Fifo : main proc(in? chan T & out! chan T).
  begin
  & x,y: var T
  & c : chan T
  |
       forever do in?x ; c!x   od
    || forever do c?y  ; out!y od
  end
```

Figure 8.3. A Tangram program for a 2-place (ripple) FIFO.

8.3.3 GCD using while and if statements

Figure 8.4 shows the code for a module that computes the greatest common divisor, the example from section 3.7. The "`do x<>y then ...od`" is a while statement and, apart from the syntactical differences, the code in figure 8.4 is identical to the code in figure 3.11 on page 39.

The module has an input channel from which it receives the two operands, and an output channel on which it sends the result.

```
  int = type [0..255]
& gcd_if : main proc (in?chan <<int,int>> & out!chan int).
  begin x,y:var int ff
  | forever do
      in?<<x,y>>
    ; do x<>y then
        if x<y then y:=y-x
               else x:=x-y
        fi
      od
    ; out!x
    od
  end
```

Figure 8.4. A Tangram for GCD using while and if statements.

8.3.4 GCD using guarded commands

Figure 8.5 shows an alternative version of GCD. This time the module has separate input channels for the two operands and its body is based on the repetition of a guarded command. The guarded repetition can be seen as a generalization of the while statement. The statement repeats until all guards are false. When at least one of the guards is true, exactly one command corresponding to such a true guard is selected (either deterministically or non-deterministically) and executed.

```
  int = type [0..255]
& gcd_gc : main proc (in1,in2?chan int & out!chan int).
  begin x,y:var int ff
  | forever do
      in1?x || in2?y
    ; do x<y then y:=y-x
      or y<x then x:=x-y
      od
    ; out!x
    od
  end
```

Figure 8.5. A Tangram program for GCD using guarded repetition.

8.4. Tangram: syntax-directed compilation

Let us now address the synthesis process. The design flow uses an intermediate format based on handshake circuits. The front-end design activity is called VLSI programming and, using syntax-directed compilation, a Tangram program is mapped into a structure of handshake components. There is a one-to-one correspondence between the Tangram program and the handshake circuit as will be clear from the following examples. The compilation process is thus fully transparent to the designer, who works entirely at the Tangram program level.

The back-end of the design flow involves a library of handshake circuits that the compiler targets as well as some tools for post-synthesis peephole optimization of the handshake circuits (i.e. replacing common structures of handshake components by more efficient equivalent ones). A number of handshake circuit libraries exist, allowing implementations using different handshake protocols (4-phase dual-rail, 4-phase bundled-data, etc.), and different implementation technologies (CMOS standard cells, FPGAs, etc.). The handshake components can be specified and designed: (i) manually, or (ii) using STGs and Petrify as explained in chapter 6, or (iii) using the lower steps in Martin's transformation-based method that is presented in the next section.

It is beyond the scope of this text to explain the details of the compilation process. We will restrict ourselves to providing a flavour of "syntax-directed compilation" by showing handshake circuits corresponding to the example Tangram programs from the previous section.

8.4.1 The 2-place shift register

As a first example of syntax-directed compilation figure 8.6 shows the handshake circuit corresponding to the Tangram program in figure 8.2.

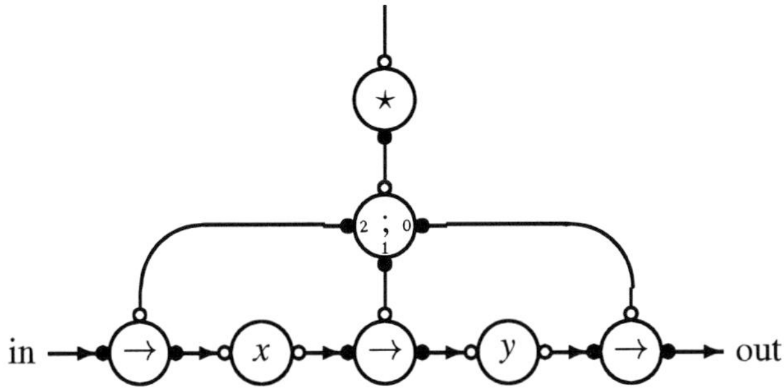

Figure 8.6. **The compiled handshake circuit for the 2-place shift register.**

Handshake components are represented by circular symbols, and the channels that connect the components are represented by arcs. The small dots on the component symbols represent ports. An open dot denotes a passive port and a solid dot denotes an active port. The arrowhead represents the direction of the data transfer. A nonput channel does not involve the transfer of data and consequently it has no direction and no arrowhead. As can be seen in figure 8.6 a handshake circuit uses a mix of push and pull channels.

The structure of the program is a forever-do statement whose body consists of three statements that are executed sequentially (because they are separated by semicolons). Each of the three statements is a kind of assignment statement: the value of variable `y` is "assigned" to output channel `out`, the value of variable `x` is assigned to variable `y`, and the value received on input chanel `in` is assigned to variable `x`. The structure of the handshake circuit is exactly the same:

- At the top is a *repeater* that implements the forever-do statement. A repeater waits for a request on its passive input port and then it performs an unbounded repetition of handshakes on its active output channel. The handshake on the input channel never completes.

- Below is a 3-way *sequencer* that implements the semicolons in the program text. The sequencer waits for a request on its passive input channel, then it performs in sequence a full handshake on each of its active output channels (in the order indicated by the numbers in the symbol) and finally it completes the handshaking on the passive input channel. In this way the sequencer activates in turn the handshake circuit constructs that correspond to the individual statements in the body of the forever-do statement.

- The bottom row of handshake components includes two *variables*, x and y, and three *transferers*, denoted by '→'. Note that variables have passive read and write ports. The transferers implement the three statements (`out!y; y:=x; in?x`) that form the body of the forever-do statement, each a form of assignment. A transferer waits for a request on its passive nonput channel and then initiates a handshake on its pull input channel. The handshake on the pull input channel is relayed to the push output channel. In this way the transferer pulls data from its input channel and pushes it onto its output channel. Finally, it completes the handshaking on the passive nonput channel.

8.4.2 The 2-place FIFO

Figure 8.7 shows the handshake circuit corresponding to the Tangram program in figure 8.3. The component labeled 'psv' in the handshake circuit of figure 8.7 is a so-called *passivator*. It relates to the internal channel `c` of the `Fifo` and implements the synchronization and communication between the active sender (`c!x`) and the active receiver (`c?y`).

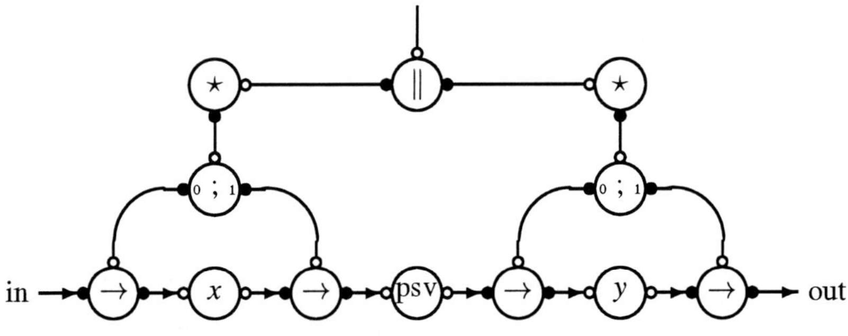

Figure 8.7. **Compiled handshake circuit for the FIFO program.**

An optimization of the handshake circuit for `Fifo` is shown in figure 8.8. The synchronization in the datapath using a passivator has been replaced by a synchronization in the control using a 'join' component. One may observe that the datapath of this handshake circuit for the FIFO design is the same as that of the shift register, shown in figure 8.2. The only difference is in the control part of the circuits.

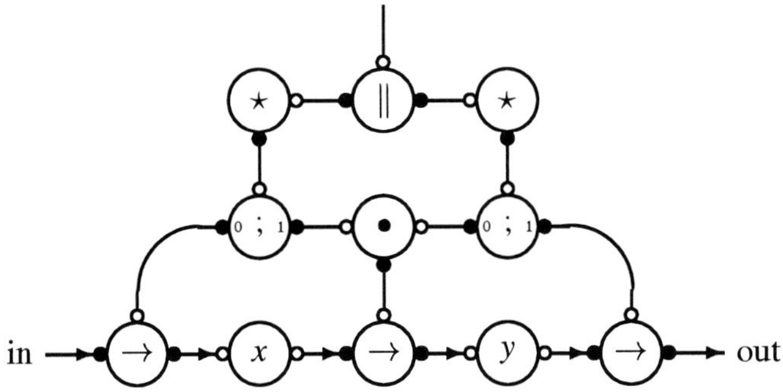

Figure 8.8. Optimized handshake circuit for the FIFO program.

8.4.3 GCD using guarded repetition

As a more complex example of syntax-directed compilation figure 8.9 shows the handshake circuit compiled from the Tangram program in figure 8.5. Compared with the previous handshake circuits, the handshake circuit for the GCD program introduces two new classes of components that are treated in more detail below.

Firstly, the circuit contains a 'bar' and a 'do' component, both of which are data-dependent control components. Secondly, the handshake circuit contains components that do not directly correspond to language constructs, but rather implement sharing: the multiplexer (denoted by 'mux'), the demultiplexer (denoted by 'dmx'), and the fork component (denoted by '•').

Warning: the Tangram fork is identical to the fork in figure 3.3 but the Tangram multiplexer and demultiplexer components are different. The Tangram multiplexer is identical to the merge in figure 3.3 and the Tangram demultiplexer is a kind of "inverse merge." Its output ports are passive and it requires the handshakes on the two outputs to be mutually exclusive.

The 'bar' and the 'do' components: The do and bar component together implement the guarded command construct with two guards, in which the do component implements the iteration part (the `do od` part, including the evaluation of the disjunction of the two guards), and the bar component implements the choice part (the `then or then` part of the command).

The do component, when activated through its passive port, first collects the disjunction of the value of all guards through a handshake on its active data port. When the value thus collected is true, it activates its active nonput port (to activate the selected command), and after completion starts a new evaluation cycle. When the value collected is false, the do component completes its operation by completing the handshake on the passive port.

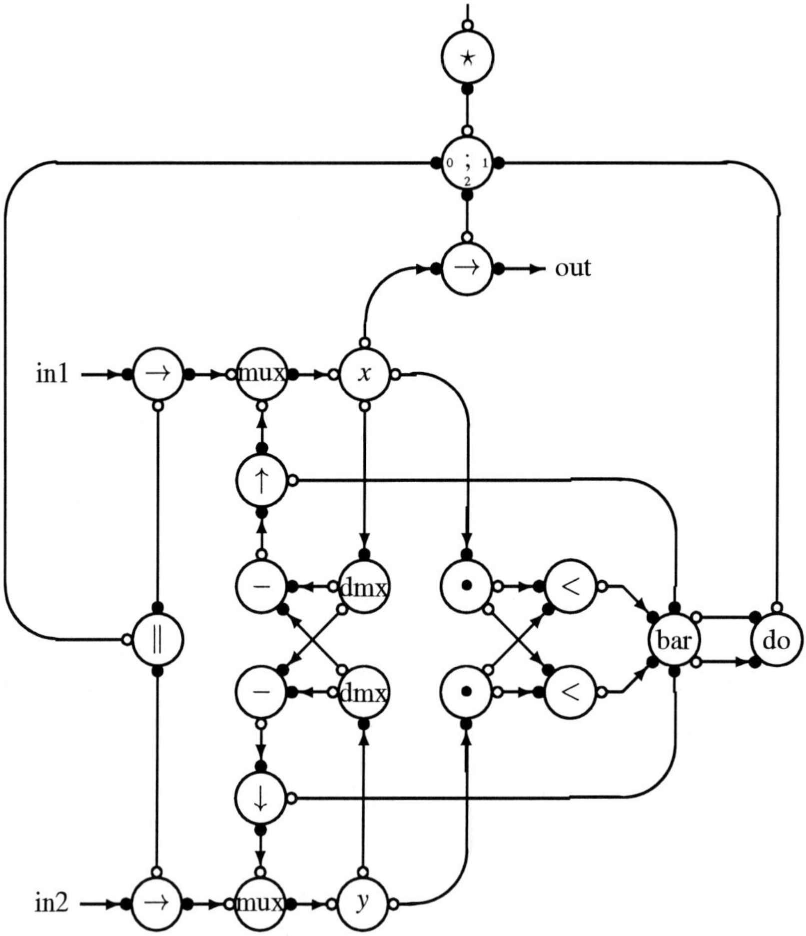

Figure 8.9. **Compiled handshake circuit for the GCD program using guarded repetition.**

The bar component can be activated either through its passive data port, or through its passive control port. (The do component, for example, sequences these two activations.) When activated through the data port, it collects the value of two guards through a handshake on the active data ports, and then sends the disjunction of these values along the passive data port, thus completing that handshake. When activated through the control port, the bar component activates an active control port of which the associated data port returned a 'true' value in the most recent data cycle. (For simplicity, this selection is typically implemented in a deterministic fashion, although this is not required at the level of the program.) One may observe that bar components can be combined in a tree or list to implement a guarded command list of arbitrary length. Furthermore, not every data cycle has to be followed by a control cycle.

The 'mux', 'demux', and 'fork' components The program for GCD in figure 8.4 has two occurrences of variable `x` in which a value is written into `x`, namely input action `in1?x` and assignment `x:=x-y`. In the handshake circuit of figure 8.9, these two write actions for Tangram variable `x` are merged by the multiplexer component so as to arrive at the write port of handshake variable `x`.

Variable `x` occurs at five different locations in the program as an expression, once in the output expression `out!x`, twice in the guard expressions `x<y` and `y<x`, and twice in the assignment expressions `x-y` and `y-x`. These five inspections of variable `x` could be implemented as five distinct read ports on the handshake variable `x`, which is shown in the handshake circuit in [135, Fig. 2.7, p.34]. In figure 8.9, a different compilation is shown, in which handshake variable `x` has three read ports:

- A read port dedicated to the occurrence in the output action.

- A read port dedicated to the guard expressions. Their evaluation is *mutually inclusive*, and hence can be combined using a synchronizing fork component.

- A read port dedicated to the assignment expressions. Their evaluation is *mutually exclusive*, and hence can be combined using a demultiplexer.

The GCD example is discussed in further detail in chapter 13.

8.5. Martin's translation process

The work of Martin and his group at Caltech has made fundamental contributions to asynchronous design and it has influenced the work of many other researchers. The methods have been used at Caltech to design several significant chips, most recently and most notably an asynchronous MIPS R3000 processor [88]. As the following presentation of the design flow hints, the design process is elaborate and sophisticated and is probably only an option to a person who has spent time with the Caltech group.

The mostly manual design process involves the following steps (semantics-preserving transformations):

(1) *Process decomposition* where each process is refined into a collection of interacting simpler processes. This step is repeated until all processes are simple enough to be dealt with in the next step in the process.

(2) *Handshake expansion* where each communication channel is replaced by explicit wires and where each communication action (e.g. send or receive)

is replaced by the signal transitions required by the protocol that is being used. For example a receive statement such as:

$$C?y$$

is replaced by a sequence of simpler statements – for example:

$$[C_{req}];\ y := data;\ C_{ack}\uparrow;\ [\neg C_{req}];\ C_{ack}\downarrow$$

which is read as: “wait for request to go high”, “read the data”, “drive acknowledge high”, “wait for request to go low”, and “drive acknowledge low”.

At this level it may be necessary to add state variables and/or to reshuffle signal transitions in order to obtain a specification that satisfies a condition similar to the CSC condition in chapter 6.

(3) *Production rule expansion* where each handshaking expansion is replaced by a set of production rules (or guarded commands), for example:

$$a \wedge b \mapsto c\uparrow \quad \text{and} \quad \neg b \wedge \neg c \mapsto c\downarrow$$

A production rule consist of a condition and an action, and the action is performed whenever the condition is true. As an aside we mention that the above two production rules express the same as the set and reset functions for the signal c on page 96. The production rules specify the behaviour of the internal signals and output signals of the process. The production rules are themselves simple concurrent processes and the guards must ensure that the signal transitions occur in program order (i.e. that the semantics of the original CHP program are maintained). This may require strengthening the guards. Furthermore, in order to obtain simpler circuit implementations, the guards may be modified and made symmetric.

(4) *Operator reduction* where production rules are grouped into clusters and where each cluster is mapped onto a basic hardware component similar to a generalized C-element. The above two production rules would be mapped into the generalized C-element shown in figure 6.17 on page 100.

8.6. Using VHDL for asynchronous design

8.6.1 Introduction

In this section we will introduce a couple of VHDL packages that provide the designer with primitives for synchronous message passing between processes – similar to the constructs found in the CSP-family of languages (send, receive and probe).

The material was developed in an M.Sc. project and used in the design of a 32-bit floating-point ALU using the IEEE floating-point number representation [110], and it has subsequently been used in a course on asynchronous circuit design. Others, including [95, 118, 149, 78], have developed related VHDL packages and approaches.

The channel packages introduced in the following support only one type of channel, using a 32-bit 4-phase bundled-data push protocol. However, as VHDL allows the overloading of procedures and functions, it is straightforward to define channels with arbitrary data types. All it takes is a little cut-and-paste editing. Providing support for protocols other than the 4-phase bundled-data push protocol will require more significant extensions to the packages.

8.6.2 VHDL versus CSP-type languages

The previous sections introduced several CSP-like hardware description languages for asynchronous design. The advantages of these languages are their support of concurrency and synchronous message passing, as well as a limited and well-defined set of language constructs that makes syntax-directed compilation a relatively simple task.

Having said this there is nothing that prevents a designer from using one of the industry standard languages VHDL (or Verilog) for the design of asynchronous circuits. In fact some of the fundamental concepts in these languages – concurrent processes and signal events – are "nice fits" with the modeling and design of asynchronous circuits. To illustrate this figure 8.10 shows how the Tangram program from figure 8.2 could be expressed in plain VHDL. In addition to demonstrating the feasibility, the figure also highlights the limitations of VHDL when it comes to modeling asynchronous circuits: most of the code expresses low-level handshaking details, and this greatly clutters the description of the function of the circuit.

VHDL obviously lacks built-in primitives for synchronous message passing on channels similar to those found in CSP-like languages. Another feature of the CSP family of languages that VHDL lacks is statement-level concurrency within a process. On the other hand there are also some advantages of using an industry standard hardware description language such as VHDL:

- It is well supported by existing CAD tool frameworks that provide simulators, pre-designed modules, mixed-mode simulation, and tools for synthesis, layout and the back annotation of timing information.

- The same simulator and test bench can be used throughout the entire design process from the first high-level specification to the final implementation in some target technology (for example a standard cell layout).

- It is possible to perform mixed-mode simulations where some entities are modeled using behavioural specifications and others are implemented using the components of the target technology.

```
library IEEE;
use IEEE.std_logic_1164.all;

  type T is std_logic_vector(7 downto 0)

  entity ShiftReg is
    port ( in_req   : in  std_logic;
           in_ack   : out std_logic;
           in_data  : in  T;
           out_req  : out std_logic;
           out_ack  : in  std_logic;
           out-data : out T );
  end ShiftReg;

  architecture behav of ShiftReg is
  begin
    process
      variable x, y: T;
    begin
      loop
        out_req <= '1' ;              --  out!y
        out_data <= y ;
        wait until out_ack = '1';
        out_req <= '0';
        wait until out_ack = '0';
        y := x;                       --  y := x
        wait until in_req = '1';      --  in?x
        x := in_data;
        in.ack <= '1';
        wait until ch_req = '0';
        ch_ack <= '0';
      end loop;
    end process;
  end behav;
```

Figure 8.10. VHDL description of the 2-place shift register FIFO stage from figure 8.2.

- Many real-world systems include both synchronous and asynchronous subsystems, and such hybrid systems can be modeled without any problems in VHDL.

8.6.3 Channel communication and design flow

The design flow presented in what follows is motivated by the advantages mentioned above. The goal is to augment VHDL with CSP-like channel communication primitives, i.e. the procedures `send(<channel>, <variable>)` and `receive(<channel>,<variable>)` and the function `probe(<channel>)`. Another goal is to enable mixed-mode simulations where one end of a channel connects to an entity whose architecture body is a circuit implementation and the other end connects to an entity whose architecture body is a behavioural description using the above communication primitives, figure 8.11(b). In this way

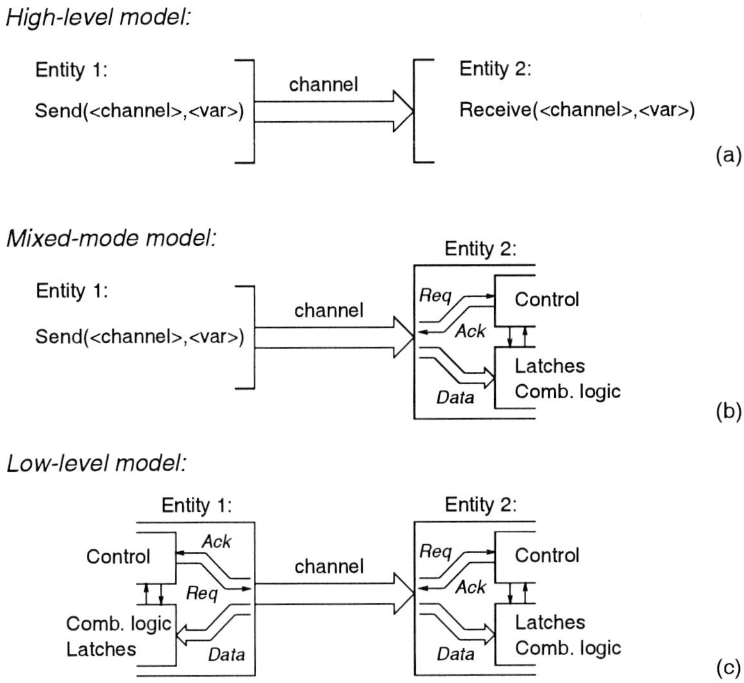

Figure 8.11. The VHDL packages for channel communication support high-level, mixed-mode and gate-level/standard cell simulations.

a *manual* top-down stepwise refinement design process is supported, where the same test bench is used throughout the entire design process from high-level specification to low-level circuit implementation, figure 8.11(a-c).

In VHDL all communication between processes takes place via signals. Channels therefore have to be declared as signals, preferably one signal per channel. Since (for a push channel) the sender drives the request and data part of a channel, and the receiver drives the acknowledge part, there are two drivers to one signal. This is allowed in VHDL if the signal is a resolved signal. Thus, it is possible to define a channel type as a record with a request, an acknowledge and a data field, and then define a resolution function for the channel type which will determine the resulting value of the channel. This type of channel, with separate request and acknowledge fields, will be called a *real channel* and is described in section 8.6.5. In simulations there will be three traces for each channel, showing the waveforms of request and acknowledge along with the data that is communicated.

A channel can also be defined with only two fields: one that describes the state of the handshaking (called the "handshake phase" or simply the "phase") and one containing the data. The type of the phase field is an enumerated type, whose values can be the handshake phases a channel can assume, as well as the values with which the sender and receiver can drive the field. This type of channel will be called an *abstract channel.* In simulations there will be two traces for each channel, and it is easy to read the phases the channel assumes and the data values that are transfered.

The procedures and definitions are organized into two VHDL-packages: one called "abstpack.vhd" that can be used for simulating high-level models and one called "realpack.vhd" that can be used at all levels of design. Full listings can be found in appendix 8.A at the end of this chapter. The design flow enabled by these packages is as follows:

- The circuit and its environment or test bench is first modelled and simulated using abstract channels. All it takes is the following statement in the top level design unit: "`usepackage work.abstpack.all`".
- The circuit is then partitioned into simpler entities. The entities still communicate using channels and the simulation still uses the abstract channel package. This step may be repeated.
- At some point the designer changes to using the real channel package by changing to: "`usepackage work.realpack.all`" in the top-level design unit. Apart from this simple change, the VHDL source code is identical.
- It is now possible to partition entities into control circuitry (that can be designed as explained in chapter 6) and data circuitry (that consist of ordinary latches and combinational circuitry). Mixed mode simulations as illustrated in figure 8.11(b) are possible. Simulation models of the control circuits may be their actual implementation in the target technology or simply an entity containing a set of concurrent signal assignments – for example the Boolean equations produced by Petrify.
- Eventually, when all entities have been partitioned into control and data, and when all leaf entities have been implemented using components of the target technology, the design is complete. Using standard technology mapping tools an implementation may be produced, and the circuit can be simulated with back annotated timing information.

Note that the same simulation test bench can be used throughout the entire design process from the high-level specification to the low-level implementation using components from the target technology.

8.6.4 The abstract channel package

An abstract channel is defined in figure 8.12 with a data type called `fp` (a 32-bit standard logic vector representing an IEEE floating-point number). The

```
type handshake_phase is
(
  u,            -- uninitialized
  idle,         -- no communication
  swait,        -- sender waiting
  rwait,        -- receiver waiting
  rcv,          -- receiving data
  rec1,         -- recovery phase 1
  rec2,         -- recovery phase 2
  req,          -- request signal
  ack,          -- acknowledge signal
  error         -- protocol error
);

subtype fp is std_logic_vector(31 downto 0);

type uchannel_fp is
  record
    phase : handshake_phase;
    data  : fp;
  end record;

type uchannel_fp_vector is array(natural range <>) of uchannel_fp;

function resolved(s : uchannel_fp_vector) return uchannel_fp;

subtype channel_fp is resolved uchannel_fp;
```

Figure 8.12. **Definition of an abstract channel.**

actual channel type is called `channel_fp`. It is necessary to define a channel for each data type used in the design. The data type can be an arbitrary type, including record types, but it is advisable to use data types that are built from `std_logic` because this is typically the type used by target component libraries (such as standard cell libraries) that are eventually used for the implementation.

The meaning of the values of the type `handshake_phase` are described in detail below:

u: Uninitialized channel. This is the default value of the drivers. As long as either the sender or receiver drive the channel with this value, the channel stays uninitialized.

idle: No communication. Both the sender and receiver drive the channel with the `idle` value.

swait: The sender is waiting to perform a communication. The sender is driving the channel with the `req` value and the receiver drives with the `idle` value.

rwait: The receiver is waiting to perform a communication. The sender is driving the channel with the `idle` value and the receiver drives with the `rwait` value. This value is used both as a driving value and as a resulting value for a channel, just like the `idle` and `u` values.

rcv: Data is transfered. The sender is driving the channel with the `req` value and the receiver drives it with the `rwait` value. After a predefined amount of time (`tpd` at the top of the package, see later in this section) the receiver changes its driving value to `ack`, and the channel changes its phase to `rec1`. In a simulation it is only possible to see the transfered value during the `rcv` phase and the `swait phase`. At all other times the data field assumes a predefined default data value.

rec1: Recovery phase. This phase is not seen in a simulation, since the channel changes to the `rec2` phase with no time delay.

rec2: Recovery phase. This phase is not seen in a simulation, since the channel changes to the `idle` phase with no time delay.

req: The sender drives the channel with this value, when it wants to perform a communication. A channel can never assume this value.

ack: The receiver drives the channel with this value when it wants to perform a communication. A channel can never assume this value.

error: Protocol error. A channel assumes this value when the resolution function detects an error. It is an error if there is more than one driver with an `rwait`, `req` or `ack` value. This could be the result if more than two drivers are connected to a channel, or if a `send` command is accidentally used instead of a `receive` command or vice versa.

Figure 8.13 shows a graphical illustration of the protocol of the abstract channel. The values in large letters are the resulting values of the channel, and the values in smaller letters below them are the driving values of the sender and receiver respectively. Both the sender and receiver are allowed to initiate a communication. This makes it possible in a simulation to see if either the

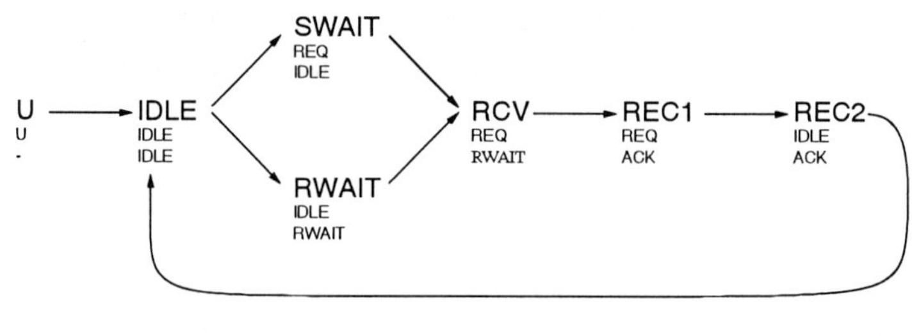

Figure 8.13. The protocol for the abstract channel. The values in large letters are the resulting resolved values of the channel, and the values in smaller letters below them are the driving values of the sender and receiver respectively.

sender or receiver is waiting to communicate. It is the procedures `send` and `receive` that follow this protocol.

Because channels with different data types are defined as separate types, the procedures `send`, `receive` and `probe` have to be defined for each of these channel types. Fortunately VHDL allows overloading of procedure names, so it is possible to make these definitions. The only differences between the definitions of the channels are the data types, the names of the channel types and the default values of the data fields in the channels. So it is very easy to copy the definitions of one channel to make a new channel type. It is not necessary to redefine the type `handshake_phase`. All these definitions are conveniently collected in a VHDL package. This package can then be referenced wherever needed. An example of such a package with only one channel type can be seen in appendix A.1. The procedures `initialize_in` and `initialize_out` are used to initialize the input and output ends of a channel. If a sender or receiver does not initialize a channel, no communications can take place on that channel.

```
library IEEE;
use IEEE.std_logic_1164.all;
use work.abstract_channels.all;

entity fp_latch is
  generic(delay : time);
  port ( d      : inout channel_fp;    -- input data channel
  port ( q      : inout channel_fp;    -- output data channel
         resetn : in std_logic        );
end fp_latch;

architecture behav of fp_latch is
begin

  process
    variable data : fp;
  begin
    initialize_in(d);
    initialize_out(q);
    wait until resetn = '1';
    loop
      receive(d, data);
      wait for delay;
      send(q, data);
    end loop;
  end process;

end behav;
```

Figure 8.14. Description of a FIFO stage.

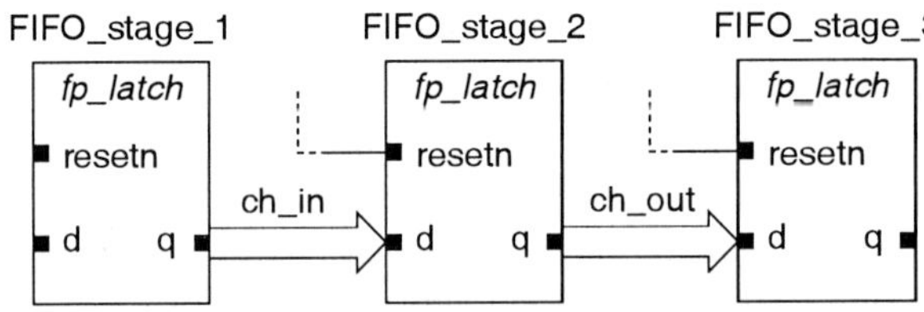

Figure 8.15. A FIFO built using the latch defined in figure 8.14.

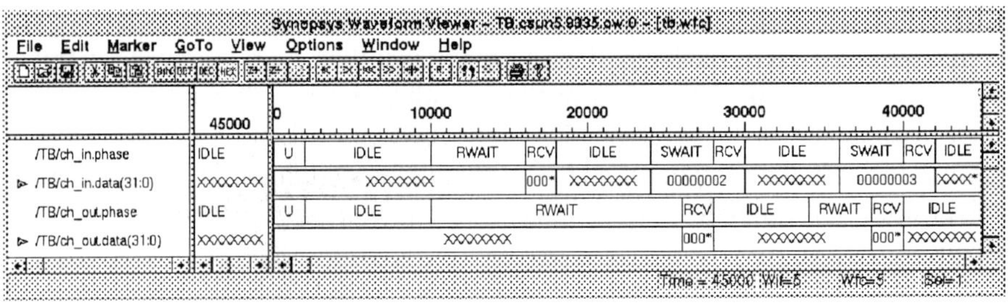

Figure 8.16. Simulation of the FIFO using the abstract channel package.

A simple example of a subcircuit is the FIFO stage `fp_latch` shown in figure 8.14. Notice that the channels in the entity have the mode `inout`, and the FIFO stage waits for the reset signal `resetn` after the initialization. In that way it waits for other subcircuits which may actually use this reset signal for initialization.

The FIFO stage uses a generic parameter `delay`. This delay is inserted for experimental reasons in order to show the different phases of the channels. Three FIFO stages are connected in a pipeline (figure 8.15) and fed with data values. The middle section has a delay that is twice as long as the other two stages. This will result in a blocked channel just before the slow FIFO stage and a starved channel just after the slow FIFO stage.

The result of this experiment can be seen in figure 8.16. The simulator used is the Synopsys VSS. It is seen that `ch_in` is predominantly in the `swait` phase, which characterizes a blocked channel, and `ch_out` is predominantly in the `rwait` phase, which characterizes a starved channel.

8.6.5 The real channel package

At some point in the design process it is time to separate communicating entities into control and data entities. This is supported by the real channel types, in which the request and acknowledge signals are separate `std_logic` signals – the type used by the target component models. The data type is the same as the abstract channel type, but the handshaking is modeled differently. A real channel type is defined in figure 8.17.

```
subtype fp is std_logic_vector(31 downto 0);

type uchannel_fp is
  record
    req  : std_logic;
    ack  : std_logic;
    data : fp;
  end record;

type uchannel_fp_vector is array(natural range <>) of
       uchannel_fp;

function resolved(s : uchannel_fp_vector) return uchannel_fp;

subtype channel_fp is resolved uchannel_fp;
```

Figure 8.17. **Definition of a real channel.**

All definitions relating to the real channels are collected in a package (similar to the abstract channel package) and use the same names for the channel types, procedures and functions. For this reason it is very simple to switch to simulating using real channels. All it takes is to change the name of the package in the `use` statements in the top level design entity. Alternatively, one can use the same name for both packages, in which case it is the last analyzed package that is used in simulations.

An example of a real channel package with only one channel type can be seen in appendix A.2. This package defines a 32-bit standard logic 4-phase bundled-data push channel. The constant `tpd` in this package is the delay from a transition on the request or acknowledge signal to the response to this transition. "Synopsys compiler directives" are inserted in several places in the package. This is because Synopsys needs to know the channel types and the resolution functions belonging to them when it generates an EDIF netlist to the floor planner, but not the procedures in the package.

Figure 8.18 shows the result of repeating the simulation experiment from the previous section, this time using the real channel package. Notice the sequence of four-phase handshakes.

Note that the data value on a channel is, at all times, whatever value the sender is driving onto the channel. An alternative would be to make the resolution function put out the default data value outside the data-validity period, but this may cause the setup and hold times of the latches to be violated. The procedure `send` provides a broad data-validity scheme, which means that it can communicate with receivers that require early, broad or late data-validity schemes on the channel. The procedure `receive` requires an early data-validity scheme,

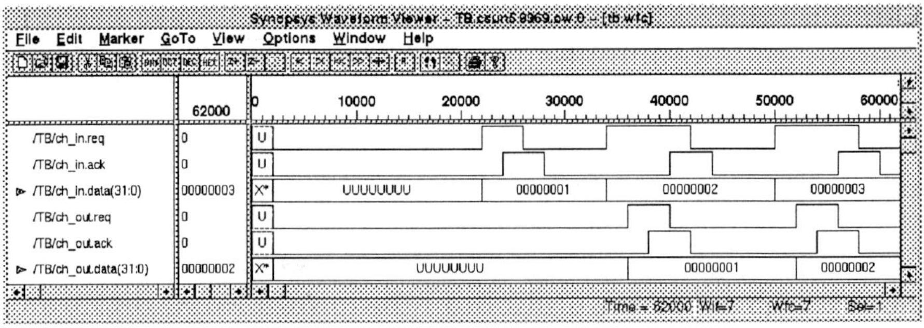

Figure 8.18. **Simulation of the FIFO using the real channel package.**

which means that it can communicate with senders that provide early or broad data-validity schemes.

The resolution functions for the real channels (and the abstract channels) can detect protocol errors. Examples of errors are more than one sender or receiver on a channel, and using a `send` command or a `receive` command at the wrong end of a channel. In such cases the channel assumes the `X` value on the request or acknowledge signals.

8.6.6 Partitioning into control and data

This section describes how to separate an entity into control and data entities. This is possible when the real channel package is used but, as explained below, this partitioning has to follow certain guidelines.

To illustrate how the partitioning is carried out, the FIFO stage in figure 8.14 in the preceding section will be separated into a latch control circuit called `latch_ctrl` and a latch called `std_logic_latch`. The VHDL code is shown in figure 8.19, and figure 8.20 is a graphical illustration of the partitioning that includes the unresolved signals `ud` and `uq` as explained below.

In VHDL a driver that drives a compound resolved signal has to drive all fields in the signal. Therefore a control circuit cannot drive only the acknowledge field in a channel. To overcome this problem a signal of the corresponding unresolved channel type has to be declared inside the partitioned entity. This is the function of the signals `ud` and `uq` of type `uchannel_fp` in figure 8.17. The control circuit then drives only the acknowledge field in this signal; this is allowed since the signal is unresolved. The rest of the fields remain uninitialized. The unresolved signal then drives the channel; this is allowed since it drives all of the fields in the channel. The resolution function for the channel should ignore the uninitialized values that the channel is driven with. Components that use the `send` and `receive` procedures also drive those fields in the

```
library IEEE;
use IEEE.std_logic_1164.all;
use work.real_channels.all;

entity fp_latch is
  port ( d      : inout channel_fp;     -- input data channel
         q      : inout channel_fp;     -- output data channel
         resetn : in std_logic        );
end fp_latch;

architecture struct of fp_latch is

  component latch_ctrl
    port ( rin, aout, resetn : in  std_logic;
           ain, rout, lt : out std_logic      );
  end component;

  component std_logic_latch
    generic (width : positive);
    port ( lt : in  std_logic;
           d  : in  std_logic_vector(width-1 downto 0);
           q  : out std_logic_vector(width-1 downto 0)  );
  end component;

  signal lt : std_logic;
  signal ud, uq : uchannel_fp;

begin

  latch_ctrl1 : latch_ctrl
    port map (d.req,q.ack,resetn,ud.ack,uq.req,lt);
  std_logic_latch1 : std_logic_latch
    generic map (width => 32)
    port map (lt,d.data,uq.data);

  d <= connect(ud);
  q <= connect(uq);

end struct;
```

Figure 8.19. **Separation of the FIFO stage into an ordinary data latch and a latch control circuit.**

channel that they do not control with uninitialized values. For example, an output to a channel drives the acknowledge field in the channel with the `U` value. The fields in a channel that are used as inputs are connected directly from the channel to the circuits that have to read those fields.

Notice in the description that the signals `ud` and `uq` do not drive `d` and `q` directly but through a function called `connect`. This function simply returns its parameter. It may seem unnecessary, but it has proved to be necessary when some of the subcircuits are described with a standard cell implementation. In a simulation a special "gate-level simulation engine" is used to simulate the

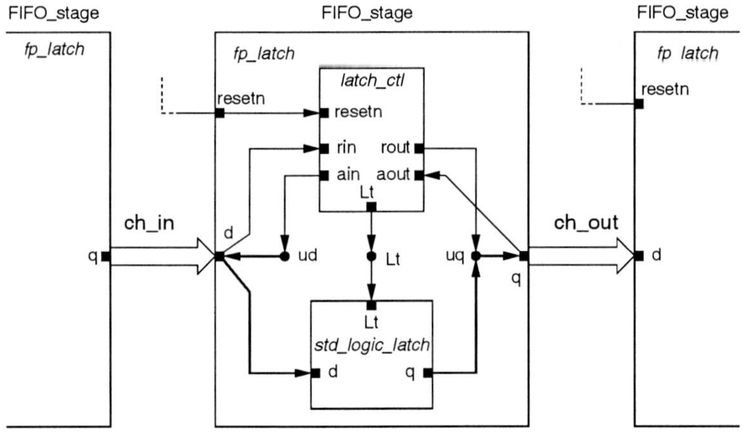

Figure 8.20. **Separation of control and data.**

standard cells [129]. During initialization it will set some of the signals to the value `X` instead of to the value `U` as it should. It has not been possible to get the channel resolution function to ignore these `X` values, because the gate-level simulation engine sets some of the values in the channel. By introducing the `connect` function, which is a behavioural description, the normal simulator takes over and evaluates the channel by means of the corresponding resolution function. It should be emphasized that it is a bug in the gate-level simulation engine that necessitates the addition of the `connect` function.

8.7. Summary

This chapter addressed languages and CAD tools for high-level modeling and synthesis of asynchronous circuits. The text focused on a few representative and influential design methods that are based languages that are similar to CSP. The reason for preferring these languages are that they support channel based communication between processes (synchronous message passing) as well as concurrency at both process and statement level – two features that are important for modeling asynchronous circuits. The text also illustrated a synthesis method known as syntax directed translation. Subsequent chapters in this book will elaborate much more on these issues.

Finally the chapter illustrated how channel based communication can be implemented in VHDL, and we provided two packages containing all the necessary procedures and functions including: `send`, `receive` and `probe`. These packages supports a *manual* top-down stepwise-refinement design flow where the same test bench can be used to simulate the design throughout the entire design process from high level specification to low level circuit implementation.

This chapter on languages and CAD-tools for asynchronous design concludes the tutorial on asynchronous circuit design and it it time to wrap up: Chapter 2 presented the fundamental concepts and theories, and provided pointers to the literature. Chapters 3 and 4 presented an RTL-like abstract view on asynchronous circuits (tokens flowing in static data-flow structures) that is very useful for understanding their operation and performance. This material is probably where this tutorial supplements the existing body of literature the most. Chapters 5 and 6 addressed the design of datapath operators and control circuits. Focus in chapter 6 was on speed-independent circuits, but this is not the only approach. In recent years there has also been great progress in synthesizing multiple-input-change fundamental-mode circuits. Chapter 7 discussed more advanced 4-phase bundled-data protocols and circuits. Finally chapter 8 addressed languages and tools for high-level modeling and synthesis of asynchronous circuits.

The tutorial deliberately made no attempts at covering of all corners of the field – the aim was to pave a road into "the world of asynchronous design". Now you are here at the end of the road; hopefully with enough background to carry on digging deeper into the literature, and equally importantly, with sufficient understanding of the characteristics of asynchronous circuits, that you can start designing your own circuits. And finally; asynchronous circuits do not represent an alternative to synchronous circuits. They have advantages in some areas and disadvantages in other areas and they should be seen as a supplement, and as such they add new dimensions to the solution space that the digital designer explores. Even today, many circuits can not be categorized as either synchronous or asynchronous, they contain elements of both.

The following chapters will introduce some recent industrial scale asynchronous chips. Additional designs are presented in [106].

Appendix: The VHDL channel packages

A.1. The abstract channel package

```
-- Abstract channel package: (4-phase bundled-data push channel, 32-bit data)

library IEEE;
use IEEE.std_logic_1164.all;

package abstract_channels is

  constant tpd : time := 2 ns;

-- Type definition for abstract handshake protocol

  type handshake_phase is
  (
    u,          -- uninitialized
    idle,       -- no communication
    swait,      -- sender waiting
    rwait,      -- receiver waiting
    rcv,        -- receiving data
    rec1,       -- recovery phase 1
    rec2,       -- recovery phase 2
    req,        -- request signal
    ack,        -- acknowledge signal
    error       -- protocol error
  );

-- Floating point channel definitions

  subtype fp is std_logic_vector(31 downto 0);

  type uchannel_fp is
    record
      phase : handshake_phase;
      data  : fp;
    end record;

  type uchannel_fp_vector is array(natural range <>) of
        uchannel_fp;

  function resolved(s : uchannel_fp_vector) return uchannel_fp;

  subtype channel_fp is resolved uchannel_fp;

  procedure initialize_in(signal ch : out channel_fp);

  procedure initialize_out(signal ch : out channel_fp);

  procedure send(signal ch : inout channel_fp; d : in fp);

  procedure receive(signal ch : inout channel_fp; d : out fp);

  function probe(signal ch : in channel_fp) return boolean;

end abstract_channels;
```

```
package body abstract_channels is

-- Resolution table for abstract handshake protocol

  type table_type is array(handshake_phase, handshake_phase) of
         handshake_phase;

  constant resolution_table : table_type := (
--------------------------------------------------------------------------
-- 2. parameter:                                              |        |
-- u     idle  swait rwait rcv   rec1  rec2  req   ack   error   |1. par:|
--------------------------------------------------------------------------
  (u,    u,    u,    u,    u,    u,    u,    u,    u,    u    ), --| u      |
  (u,    idle, swait,rwait,rcv,  rec1, rec2, swait,rec2, error), --| idle   |
  (u,    swait,error,rcv,  error,error,rec1, error,rec1, error), --| swait  |
  (u,    rwait,rcv,  error,error,error,error,rcv,  error,error), --| rwait  |
  (u,    rcv,  error,error,error,error,error,error,error,error), --| rcv    |
  (u,    rec1, error,error,error,error,error,error,error,error), --| rec1   |
  (u,    rec2, rec1, error,error,error,error,rec1, error,error), --| rec2   |
  (u,    error,error,error,error,error,error,error,error,error), --| req    |
  (u,    error,error,error,error,error,error,error,error,error), --| ack    |
  (u,    error,error,error,error,error,error,error,error,error));--| error  |

-- Fp channel

  constant default_data_fp : fp := "XXXXXXXXXXXXXXXXXXXXXXXXXXXXXXXX";

  function resolved(s : uchannel_fp_vector) return uchannel_fp is
    variable result : uchannel_fp := (idle, default_data_fp);
  begin
    for i in s'range loop
      result.phase := resolution_table(result.phase, s(i).phase);
      if (s(i).phase = req) or (s(i).phase = swait) or
         (s(i).phase = rcv) then
           result.data := s(i).data;
      end if;
    end loop;
    if not((result.phase = swait) or (result.phase = rcv)) then
      result.data := default_data_fp;
    end if;
    return result;
  end resolved;

  procedure initialize_in(signal ch : out channel_fp) is
  begin
    ch.phase <= idle after tpd;
  end initialize_in;

  procedure initialize_out(signal ch : out channel_fp) is
  begin
    ch.phase <= idle after tpd;
  end initialize_out;

  procedure send(signal ch : inout channel_fp; d : in fp) is
  begin
    if not((ch.phase = idle) or (ch.phase = rwait)) then
      wait until (ch.phase = idle) or (ch.phase = rwait);
```

```
    end if;
    ch <= (req, d);
    wait until ch.phase = rec1;
    ch.phase <= idle;
  end send;

  procedure receive(signal ch : inout channel_fp; d : out fp) is
  begin
    if not((ch.phase = idle) or (ch.phase = swait)) then
      wait until (ch.phase = idle) or (ch.phase = swait);
    end if;
    ch.phase <= rwait;
    wait until ch.phase = rcv;
    wait for tpd;
    d := ch.data;
    ch.phase <= ack;
    wait until ch.phase = rec2;
    ch.phase <= idle;
  end receive;

  function probe(signal ch : in channel_fp) return boolean is
  begin
    return (ch.phase = swait);
  end probe;

end abstract_channels;
```

A.2. The real channel package

```
-- Low-level channel package (4-phase bundled-data push channel, 32-bit data)

library IEEE;
use IEEE.std_logic_1164.all;

package real_channels is

  -- synopsys synthesis_off
  constant tpd : time := 2 ns;
  -- synopsys synthesis_on

-- Floating point channel definitions

  subtype fp is std_logic_vector(31 downto 0);

  type uchannel_fp is
    record
      req  : std_logic;
      ack  : std_logic;
      data : fp;
    end record;

  type uchannel_fp_vector is array(natural range <>) of
        uchannel_fp;

  function resolved(s : uchannel_fp_vector) return uchannel_fp;
```

```
  subtype channel_fp is resolved uchannel_fp;

  -- synopsys synthesis_off
  procedure initialize_in(signal ch : out channel_fp);

  procedure initialize_out(signal ch : out channel_fp);

  procedure send(signal ch : inout channel_fp; d : in fp);

  procedure receive(signal ch : inout channel_fp; d : out fp);

  function probe(signal ch : in uchannel_fp) return boolean;
  -- synopsys synthesis_on

  function connect(signal ch : in uchannel_fp) return channel_fp;

end real_channels;

package body real_channels is

-- Resolution table for 4-phase handshake protocol

  -- synopsys synthesis_off
  type stdlogic_table is array(std_logic, std_logic) of std_logic;

  constant resolution_table : stdlogic_table := (
  --  -------------------------------------------------------------
  --  | 2. parameter:                                      |       |
  --  |  U    X    0    1    Z    W    L    H    -         |1. par:|
  --  -------------------------------------------------------------
      ( 'U', 'X', '0', '1', 'X', 'X', 'X', 'X', 'X' ),  -- |   U   |
      ( 'X', 'X', 'X', 'X', 'X', 'X', 'X', 'X', 'X' ),  -- |   X   |
      ( '0', 'X', 'X', 'X', 'X', 'X', 'X', 'X', 'X' ),  -- |   0   |
      ( '1', 'X', 'X', 'X', 'X', 'X', 'X', 'X', 'X' ),  -- |   1   |
      ( 'X', 'X', 'X', 'X', 'X', 'X', 'X', 'X', 'X' ),  -- |   Z   |
      ( 'X', 'X', 'X', 'X', 'X', 'X', 'X', 'X', 'X' ),  -- |   W   |
      ( 'X', 'X', 'X', 'X', 'X', 'X', 'X', 'X', 'X' ),  -- |   L   |
      ( 'X', 'X', 'X', 'X', 'X', 'X', 'X', 'X', 'X' ),  -- |   H   |
      ( 'X', 'X', 'X', 'X', 'X', 'X', 'X', 'X', 'X' )); -- |   -   |
  -- synopsys synthesis_on

-- Fp channel

  -- synopsys synthesis_off
  constant default_data_fp : fp := "XXXXXXXXXXXXXXXXXXXXXXXXXXXXXXXX";
  -- synopsys synthesis_on

  function resolved(s : uchannel_fp_vector) return uchannel_fp is
  -- pragma resolution_method three_state
  -- synopsys synthesis_off
    variable result : uchannel_fp := ('U','U',default_data_fp);
  -- synopsys synthesis_on
  begin
  -- synopsys synthesis_off
    for i in s'range loop
      result.req := resolution_table(result.req,s(i).req);
      result.ack := resolution_table(result.ack,s(i).ack);
```

```
      if (s(i).req = '1') or (s(i).req = '0') then
        result.data := s(i).data;
      end if;
    end loop;
    if not((result.req = '1') or (result.req = '0')) then
      result.data := default_data_fp;
    end if;
    return result;
  -- synopsys synthesis_on
  end resolved;

  -- synopsys synthesis_off
  procedure initialize_in(signal ch : out channel_fp) is
  begin
    ch.ack <= '0' after tpd;
  end initialize_in;

  procedure initialize_out(signal ch : out channel_fp) is
  begin
    ch.req <= '0' after tpd;
  end initialize_out;

  procedure send(signal ch : inout channel_fp; d : in fp) is
  begin
    if ch.ack /= '0' then
      wait until ch.ack = '0';
    end if;
    ch.req <= '1' after tpd;
    ch.data <= d after tpd;
    wait until ch.ack = '1';
    ch.req <= '0' after tpd;
  end send;

  procedure receive(signal ch : inout channel_fp; d : out fp) is
  begin
    if ch.req /= '1' then
      wait until ch.req = '1';
    end if;
    wait for tpd;
    d := ch.data;
    ch.ack <= '1';
    wait until ch.req = '0';
    ch.ack <= '0' after tpd;
  end receive;

  function probe(signal ch : in uchannel_fp) return boolean is
  begin
    return (ch.req = '1');
  end probe;
  -- synopsys synthesis_on

  function connect(signal ch : in uchannel_fp) return channel_fp is
  begin
    return ch;
  end connect;

end real_channels;
```

II

BALSA - AN ASYNCHRONOUS HARDWARE SYNTHESIS SYSTEM

Author: Doug Edwards, Andrew Bardsley

Department of Computer Science

The University of Manchester

{doug,bardsley}@cs.man.ac.uk

Abstract Balsa is a system for describing and synthesising asynchronous circuits based on syntax-directed compilation into communicating handshake circuits. In these chapters, the basic Balsa design flow is described and several simple circuit examples are used to illustrate the Balsa language in an informal tutorial style. The section concludes with a walk-through of a major design exercise – a 4 channel DMA controller described entirely in Balsa.

Keywords: asynchronous circuits, high-level synthesis

Chapter 9

AN INTRODUCTION TO BALSA

9.1. Overview

Balsa is both a framework for synthesising asynchronous hardware systems and a language for describing such systems. The approach adopted is that of syntax-directed compilation into communicating handshaking components and closely follows the Tangram system ([141, 135] and Chapter 13 on page 221) of Philips. The advantage of this approach is that the compilation is transparent: there is a one-to-one mapping between the language constructs in the specification and the intermediate handshake circuits that are produced. It is relatively easy for an experienced user to envisage the micro-architecture of the circuit that results from the original description. Incremental changes made at the language level result in predictable changes at the circuit implementation level. This is important if optimisations and design trade-offs are to be made easily and contrasts with synchronous VHDL synthesis in which small changes in the specification may make radical alterations to the resulting circuit.

It is important to understand what Balsa offers the designer and what obligations are still placed upon the designer. The tight "edit description – synthesise – simulate – revise description" loop made possible by the fast compilation process makes it very easy for the design space of a system to be explored and prototypes rapidly evaluated. However, there is no substitute for creativity. Poor designs may be created as easily as elegant designs and some experience in designing asynchronous circuits is required before even a good designer of conventional clocked circuits will best be able to exploit the system. Be warned that although Balsa guarantees correct-by-construction circuits, it does not guarantee correct systems. In particular, it is quite feasible, as in any asynchronous system, to describe an elegant circuit which will exhibit deadlock. Furthermore, post-layout simulation is still required in order to check that when the instantiated circuit has been placed and routed by conventional CAD tools, it meets basic timing requirements. On the other hand, a choice of implementation libraries is available allowing the designer to trade

J. Sparsø and S. Furber (eds.), Principles of Asynchronous Circuit Design, 155–171.

the greater process portability of a delay-insensitive implementation against, perhaps, smaller circuit area which may require a larger post-layout validation effort.

Although Balsa has evolved from a research environment, it is not a toy system unsuited for large-scale designs; Balsa has been used to synthesise the 32 channel DMA controller [11] for the Amulet3i asynchronous microprocessor macro-cell [48]. The controller has a complex specification and the resulting implementation occupies $2mm^2$ on a $0.35\mu m$ 3-layer metal process. Balsa is at the time of writing being used to synthesise a complete Amulet core as part of the EU funded G3 smartcard project [46].

As noted earlier, Balsa is very similar to Tangram. It is a less mature package lacking some useful tools contained within the Tangram package such as the power performance analyser. However, Balsa is freely available whereas Tangram is not generally available outside Philips. As far as the expressiveness of the languages is concerned, Balsa adds powerful parameterisation using recursive expansion definition facilities whereas Tangram allows more flexibility in interacting with non delay-insensitive external interfaces. Balsa has deliberately chosen not to add such features to ensure that its channels-only delay-insensitive model is not compromised.

The reader should be aware that not all aspects of the Balsa language or its syntax are explored in the material that follows: a more detailed introduction is available in the Balsa User Guide available from [7]. The Balsa system is freely available from the same site. The system is still evolving: the description here refers to Balsa release 3.1.0.

9.2. Basic concepts

A circuit described in Balsa is compiled into a communicating network composed from a small (about 40) set of handshake components. The components are connected by *channels* over which atomic *communications* or *handshakes* take place. Channels may have datapaths associated with them (in which case a handshake involves the transfer of data), or may be purely control (in which case the handshake acts as a synchronisation or rendezvous point).

Each channel connects exactly one *passive* port of a handshake component to one *active* port of another handshake component. An active port is a port which initiates a communication. A passive port responds (when it is ready) to the *request* from the active port by an *acknowledge* signal.

Data channels may be *push* channels or *pull* channels. In a push channel, the direction of the data flow is from the active port to the passive port. This is similar to the communication style of micropipelines. Data validity is signalled by request and released on acknowledge. In a pull channel, the direction of data flow is from the passive port to the active port. The active port requests a transfer, data validity is signalled by an acknowledge from the passive port. An example of a circuit composed from handshake components is shown in figure 9.1. Active ports are denoted by filled bubbles on a handshake component and passive ports are denoted by open bubbles.

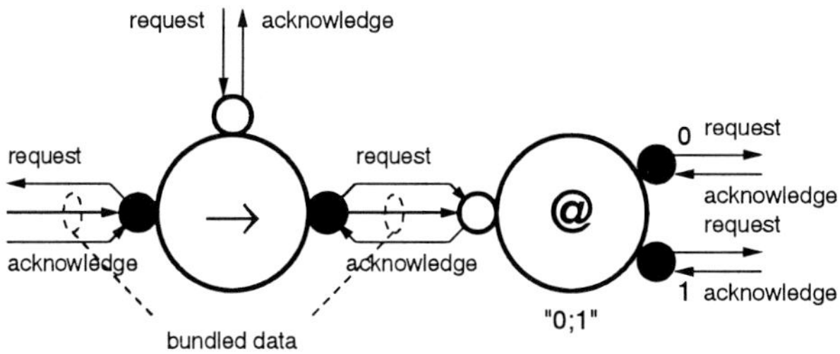

Figure 9.1. Two connected handshake components.

Here, a *Fetch* component, or *Transferrer*, denoted by "$\rightarrow$ ") and a *Case* component (denoted by "@") are connected by an internal data-bearing channel. Circuit action is activated by a request to the Transferrer which in turn issues a request to the environment on its active pull input port (on the left of the diagram). The environment supplies the demanded data indicating its validity by the acknowledgement signal. The Transferrer then presents a handshake request and data to the Case component on its active push output port which the Case component receives on a passive port. Depending on the data value, the Case component issues a handshake to its environment on either the top right or bottom right port. Finally, when the acknowledgement is received by the Case component, an acknowledgement is returned along the original channel and terminating this handshake. The circuit is ready to operate once more.

Data follows the direction of the request in this example and the acknowledgement to that request flows in the opposite direction. In this figure, individual physical request, acknowledgement and data wires are explicitly shown. Data is carried on separate wires from the signalling (it is "bundled" with the control) although this is not necessarily true for other data/signalling encoding schemes.

The bundled-data scheme illustrated in figure 9.1 is not the only implementation possible. Methodologies exist to implement channel connections with delay-insensitive signalling where timing relationships between individual wires of an implemented channel do not affect the functionality of the circuit. Handshake circuits can be implemented using these methodologies which are robust to naïve realisations, process variations and interconnect delay properties. Future releases of Balsa will include several alternative back-ends. A more detailed discussion of handshake protocols can be found in section 2.1 on page 9 and section 7.1 on page 115.

Normally, handshake circuit diagrams are not shown at the level of detail of figure 9.1, a channel usually being shown as a single arc with the direction of data being denoted by an arrow head on the arc. Similarly, control only channels, comprising only request/acknowledge wires, are indicated by an arc without an arrowhead. The circuit complexity of handshake circuits is often low: for example, a Transferrer may be implemented using only wires. An example of a handshake circuit for a modulo-10 counter (see page 185) is shown in figure 9.2. The corresponding gate-level implementation is shown in figure 9.3.

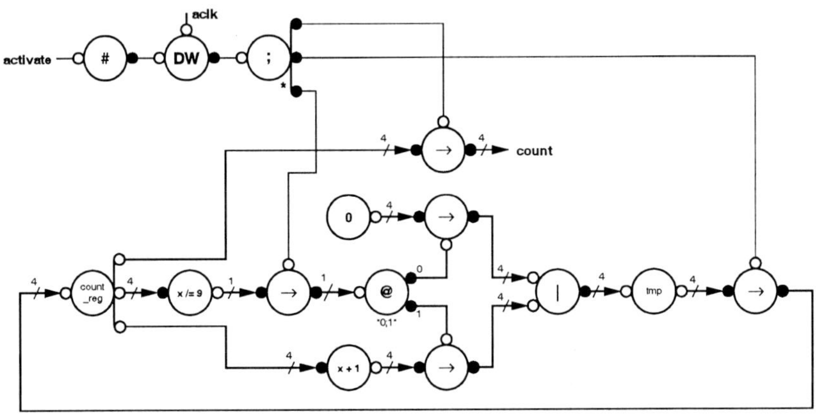

Figure 9.2. Handshake circuit of a modulo-10 counter.

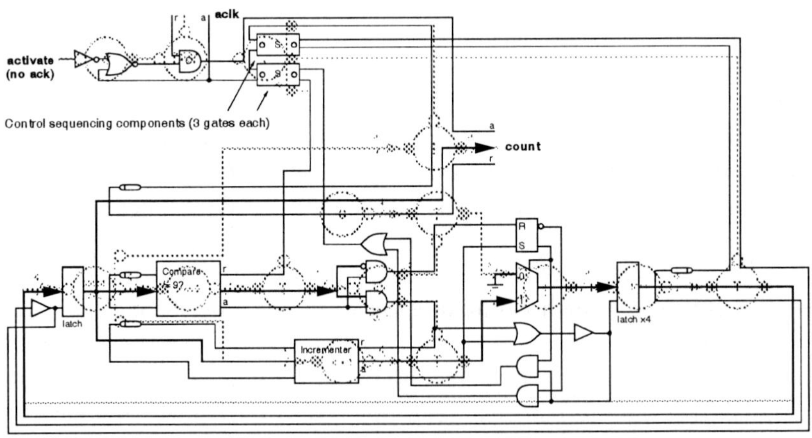

Figure 9.3. Gate-level circuit of a modulo-10 counter.

9.3. Tool set and design flow

An overview of the Balsa design flow is shown in figure 9.4. Behavioural simulation is provided by LARD [38], a language developed within the Amulet group for modelling asynchronous systems. However, the target CAD system can also be used to perform more accurate simulations and to validate the design. Most of the Balsa tools are concerned with manipulating the Breeze handshake intermediate files produced by compiling Balsa descriptions. Breeze files can be used by back-end tools to provide implementations for Balsa descriptions, but also contain procedure and type definitions passed on from Balsa source files allowing Breeze to be used as the package description format for Balsa.

The Balsa system comprises the following collection of tools:

- *balsa-c*: the compiler for the Balsa language. The compiler produces *Breeze* from Balsa descriptions.
- *balsa-netlist*: produces a netlist, currently EDIF, Compass or Verilog, from a Breeze description, performing technology mapping and handshake expansion.
- *breeze2ps*: a tool which produces a PostScript file of the handshake circuit graph.
- *breeze2lard*: a translator that converts a *Breeze* file to a LARD behavioural model.
- *breeze-cost*: a tool which gives an area cost estimate of the circuit.
- *balsa-md*: a tool for generating Makefiles for make(1).
- *balsa-mgr*: a graphical front-end to *balsa-md* with project management facilities.

The interfaces between the Balsa and target CAD systems are handled by the following scripts:

- *balsa-pv*: uses powerview tools to produce an EDIF file from a top-level powerview schematic which incorporates Balsa generated circuits.
- *balsa-xi*: produces a Xilinx download file from an EDIF description of a compiled circuit.
- *balsa-ihdl*: an interface to the Cadence Verilog-XL environment.

9.4. Getting started

In this section, simple buffer circuits are described in Balsa introducing the basic elements of a Balsa description.

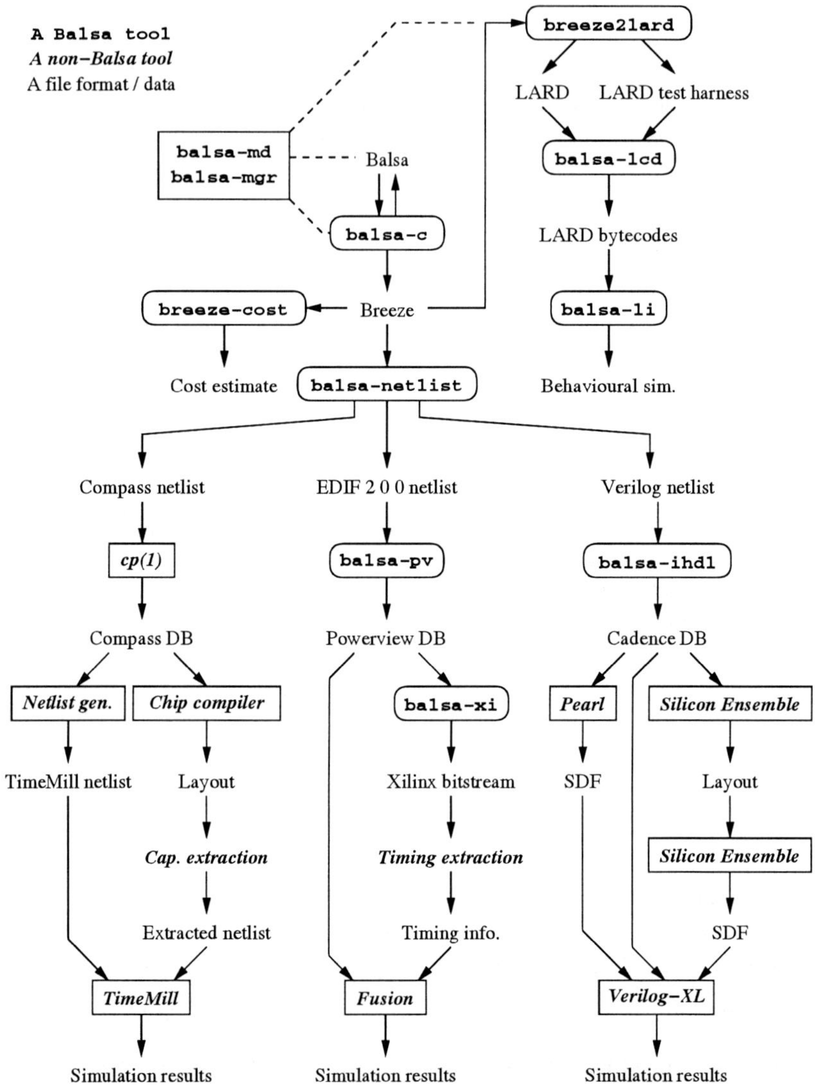

Figure 9.4. Design flow.

9.4.1 A single-place buffer

This buffer circuit is the HDL equivalent of the "*hello, world*" program. Its Balsa description is:

```
import [balsa.types.basic]
-- a single line comment
-- buffer1a: A single place buffer
procedure buffer1 (input i : byte; output o : byte) is
  variable x : byte
begin
  loop
    i -> x  -- Input  communication
    ;       -- sequence the two communications
    o <- x  -- Output communication
  end
end
```

Commentary on the code

This Balsa description builds a single-place buffer, 8-bits wide. The circuit requests a byte from the environment which, when ready, transfers the data to the register. The circuit signals to the environment on its output channel that data is available and the environment reads it when it chooses. This small program introduces:

comments: Balsa supports both multi-line comments and single-line comments.

modular compilation: Balsa supports modular compilation. The `import` statement in this example includes the definition of some standard data types such as `byte`, `nibble`, etc. The search path given in the `import` statement is a dot-separated directory path similar to that of Java (although multi-file packages are not implemented). The `import` statement may be used to include other precompiled Balsa programs thereby acting as a library mechanism. Any `import` statements must precede other declarations in the files.

procedures: The `procedure` declaration introduces an object that looks similar to a procedure definition in a conventional programming language. A Balsa procedure is a process. The parameters of the procedure define the interface to the environment outside the circuit block. In this case, the module has an 8-bit input and an 8-bit output. The body of the procedure definition defines an algorithmic behaviour for the circuit; it also implies a structural implementation. In this example, a variable `x` (of type `byte`) is declared implying that an 8-bit wide storage element will be appear in the synthesised circuit.

The behaviour of the circuit is obvious from the code: 8-bit values are transferred from the environment to the storage variable, x, and then sequentially output from the variable to the environment. This sequence of events is continually repeated (loop ... end).

channel communication: the communication operators "->" and "<-" are channel assignments and imply a communication or handshake over the channel. Because of the sequencing explicit in the description, the variable x will only accept a new value when it is ready; the value will only be passed out to the environment when requested. Note that the channel is always on the left-hand side of the operator and the corresponding variable or expression on the right-hand side.

sequencing: The ";" operator separating the two assignments is not merely a syntactic statement separator, it explicitly denotes sequentiality. The contents of x are transferred to the output port after the input transfer has completed. Because a ";" connects two sequenced statements or blocks, it is an error to place a ";" after the last statement in a block.

repetition The loop ... end construct causes infinite repetition of the code contained within its body. Procedures without loop ... end are permitted and will terminate, allowing procedure calls to be sequenced if required.

Compiling the circuit

```
balsa-c buffer1a
```

The compiler produces an output file *buffer1a.breeze*. This is a file in an intermediate format which can be imported back into other Balsa source files (thereby providing a simple library mechanism). Breeze is a textual format file designed for ease of parsing and it is therefore somewhat opaque. A primitive graphical representation of the compiled circuit in terms of handshake components can be produced (as *buffer1a.ps*) by:

```
breeze2ps buffer1a
```

The synthesised circuit

The resulting handshake circuit is shown in figure 9.5. This is not actually taken from the output of *breeze2ps*, but has been redrawn to make the diagram more readable. Although it is not necessary to understand the exact operation of the compiled circuit, a knowledge of the structure is helpful to gain an understanding of how best to describe circuits which can be synthesised efficiently using Balsa. A brief description of the operation of the circuit is given

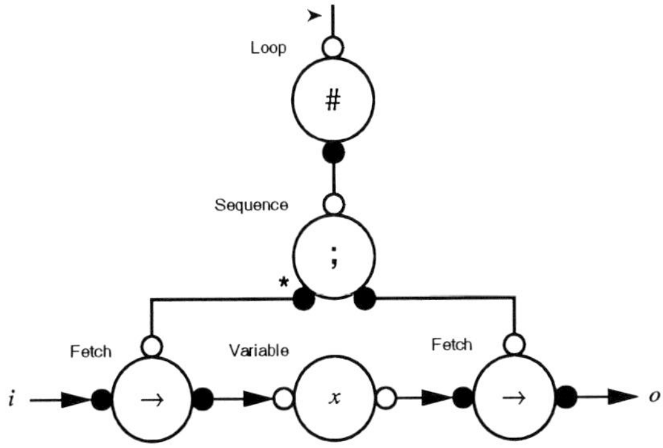

Figure 9.5. **Handshake circuit for a single-place buffer.**

below. The circuit has been annotated with the names of the various handshake components.

The port at the top, denoted by ">", is an *activation* port generating a handshake enclosing the behaviour of the circuit. It can be thought of as a reset signal which, when de-asserted, initiates the operation of the circuit. All compiled Balsa programs contain an activation port.

The activation port starts the operation of the *Repeater* ("#") which initiates a handshake with the *Sequencer*. The Repeater corresponds directly to the `loop...` end construct, and the Sequencer to the "`;`" operator. The Sequencer first issues a handshake to the left-hand *Fetch* component, causing data to be moved to the storage element in the *Variable* element. The Sequencer then handshakes with the right-hand Fetch component, causing data to be read from the Variable element. When these operations are complete, the Sequencer completes its handshake with the Repeater which starts the cycle again.

9.4.2 Two-place buffers

1st design

Having built a single-place buffer, an obvious goal is a pipeline of single buffer stages. Initially consider a two-place buffer; there are a number of ways we might describe this. One choice is to define a circuit with two storage elements:

```
-- buffer2a: Sequential 2-place buffer with assignment
--           between variables
import [balsa.types.basic]

procedure buffer2 (input i : byte; output o : byte) is
  variable x1, x2 : byte
begin
```

```
  loop
    i -> x1;     -- Input communication
    x2 := x1;    -- Implied communication
    o <- x2      -- Output communication
  end
end
```

In this example in we explicitly introduce two storage elements, x1 and x2. The contents of the variable x1 are caused to be transferred to the variable x2 by means of the assignment operator ":=". However, transfer is still effected by means of a handshaking communication channel. This assignment operator is merely a way of concealing the channel for convenience.

2nd design

The implicit channel can be made explicit as shown in *buffer2b.balsa*:

```
-- buffer2b: Sequential version with an explicit
--           internal channel
import [balsa.types.basic]

procedure buffer2 (input i:byte; output o:byte) is
  variable x1, x2 : byte
  channel chan: byte
begin
  loop
    i -> x1;                      -- Input  communication
    chan <- x1 || chan -> x2;     -- Transfer x1 to x2
    o <- x2                       -- Output communication
  end
end
```

The channel which, in the previous example, was concealed behind the use of the ":=" assignment operator, has been made explicit. The handshake circuit produced (after some simple optimisations) is identical to *buffer2a*. The "||" operator is explained in the next example.

It is important to understand the significance of the operation of the circuits produced by *buffer2a* and *buffer2b*. Remember that ";" is more than a syntactic separator: it is an operator denoting sequence. Thus, first the input, i, is transferred to x1. When this operation is complete, x1 is transferred to x2 and finally the contents of x2 are written to the environment on port o. Only after this sequence of operations is complete can new data from the environment be read into x1 again.

9.4.3 Parallel composition and module reuse

The operation above is unnecessarily constrained: there is no reason why the circuit cannot be reading a new value into x1 at the same time that x2 is writing out its data to the environment. The program in *buffer2c* achieves this optimisation.

```
-- buffer2c: a 2-place buffer using parallel composition
import [buffer1a]

procedure buffer2 (input i : byte; output o : byte) is
  channel c : byte
begin
  buffer1 (i, c) ||
  buffer1 (c, o)
end
```

Commentary on the code

In the program above, a 2-place buffer is composed from 2 single-place buffers. The output of the first buffer is connected to the input of the second buffer by their respective output and input ports. However, apart from communications across the common channel, the operation of the two buffers is independent.

The deceptively simple program above illustrates a number of new features of the Balsa language:

modular compilation: The *buffer1a* circuit is included by the import mechanism described earlier. The circuit must have been compiled previously. The Makefile generation program `balsa-md` (see page 166) can be used to generate a Makefile which will automatically take care of such dependencies.

connectivity by naming: The output of the first buffer is connected to the input of the second buffer because of the common channel name, `c`, in the parameter list in the instantiation of the buffers.

parallel composition: The "||" operator specifies that the two units which it connects should operate in parallel. This does not mean that the two units may operate totally independently: in this example, the output of one buffer writes to the input of the other buffer, creating a point of synchronisation. Note also that the parallelism referred to is a temporal parallelism. The two buffers are physically connected in series.

9.4.4 Placing multiple structures

If we wish to extend the number of places in the buffer, the previous technique of explicitly enumerating every buffer becomes tedious. What is required is a means of parameterising the buffer length (though in any real hardware implementation the number of buffers cannot be variable and must be known before-hand). One technique, shown in *buffer_n*, is to use the `for` construct together with compile-time constants:

```
-- buffer_n: an n-place parameterised buffer
import [buffer1a]
constant n = 8

procedure buffer_n (input i:byte; output o:byte)
is
  array 1 .. n-1 of channel c : byte
begin
  buffer1 (i, c[1]) ||          -- First buffer
  buffer1 (c[n-1], o) ||        -- Last buffer
  for || i in 1 .. n-2 then     -- Buffer i
    buffer1 (c[i], c[i+1])
  end
end
```

Commentary on the Code

constants: the value of an expression (of any type) may be bound to a name. The value of the expression is evaluated at compile time and the type of the name when used will be the same as the original expression in the constant declaration. Numbers can be given in decimal (starting with one of 1..9), hexadecimal (0x prefix), octal (0 prefix) and binary (0b prefix).

arrayed channels: procedure ports and locally declared channels may be arrayed. Each channel can be referred to by a numeric or enumerated index, but from the point of view of handshaking, each channel is distinct and no indexed channel has any relationship with any other such channel other than the name they share. Arraying is not part of a channel's type.

for loops: a `for` loop allows iteration over the instantiation of a subcircuit. The composition of the circuits may either be a parallel composition – as in the example above – or sequential. In the latter case, "`;`" should be substituted for "`||`" in the loop specifier. The iteration range of the loop must be resolvable at compile time.

A more flexible approach uses parameterised procedures and is discussed later in chapter 11 on page 193.

9.5. Ancillary Balsa tools

9.5.1 Makefile generation

Makefiles are commonly used in Unix by the utility make(1) to specify and control the processes by which complicated programs are compiled. Specifying the dependencies involved is often tedious and error prone. The Balsa system has a utility, `balsa-md`, to generate the Makefile for a given program automatically. The generated Makefile knows not only how to compile a Balsa module with multiple imports, but also how to generate and run test-harnesses for the simulation environment, LARD, used by Balsa. `Balsa-mgr` provides a convenient, intuitive, GUI front-end to `balsa-md` and considerably simplifies project management, in particular the handling of multiple test harnesses. However, since a textual description of any GUI is tedious, `balsa-mgr` will not be discussed further and only the facilities to which the underlying `balsa-md` provides a gateway will be described in the examples that follow. The interface presented by `balsa-mgr` is shown in figure 9.6.

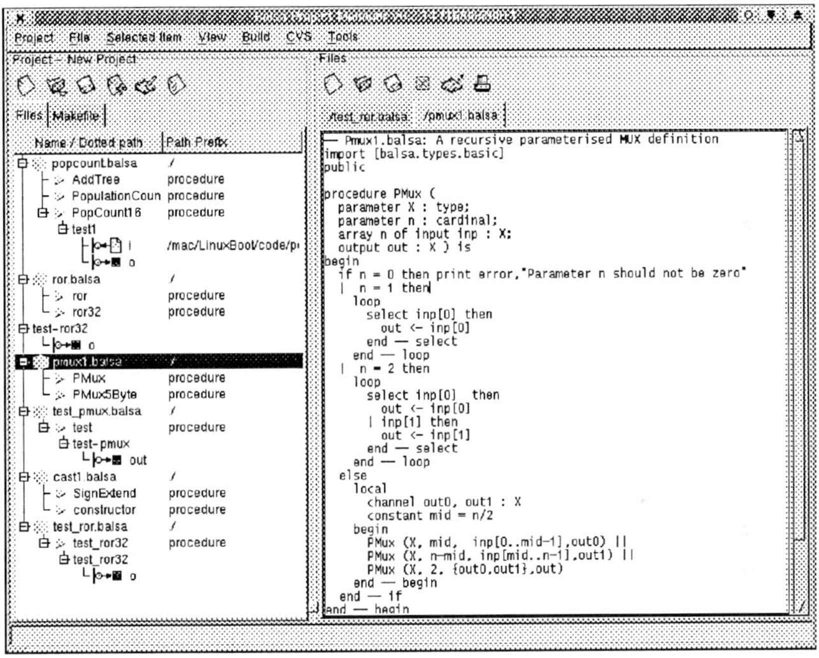

Figure 9.6. balsa-mgr IDE.

9.5.2 Estimating area cost

The area cost of a circuit may be estimated by executing the Makefile rule `cost`. For example, an extract of the output produced for the 2-place buffer is shown below:

```
Part: buffer2
(0 (component "$BrzFetch" (8) (10 2 9)))
(0 (component "$BrzFetch" (8) (8 6 7)))
(0 (component "$BrzFetch" (8) (5 4 3)))
```

```
(20.75 (component "$BrzLoop" () (1 11)))
(99.0 (component "$BrzSequence" (3) (11 (10 8 5))))
(198.0 (component "$BrzVariable" (8 1 "x1[0..7]") (9 (6))))
(198.0 (component "$BrzVariable" (8 1 "x2[0..7]") (7 (4))))

Total cost: 515.75
```

The exact format of the report produced is somewhat obscure. Each line corresponds to a handshake component. Its area cost is the first number on the line. The parameters after the component name correspond to the width of the various channels of that component and the internal channel names. The area reported is proportional to the cost of implementing the circuit in a particular silicon process and is of most use in comparing different circuit descriptions.

9.5.3 Viewing the handshake circuit graph

A PostScript view of the handshake circuit graph can be produced by running the rule `make ps`. A (flattened) view of the handshake circuit graph for the example *buffer.2c* is shown in figure 9.7.

The two single-place buffers from which the circuit is composed are recognisable in the circuit. Apart from minor differences in the labelling of the handshake component symbols, the circuit is identical to that shown in figure 8.6 discussed in section 8.4 on page 128 and the same optimisations have been (automatically) applied.

9.5.4 Simulation

Ignoring the various simulation possibilities available once the design has been converted to a silicon layout, there are three strategies for evaluating and simulating the design from Balsa:

1 Default LARD test harness.

   The command `make sim` will generate a LARD test harness and run it. The test harness reads data from a file for each input port of the module under test. Data sent to output channels appears on the standard output. This method needs no knowledge of LARD at all.

2 Balsa test harness.

   If a more sophisticated test sequence is required, Balsa is a sufficiently flexible language in its own right to be able to specify most test sequences. A default LARD test harness can then be generated for the Balsa test harness. Again no detailed knowledge of LARD is required.

3 Custom LARD test harness.

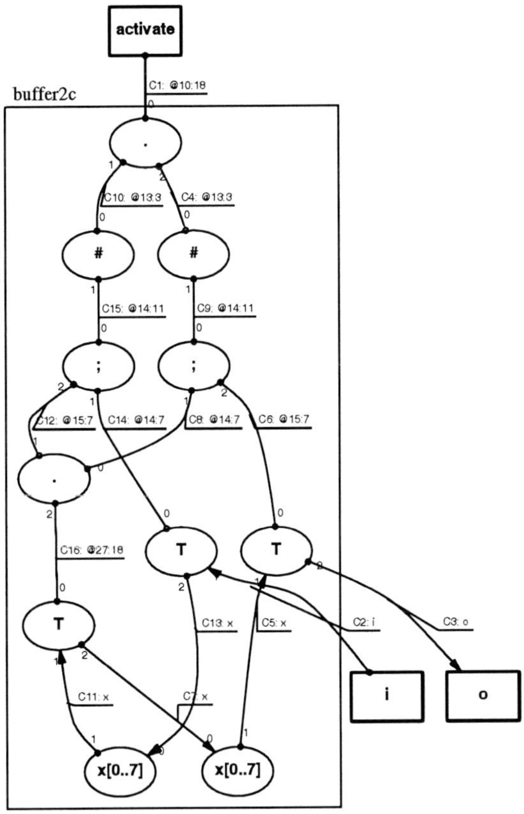

Figure 9.7. Flattened view of buffer2c.

For some applications, it may be necessary to write a custom test harness in LARD. The Makefile generated test harness may be used as template.

The default test harness exercises the target Balsa block by repeatedly handshaking on all external channels; input data channels receive the value 0 on each handshake, although it is possible to associate an input channel with a data file.

Simulating buffer2c

A simulation can be generated by invoking the appropriate simulation rule from the Makefile, producing the following output:

```
0: chan 'i': writing 0
6: chan 'i': writing 0
15: chan 'o': reading 0
19: chan 'i': writing 0
28: chan 'o': reading 0
32: chan 'i': writing 0
41: chan 'o': reading 0
45: chan 'i': writing 0
54: chan 'o': reading 0
58: chan 'i': writing 0
67: chan 'o': reading 0
71: chan 'i': writing 0
80: chan 'o': reading 0
```

The simulation runs forever unless terminated (by Ctrl-C). The numbers reported on the left hand side of each channel activity line are simulation times. LARD uses a unit delay model so these values should be treated with caution.

Simulation data file

This particular simulation stimulus is not very informative. A better strategy is to arrange for the data on the input channel `i` to be externally defined. In the next example, a file contains the following set of test data (in a variety of number representations):

```
1
0x10
022
0b011101
5
```

The Makefile can be forced to generate a rule for running a simulation from this stimulus file. If the simulation is now run, the following output is produced:

```
3: chan 'i': writing 1
15: chan 'o': reading 1
16: chan 'i': writing 16
28: chan 'o': reading 16
29: chan 'i': writing 18
41: chan 'o': reading 18
42: chan 'i': writing 29
54: chan 'o': reading 29
55: chan 'i': writing 5
67: chan 'o': reading 5
Program terminated
```

Channel viewer

In the previous examples, the output of the simulation is textual appearing on the standard output. LARD has a graphical interface which displays the handshakes and data values associated with the internal and external channels. Assuming the building of a test harness rule has been specified to `balsa-md`, the channel viewer can be invoked causing two windows to appear on the screen: the LARD interpreter control window and the channel viewer window itself.

Starting the simulation will cause a trace of the various channels in the design to appear in the channel view window. For each channel the request and acknowledge signals and data values are displayed.

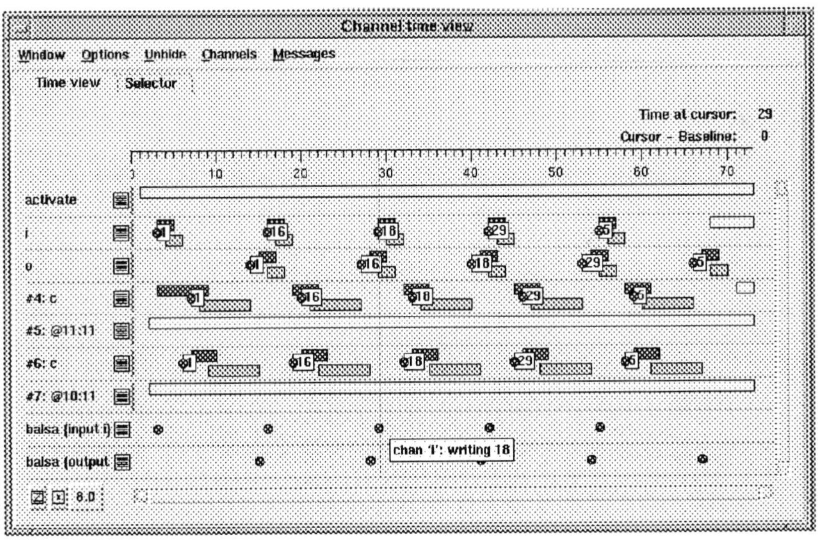

Figure 9.8. Channel viewer window.

Chapter 10

THE BALSA LANGUAGE

In this chapter, a tutorial overview of the language is given together with several small designs which illustrate various aspects of the language.

10.1. Data types

Balsa is strongly typed with data types based on bit vectors. The results of expressions must be guaranteed to fit within the range of the underlying bit vector representation. Types are either anonymous or named. Type equivalence for anonymous types is checked on the basis of the size and properties of the type, whereas type equivalence for named types is checked against the point of declaration.

There are two classes of anonymous types: numeric types which are declared with the `bits` keyword, and arrays of other types. Numeric types can be either signed or unsigned. Signedness has an effect on expression operators and casting. Only numeric types and arrays of other types may be used without first binding a name to those types. Balsa has three separate name spaces: one for procedure and function names, a second for variable and channel names and a third for type declarations.

Numeric types

Numeric types support the number range $[0, 2^n - 1]$ for n-bit unsigned numbers or $[-2^{n-1}, 2^{n-1} - 1]$ for n-bit signed numbers. Named numeric types are just aliases of the same range. An example of a numeric type declaration is:

```
type word is 16 bits
```

This defines a new type `word` which is unsigned (there is no unsigned keyword) covering the range $[0, 2^{16} - 1]$. Alternatively, a signed type could have been declared as:

```
type sword is 16 signed bits
```

which defines a new type `sword` covering the range $[-2^{15}, 2^{15} - 1]$.

J. Sparsø and S. Furber (eds.), Principles of Asynchronous Circuit Design, 173–191.

The only predefined type is `bit`. However the standard Balsa distribution comes with with a set of library declarations for such types as `byte`, `nibble`, `boolean` and `cardinal` as well as the constants `true` and `false`.

Enumerated types

Enumerated types consist of named numeric values. The named values are given values starting at zero and incrementing by one from left to right. Elements with explicit values reset the counter and many names can be given to the same value, for example:

```
type Colour is enumeration
  Black, Brown, Red, Orange, Yellow, Green, Blue,
  Violet, Purple=Violet, Grey, Gray=Grey, White
end
```

The value of the `Violet` element of `Colour` is 7, as is `Purple`. Both `Grey` and `Gray` have value 8. The total number of elements is 10. An enumeration can be padded to a fixed size by use of the `over` keyword:

```
type SillyExample is enumeration
  e1=1, e2 over 4 bits
end
```

Here 2 bits are sufficient to specify the 3 possible values of the enumeration (0 is not bound to a name, `e1` has the value 1 and `e2` has the value 2). The `over` keyword ensures that the representation of the enumerated type is actually 4 bits. Enumerated types must be bound to names by a type declaration before use.

Constants

Constant values can be defined in terms of an expression resolvable at compile time. Constants may be declared in terms of a predefined type otherwise they default to a numeric type. Examples are:

```
constant minx = 5
constant maxx = minx + 10
constant hue = Red : Colour
constant colour = Colour'Green   -- explicit enumeration element
```

Record types

Records are bit-wise compositions of named elements of possibly different (pre-declared) types with the first element occupying the least significant bit positions, e.g.:

```
type Resistor is record
```

```
  FirstBand, SecondBand, Multiplier : Colour;
  Tolerance : ToleranceColour
end
```

`Resistor` has four elements: `FirstBand`, `SecondBand`, `Multiplier` of type `Colour` and `Tolerance` of type `ToleranceColour` (both types must have been declared previously). `FirstBand` is the first element and so represents the least significant portion of the bit-wise value of a type `Resistor`. Selection of elements within the record structure is accomplished with the usual dot notation. Thus if `R15` is a variable of type `Resistor`, the value of its `SecondBand` can extracted by `R15.SecondBand`. As with enumerations, record types can be padded using the `over` notation.

Array types

Arrays are numerically indexed compositions of same-typed values. An example of the declaration of an array type is:

```
type RegBank_t : array 0..7 of byte
```

This introduces a new type `RegBank_t` which is an array type of 8 elements indexed across the range [0, 7], each element being of type `byte`. The ordering of the range specifier is irrelevant; `array 0..7` is equivalent to `array 7..0`. In general a single expression, `expr`, can be used to specify the array size: this is equivalent to a range of `0..expr-1`. Anonymous array types are allowed in Balsa, so that variables can be declared as an array without first defining the array type:

```
variable RegBank : array 0..7 of byte
```

Arbitrary ranges of elements within an array can be accessed by an array slicing mechanism e.g. `a[5..7]` extracts elements `a5`, `a6`, and `a7`. As with all range specifiers, the ordering of the range is irrelevant. In general Balsa packs all composite typed structures in a least significant to most significant, left to right manner. Array slices always return values which are based at index 0.

Arrays can be constructed by a tupling mechanism or by concatenation of other arrays of the same base type:

```
variable a, b, c, d, e ,f: byte
variable z2 : array 2 of byte
variable z4 : array 4 of byte
variable z6 : array 6 of byte

z4:= {a,b,c,d}              -- array construction
z6:= z4 @ {e, f}            -- array concatenation
z2:= (z4 @ {e, f}) [3..4]   -- element extraction by array slicing
```

In the last example, the first element of z2 is set to d and the second element is set to e. Array slicing is useful for extracting arbitrary bitfields from other datatypes.

Arrayed channels

Channels may be arrayed, that is they may consist of several distinct channels which can be referred to by a numeric or enumerated index. This is similar to the way in which variables can have an array type except that each channel is distinct for the purposes of handshaking and each indexed channel has no relationship to the other channels in the array other than the single name they share. The syntax for arrayed channels is different from that of array typed variables making it easier to disambiguate arrays from arrayed channels. As an example:

```
array 4 of channel XYZ : array 4 of byte
```

declares 4 channels, `XYZ[0]` to `XYZ[3]`, each channel is a 32-bit wide type `array 0..3 of byte`. An example of the use of arrayed channels is shown in section 9.4.4 on page 165.

10.2. Data typing issues

As stated previously, Balsa is strongly typed: the left and right sides of assignments are expected to have the same type. The only form of implicit type-casting is the promotion of numeric literals and constants to a wider numeric type. In particular, care must be taken to ensure that the result of an arithmetic operation will always be compatible with the declared result type. Consider the assignment statement `x := x + 1`. This is not a valid Balsa statement because potentially the result is one bit wider than the width of the variable x. If the carry-out from the addition is to be ignored, the user must explicitly force the truncation by means of a cast.

Casts

If the variable x was declared as 32 bits, the correct form of the assignment above is:

```
x := (x + 1 as 32 bits)
```

The keyword `as` indicates the cast operation. The parentheses are a necessary part of the syntax. If the carry out of the addition of two 32-bit numbers is required, a record type can be used to hold the composite result:

```
type AddResult is record
  Result : 32 bits;
```

```
  Carry : bit;
end
variable r : AddResult

r := (a + b as AddResult)
```

The expression `r.Carry` accesses the required carry bit, `r.Result` yields the 32-bit addition result.

Casts are required when extracting bit fields. Here is an example from the instruction decoder of a simple microprocessor. The bottom 5 bits of the 16-bit instruction word contain an 5-bit signed immediate. It is required to extract the immediate field and sign-extend it to 16 bits:

```
type Word is 16 signed bits
type Imm5 is 5 signed bits

variable Instr : 16 bits -- bottom 5 bits contain an immediate
variable Imm16 : Word

Imm16 := (((Instr as array 16 of bit) [0..4] as Imm5) as Word)
```

First, the instruction word, `Instr`, is cast into an array of bits from which an arbitrary sub-range can be extracted:

```
(Instr as array 16 of bit)
```

Next the bottom (least significant) 5 bits must be extracted:

```
(Instr as array 16 of bit) [0..4]
```

The extracted 5 bits must now be cast back into a 5-bit signed number:

```
((Instr as array 16 of bit) [0..4] as Imm5)
```

The 5-bit signed number is then signed extended to the 16-bit immediate value:

```
(((Instr as array 16 of bit) [0..4] as Imm5) as Word)
```

The double cast is required because a straightforward cast from `5 bits` to the variable `Imm16` of type `Word` would have merely zero filled the topmost bit positions even though `Word` is a signed type. However, a cast from a signed numeric type to another (wider) signed numeric type will sign extend the narrower value into the width of the wider target type.

Extracting bits from a field is a fairly common operation in many hardware designs. In general, the original datatype has to be cast into an array first before bitfield extraction. The *smash* operator "#" provides a convenient shorthand for casting an object into an array of bits. Thus the sign extension example above is more simply written:

```
((#Instr [0..4] as Imm5) as Word)
```

Table 10.1. Balsa commands.

Command	Notes
`sync`	control only (dataless) handshake
`<-`	handshake data transfer from an expression to an output port
`->`	handshake data transfer to a variable from an input port
`:=`	assigns a value to a variable
`;`	sequence operator
`||`	parallel composition operator
`continue`	a no-op
`halt`	causes deadlock (useful in simulation)
`loop ... end`	repeat forever
`while ... else ... end`	conditional loop
`for ... end`	structural (not temporal) iteration
`if ... then ... else ... end`	conditional execution, may have multiple guarded commands
`case ... end`	conditional execution based on constant expressions
`select`	non-arbitrated choice operator
`arbitrate`	arbitrated choice operator
`print`	compile time printing of diagnostics

Auto-assignment

Statements of the form:

```
x := f(x)
```

are allowed in Balsa. However, the implementation generates an auxiliary variable which is then assigned back to the variable visible to the programmer – the variable is enclosed within a single handshake and cannot be read from and written to simultaneously. Since auto-assignment generates twice as many variables as might be suspected, it is probably better practice to avoid the auto-assignment, explicitly introduce the extra variable and then rewrite the program to hide the sequential update thereby avoiding any time penalty. An example of this approach is given in *count10b* on page 184.

10.3. Control flow and commands

Balsa's command set is listed in table 10.1.

Dataless handshakes

`sync` ⟨`Channel`⟩ – awaits a handshake on the named channel. Circuit action does not proceed until the handshake is completed.

Channel communications

Data can be transferred between a variable and a channel, between channels or from a channel to a command code block as shown below:

⟨`Channel`⟩ `<-` ⟨`Variable`⟩ – transfers data from a variable to the named channel. This may either be an internal channel local to a procedure or an output port listed in the procedure declaration.

⟨`Channel`⟩ `->` ⟨`Variable`⟩ – transfers data from the channel connected to a variable. The channel may either be an internal channel local to a procedure or an input port listed in the procedure declaration.

⟨`Channel1`⟩ `->` ⟨`Channel2`⟩ – transfers data between channels.

⟨`Channel`⟩ `-> then` ⟨`Command`⟩ – allows the data to be accessed throughout the command block. However, the handshake on the channel is not completed and thus the data not released until the command block itself has terminated.

Variable assignment

⟨`Variable`⟩ `:=` ⟨`Expression`⟩ – transfers the result of an expression into a variable. The result type of the expression and that of the variable must agree.

Sequential composition

⟨`Command1`⟩ `;` ⟨`Command2`⟩ – the two commands execute sequentially. The first must terminate before the second commences.

Parallel composition

⟨`Command1`⟩ `||` ⟨`Command2`⟩ – composes two commands such that they operate concurrently and independently. Both commands must complete before the circuit action proceeds. Beware of inadvertently introducing dependencies between the two commands so that neither can proceed until the other has completed. The "`||`" operator binds tighter than "`;`". If that is not what is intended, then commands may be grouped in blocks as shown below

`[` ⟨`Command1`⟩ `;` ⟨`Command2`⟩ `]` `||` ⟨`Command3`⟩

Note the use of square brackets to group commands rather than parentheses. Alternatively, the keywords `begin` ... `end` may be used and are mandatory if variables local to a block are to be declared.

Continue and halt commands

`continue` is effectively a no-op. The command `halt` causes a process thread to deadlock.

Looping constructs

The loop command causes an infinite repetition of a block of code. Finite loops may be constructed using the while construct. The simplest example of its use is:

```
while ⟨Condition⟩ then ⟨Command⟩ end
```

However, multiple guards are allowed, so a more general form of the construct is:

```
while
  ⟨Condition1⟩ then ⟨Command1⟩
| ⟨Condition1⟩ then ⟨Command2⟩
| ⟨Condition3⟩ then ⟨Command3⟩
else
  ⟨Command4⟩
end
```

The ability of the while construct to take an else clause is a minor convenience. The code sequence above could have been written, with only a small difference in the resultant handshake circuit implementation, as:

```
while
  ⟨Condition1⟩ then ⟨Command1⟩
| ⟨Condition2⟩ then ⟨Command2⟩
| ⟨Condition3⟩ then ⟨Command3⟩
end;
⟨Command4⟩
```

If more than one guard is satisfied, the particular command that is executed is unspecified.

Structural iteration

Balsa has a structural looping construct. In many programming languages it is a matter of convenience or style as to whether a loop is written in terms of a for loop or a while loop. This is not so in Balsa. The for loop is similar to VHDL's for ... generate command and is used for iteratively laying out repetitive structures. An example of its use was given earlier in section 9.4.4 on page 165. An illustration of the inappropriate use of the for command is given in the example *count10e* on page 189. Structures may be iteratively instantiated to operate either sequentially or concurrently with one another depending on whether for ; or for || is employed.

Conditional execution

Balsa has two constructs to achieve conditional execution. Balsa's case statement is similar to that found in conventional programming languages. A single guard may match more than one value of the guard expression.

```
case x+y of
  1 .. 4, 11  then o <- x
| 5 .. 10 then o <- y
  else o <- z
end
```

An `if` ... `then` ... `else` statement allows conditional execution based on the evaluation of expressions at run-time. Its syntax is similar to that of the `while` loop. Note the sequencing implicit in nested `if` statements, such as that shown below:

```
if ⟨Condition1⟩ then
  ⟨Command1⟩
else
  if ⟨Condition2⟩ then
    ⟨Command2⟩
  end
end
```

The test for `Condition2` is made after the test for `Condition1`. If it is known that the two conditions are mutually exclusive, the expression may be written:

```
if ⟨Condition1⟩ then ⟨Command1⟩
|  ⟨Condition2⟩ then ⟨Command2⟩
end
```

The "|" separator causes `Condition1` and `Condition2` to be evaluated in parallel. The result is undefined if more than one guard (condition) is satisfied.

10.4. Binary/unary operators

Balsa's binary/unary operators are shown in order of decreasing precedence in table 10.2.

10.5. Program structure

File structure

A typical design will consist of several files containing procedure/type/constant declarations which come together in a top-level procedure that constitutes the overall design. This top-level procedure would typically be at the end of a file which imports all the other relevant design files. This importing feature forms a simple but effective way of allowing component reuse and maps simply onto the notion of the imported procedures being either pre-compiled handshake circuits or existing (possibly hand crafted) library components. Declarations have a syntactically defined order (left to right, top to bottom) with each declaration having its scope defined from the point of declaration to the end of

Table 10.2. Balsa binary/unary operators.

Symbol	Operation	Valid types	Notes
.	record indexing	record	
#	smash	any	takes value from any type and reduces it to an array of bits
[]	array indexing	array	non-constant index possible (can generate lots of hardware)
not,log, - (unary)	unary operators	numeric	`log` only works on constants and returns the ceiling: e.g. `log 15` returns 4 "-" returns a result 1 bit wider than the argument
^	exponentiation	numeric	
*, /, %	multiply, divide, remainder	numeric	only applicable to constants
+,-	add, subtract	numeric	results are 1 or 2 bits longer than the largest argument
@	concatenation	arrays	
<,>,<=,>=	inequalities	numeric, enumerations	
=, /=	equals, not equals	all types	comparison is by sign extended value for signed numeric types
and	bitwise AND	numeric	Balsa uses type `1 bits` for `if/while` guards so bitwise and logical operators are the same.
or, xor	bitwise OR,XOR	numeric	

the current (or importing) file. Thus Balsa has the same simple "declare before use" rule of C and Modula, though without any facility for prototypes.

Declarations

Declarations introduce new type, constant or procedure names into the global name spaces from the point of declaration until the end of the enclosing block (or file in the case of top-level declarations). There are three disjoint name spaces: one for types, one for procedures and a third for all other declarations. At the top level, only constants are in this last category. However, variables and channels may be included in procedure local declarations. Where a declaration within an enclosed/inner block has the same name as one previously made in an outer/enclosing context, the local declaration will hide the outer declaration for the remainder of that inner block.

Procedures

Procedures form the bulk of a Balsa description. Each procedure has a name, a set of ports and an accompanying behavioural description. The `sync` keyword introduces dataless channels. Both dataless and data bearing channels can be members of "arrayed channels". Arrayed channels allow numeric/enumerated indexing of otherwise functionally separate channels. Procedures can also carry a list of local declarations which may include other procedures, types and constants.

Shared procedures

Normally, each call to a procedure generates separate hardware to instantiate that procedure. A procedure may be shared, in which case calls to that procedure access common hardware thereby avoiding duplication of the circuit at the cost of some multiplexing to allow sharing to occur. The use of shared procedures is discussed further on page 187.

Functions

In many programming languages, functions can be thought of as procedures without side effects that return a result. However, in Balsa there is a fundamental difference between functions and procedures. Parameters to a procedure define handshaking channels that interface to the circuit block defined by the procedure. Function parameters, on the other hand, are just expression aliases returning values. An example of the use of function definitions can be found in the arbiter tree design on page 202.

10.6. Example circuits

In this section, various designs of counter are described in Balsa. In flavour, they resemble the specifications of conventional synchronous counters, since these designs are more familiar to newcomers to asynchronous systems. More sophisticated systolic counters, better suited to an asynchronous approach, are described by van Berkel [14].

In this design, the role of the clock which updates the state of the counter is taken by a dataless sync channel, named `aclk`. The counter issues a handshake request over the sync channel, the environment responds with an acknowledgement completing the handshake and the counter state is updated.

A modulo-16 counter

```
-- count16a.balsa: modulo 16 counter
import [balsa.types.basic]
```

```
procedure count16 (sync aclk; output count : nibble) is
  variable count_reg : nibble
begin
  loop
    sync aclk ;
    count <- count_reg ;
    count_reg := (count_reg + 1 as nibble)
  end
end
```

This counter interfaces to its environment by means of two channels: the dataless `aclk` channel and the output channel `count` which carries the current count value. The internal register which implements the variable `count_reg` and the output channel are of the predefined type `nibble` (4-bits wide). After `count_reg` is incremented, the result must be cast back to type `nibble`. For simplicity, issues of initialisation/reset have been ignored. A LARD simulation of this circuit will give a harmless warning when uninitialised variables are accessed.

Removing auto-assignment

The auto-assignment statement in the example above, although concise and expressive, hides the fact that, in most back-ends, an auxiliary variable is created so that the update can be carried out in a race-free manner. By making this auxiliary variable explicit, advantage may be taken of its visibility to overlap its update with other activity as shown in the example below.

```
-- count16b.balsa: write-back overlaps output assignment
import [balsa.types.basic]

procedure count16 (sync aclk; output count : nibble) is
  variable count_reg, tmp : nibble
begin
  loop
    sync aclk;
    tmp := (count_reg + 1 as nibble) ||
    count <- count_reg;
    count_reg := tmp
  end
end
```

In this example, the transfer of the count register to the output channel is overlapped with the incrementing of the auxiliary shadow register. There is some slight area overhead involved in parallelisation and any potential speed-up may be minimal in this case, but the principle of making trade-offs at the level of the source code is illustrated.

A modulo-10 counter

The basic counter description above can easily be modified to produce a modulo-10 counter. A simple test is required to detect when the internal register reaches its maximum value and then to reset it to zero.

```
-- count10a.balsa: an asynchronous decade counter
import [balsa.types.basic]

type C_size is nibble
constant max_count = 9

procedure count10(sync aclk; output count: C_size) is
  variable count_reg : C_size
  variable tmp : C_size
begin
  loop
    sync aclk;
    if count_reg /= max_count then
      tmp := (count_reg + 1 as C_size)
    else
      tmp := 0
    end || count <- count_reg ;
    count_reg := tmp
  end -- loop
end -- begin
```

A loadable up/down decade counter

This example describes a loadable up/down decade counter. It introduces many of the language features discussed earlier in the chapter. The counter requires two control bits, one to determine the direction of count, and the other to determine whether the counter should load or {inc,dec}rement on the next operation. The are several valid design options; in this example, *count10b* below, the control bits and the data to be loaded are bundled together in a single channel, in_sigs.

```
-- count10b.balsa: an aysnchronous up/down decade counter
import [balsa.types.basic]

type C_size is nibble
constant max_count = 9

type dir is enumeration down, up end
type mode is enumeration load, count end
type In_bundle is record
  data : C_size ;
  mode : mode;
  dir : dir
```

```
end

procedure updown10 (
  input in_sigs: In_bundle;
  output count: C_size
) is
  variable count_reg : C_size
  variable tmp : In_bundle
begin
  loop
    in_sigs -> tmp; -- read control+data bundle
    if  tmp.mode = count then
      case tmp.dir of
        down then  -- counting down
        if count_reg /= 0 then
          tmp.data := (count_reg - 1 as C_size)
        else
          tmp.data := max_count
        end -- if
      | up then  -- counting up
        if count_reg /= max_count then
              tmp.data := (count_reg + 1 as C_size)
        else
          tmp.data := 0
        end -- if
      end -- case tmp.dir
    end; -- if
    count <- tmp.data || count_reg:= tmp.data
  end -- loop
end
```

The example above illustrates the use of `if ... then ... else` and `case` control constructs as well the use of record structures and enumerated types. The use of symbolic values within enumerated types makes the code more readable. Test harnesses which can be generated automatically by the Balsa system can also read the symbolic enumerated values. For example, here is a test file which initialises the counter to 8, counts up, testing that the counter wraps round to zero and then counts down allowing the user to check that the counter correctly wraps to 9.

```
{8, load, up}        load  counter with 8
{0, count, up}       count to 9
{0, count, up}       count & wrap to 0
{0, count, up}       count to 1
{0, count, down}     count down to 0
{0, count, down}     count down to 9
{1, load, down}      load counter with 1
{0, count, down}     count down to 0
{0, count, down}     count down & wrap to 9
```

Sharing hardware

In Balsa, every statement instantiates hardware in the resulting circuit. It is therefore worth examining descriptions to see if there are any repeated constructs that could either be moved to a common point in the code or replaced by shared procedures. In *count10b* above, the description instantiates two adders: one used for incrementing and the other for decrementing. Since these two units are not used concurrently, area can be saved by sharing a single adder (which adds either $+1$ or -1 depending in the direction of count) described by a shared procedure. The code below illustrates how *count10b* can be rewritten to use a shared procedure. The shared procedure `add_sub` computes the next count value by adding the current count value to a variable, `inc`, which can take values of $+1$ or -1. Note that to accommodate these values, `inc` must be declared as `2 signed bits`.

The area advantage of the approach is shown by examining the cost of the circuit reported by *breeze-cost*: *count10b* has a cost of 2141 units, whereas the shared procedure version has a cost of only 1760. The relative advantage, of course, becomes more pronounced as the size of the counter increases.

```
-- count10c.balsa: introducing shared procedures
import [balsa.types.basic]

type C_size is nibble
constant max_count = 9

type dir is enumeration down, up end
type mode is enumeration load, count end
type inc is 2 signed bits

type In_bundle is record
  data : C_size ;
  mode :  mode;
  dir : dir
end

procedure updown10 (
  input in_sigs: In_bundle;
  output count: C_size
) is
  variable count_reg : C_size
  variable tmp : In_bundle
  variable inc : inc

  shared add_sub is
  begin
    tmp.data:= (count_reg + inc as C_size)
  end
```

```
begin
  loop
    in_sigs -> tmp; -- read control+data bundle
    if  tmp.mode = count then
      case tmp.dir of
      down then  -- counting down
        if count_reg /= 0 then
          inc:= -1;
          add_sub()
        else
          tmp.data := max_count
        end -- if
      | up then  -- counting up
        if count_reg /= max_count then
          inc := +1;
          add_sub()
        else
          tmp.data := 0
        end -- if
      end -- case tmp.dir
    end; -- if
    count <- tmp.data || count_reg:= tmp.data
  end -- loop
end
```

In order to guarantee the correctness of implementations, there are a number of minor restrictions on the use of shared procedures:

- shared procedures cannot have any arguments;
- shared procedures cannot use local channels;
- shared procedures using elements of the channel referenced by a `select` statement (see section 10.7 on page 190) must be declared as local within the body of that `select` block.

"while" loop description

An alternative description of the modulo-10 counter employs the `while` construct:

```
-- count10d.balsa: mod-10 counter alternative implementation
import [balsa.types.basic]

type C_size is nibble
constant max_count = 10

procedure count10(sync aclk; output count : C_size) is
  variable count_reg : C_size
begin
```

```
    loop
      while count_reg < max_count then
        sync aclk;
        count <- count_reg;
        count_reg:= (count_reg + 1 as C_size)
      end; -- while
      count_reg:= 0
    end -- loop
  end
```

Structural "for" loops

`for` loops are a potential pitfall for beginners to Balsa. In many programming languages, `while` loops and `for` loops can be used interchangeably. This is not the case in Balsa: a `for` loop implements structural iteration, in other words, separate hardware is instantiated for each pass through the loop. The following description, which superficially appears very similar to the `while` loop example of *count10d* previously, appears to be correct: it compiles without problems and a LARD simulation appears to give the correct behaviour. However, examination of the cost reveals an area cost of 11577, a large increase. It is important to understand why this is the case. The `for` loop is unrolled at compile time and 10 instances of the circuit to increment the counter are created. Each instance of the loop is activated sequentially. The PostScript plot of the handshake circuit graph is rather unreadable; setting `max_count` to 3 produces a more readable plot.

```
-- count10e.balsa: beware the "for" construct
import [balsa.types.basic]

type C_size is nibble
constant max_count = 10

procedure count10(sync aclk; output count: C_size) is
  variable count_reg : C_size
begin
  loop
    for ; i in 1 .. max_count then
      sync aclk;
      count <- count_reg;
      count_reg:= (count_reg + 1 as C_size)
    end; -- for ; i
    count_reg:= 0
  end -- loop
end -- begin
```

If, instead of using the sequential `for` construct, the parallel `for` construct had been employed (`for ||` ...), the compiler would give an error message complaining about read/write conflicts from parallel threads. In this case, all instances of the counter circuit would attempt to update the counter register at the same time, leading to possible conflicts. A reader who understands the resulting potential handshake circuit is well on the way to a good understanding of the methodology.

10.7. Selecting channels

The asynchronous circuit described below merges two input channels into a single output channel; it may be thought of as a self-selecting multiplexer. The select statement chooses between the two input channels a and b by waiting for data on either channel to arrive. When a handshake on either a or b commences data is held valid on the input, and the handshake not completed, until the end of the select ... end block.

This circuit is an example of handshake *enclosure* and avoids the need for an internal latch to be created to store the data from the input channel; a possible disadvantage is that, because of the delayed completion of the handshake, the input is not released immediately to continue processing independently. In this example, data is transferred to the output channel and the input handshake will complete as soon as data has been removed from the output channel. An example of a more extended enclosure can be found in the code for the population counter on page 197.

```
-- mux1.balsa: unbuffered Merge
import [balsa.types.basic]

procedure mux (input a, b :byte; output c :byte) is
begin
  loop
    select a then c <- a    -- channel behaves like a variable
    |      b then c <- b    -- ditto
    end -- select
  end -- loop
end
```

Because of the enclosed nature of the handshake associated with select, inputs a and b should be mutually exclusive for the duration of the block of code enclosed by the select. In many cases, this is not a difficult obligation to satisfy. However, if a and b are truly independent, select can be replaced by arbitrate which allows an arbitrated choice to be made. Arbiters are relatively expensive in terms of speed and may not be possible to implement in some technologies. Further, the loss of determinism in circuits with arbitration can also introduce testing and design correctness verification problems. Designers should therefore not use arbiters unnecessarily.

```
-- mux2.balsa: unbuffered arbitrated Merge.
import [balsa.types.basic]

procedure mux (input a, b :byte; output c :byte) is
begin
  loop
    arbitrate a then c <- a   -- channel behaves like a variable
    |         b then c <- b   -- ditto
    end -- arbitrate
  end -- loop
end
```

Chapter 11

BUILDING LIBRARY COMPONENTS

11.1. Parameterised descriptions

Parameterised procedures allow designers to develop a library of commonly used components and then to instantiate those structures later with varying parameters. A simple example is the specification of a buffer as a library part without knowing the width of the buffer. Similarly, a pipeline of buffers can be defined in the library without requiring any knowledge of the depth chosen for the pipeline when it is instantiated.

11.1.1 A variable width buffer definition

The example *pbuffer1* below defines a single place buffer with a parameterised width.

```
-- pbuffer1.balsa - parameterised buffer example
import [balsa.types.basic]

procedure Buffer (
  parameter X : type ;
  input i : X; output o : X
) is
  variable x : X
begin
  loop
    i -> x ;
    o <- x
  end
end

-- now define a byte wide buffer
procedure Buffer8 is Buffer (byte)

-- now use the definition
procedure test1 (input a : byte ; output b : byte) is
begin
  Buffer8 (a, b)
```


J. Sparsø and S. Furber (eds.), Principles of Asynchronous Circuit Design, 193–204.

```
end

-- alternatively
procedure test2 (input a : byte ; output b : byte) is
begin
  Buffer (byte, a, b)
end
```

The definition of the single place buffer given earlier on page 161 is modified by the addition of the parameter declaration which defines `X` to be of type `type`. In other words `X` is identified as being a type to be refined later. Once a parameter type has been declared, it can be used in later declarations and statements: for example, input channel `i` is defined as being of type `X`. No hardware is generated for the parameterised procedure definition itself.

Having defined the procedure, it can be used in other procedure definitions. `Buffer8` defines a byte wide buffer that can be instantiated as required as shown, for example, in procedure `test1`. Alternatively, a concrete realisation of the parameterised procedure can be used directly as shown in procedure `test2`.

11.1.2 Pipelines of variable width and depth

The next example illustrates how multiple parameters to a procedure may be specified. The parameterised buffer element is included in a pipeline whose depth is also parameterised.

```
-- pbuffer2.balsa - parameterised pipeline example
import [balsa.types.basic]
import [pbuffer1]

-- BufferN: an n-place parameterised, variable width buffer
procedure BufferN (
  parameter n : cardinal ;
  parameter X : type ;
  input i : X ;
  output o : X
) is
begin
  if n = 1 then -- single place pipeline
    Buffer(X, i, o)
  | n >= 2 then -- parallel evaluation
    local array 1 .. n-1 of channel c : X
    begin
      Buffer(x, i, c[1])    ||  -- first buffer
      Buffer(x, c[n-1], o) ||  -- last buffer
      for || i in 1 .. n-2 then
        Buffer(X, c[i], c[i+1])
      end -- for || i
    end
```

```
    else print error, "zero length pipeline specified"
    end -- if
  end

  -- Now define a 4 deep, byte-wide pipeline.
  procedure Buffer4 is BufferN(4, byte)
```

Buffer is the single place parameterised width buffer of the previous example and this is reused by means of the library statement `import[pbuffer1]`. In this code, *BufferN* is defined in a very similar manner to the example in section 9.4.4 on page 165 except that the number of stages in the pipeline, `n`, is not a constant but is a parameter to the definition of type cardinal. Note that this definition includes some error checking. If an attempt is made to build a zero length pipeline during a definition, an error message is printed.

11.2. Recursive definitions

Balsa allows a form of recursion in definitions (as long as the resulting structures can be statically determined at compile time). Many structures can be described elegantly using this technique which forms a natural extension to the powerful parameterisation mechanism. The remainder of this chapter illustrates recursive parameterisation by means of some interesting (and useful) examples.

11.2.1 An n-way multiplexer

An n-way multiplexer can be constructed from a tree of 2-way multiplexers. A recursive definition suggests itself as the natural specification technique: an n-way multiplexer can be split into two n/2-way multiplexers connected by internal channels to a 2-way multiplexer as shown in figure 11.1 on page 196.

```
--- Pmux1.balsa: A recursive parameterised MUX definition
import [balsa.types.basic]

procedure PMux (
  parameter X : type;
  parameter n : cardinal;
  array n of input inp : X;  -- note use of arrayed port
  output out : X
) is
begin
  if n = 0 then print error,"Parameter n should not be zero"
  |  n = 1 then
    loop
      select inp[0] then
        out <- inp[0]
      end -- select
    end -- loop
```

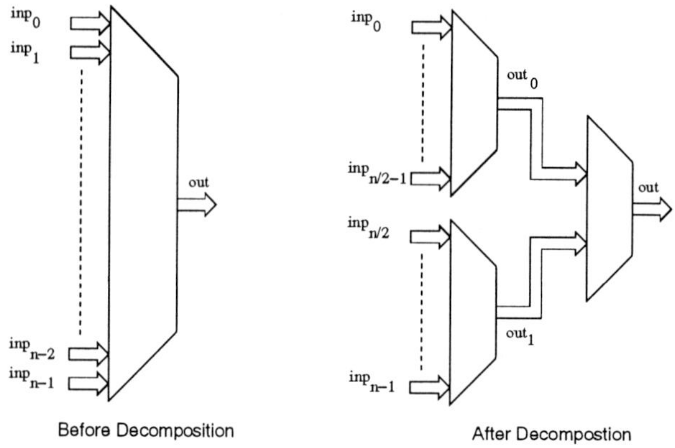

Figure 11.1. Decomposition of an n-way multiplexer.

```
  | n = 2 then
    loop
      select inp[0] then
        out <- inp[0]
      | inp[1] then
        out <- inp[1]
      end -- select
    end -- loop
  else
    local  -- local block with local definitions
      channel out0, out1 : X
      constant mid = n/2
    begin
      PMux (X, mid,   inp[0..mid-1], out0) ||
      PMux (X, n-mid, inp[mid..n-1], out1) ||
      PMux (X, 2, {out0,out1},out)
    end
  end -- if
end

-- Here is a 5-way multiplexer
procedure PMux5Byte is PMux(byte, 5)
```

Commentary on the code

The multiplexer is parameterised in terms of the type of the inputs and the number of channels n. The code is straightforward. A multiplexer of size greater than 2 is decomposed into two multiplexers half the size connected by internal channels to a 2-to-1 multiplexer. Notice how the arrayed channels, out0 and out1 are specified as a tuple. The recursive decomposition stops when the number of inputs is 2 or 1 (specifying a multiplexer with zero inputs generates an error). A 1-input multiplexer makes no choice of inputs.

A Balsa test harness

The code below illustrates how a simple Balsa program can be used as a test harness to generate test values for the multiplexer. The test program is actually rather naïve.

```
-- test_pmux.balsa - A test-harness for Pmux1
import [balsa.types.basic]
import [pmux1]

procedure test (output out : byte) is
  type ttype is sizeof byte + 1 bits
  array 5 of channel inp : byte
  variable i : ttype
begin
  begin
    i:= 1;
    while i <= 0x80 then
      inp[0] <- (i as byte);
      inp[1] <- (i+1 as byte);
      inp[2] <- (i+2 as byte);
      inp[3] <- (i+3 as byte);
      inp[4] <- (i+4 as byte);
      i:= (i + i as ttype)
    end
  end ||  PMux5Byte(inp, out)
end
```

11.2.2 A population counter

This next example counts the number of bits set in a word. It comes from the requirement in an Amulet processor to know the number of registers to be restored/saved during LDM/STM (Load/Store Multiple) instructions.

The approach taken is to partition the problem into two parts. Initially, adjacent bits are added together to form an array of 2-bit channels representing the numbers of bits that are set in each of the adjacent pairs. The array of 2-bit numbers are then added in a recursively defined tree of adders (`procedure AddTree`). The structure of the bit-counter is shown in figure 11.2.

```
-- popcount: count the number of bits set in a word
import [balsa.types.basic]

procedure AddTree (
  parameter inputCount : cardinal;
  parameter inputSize : cardinal;
  parameter outputSize : cardinal;
```

```
  array inputCount of input i : inputSize bits;
  output o : outputSize bits
) is
begin
  if inputCount = 1 then
    select i[0] then o <- (i[0] as outputSize) end
  | inputCount = 2 then
    select i[0], i[1] then
      o <- (i[0] + i[1] as outputSize bits)
    end -- select
  else
    local
      constant lowHalfInputCount = inputCount / 2
      constant highHalfInputCount = inputCount - lowHalfInputCount

      channel lowO, highO : outputSize - 1 bits
    begin
      AddTree (lowHalfInputCount, inputSize, outputSize - 1,
        i[0..lowHalfInputCount-1], lowO) ||
      AddTree (highHalfInputCount, inputSize, outputSize - 1,
        i[lowHalfInputCount..inputCount-1], highO) ||
      AddTree (2, outputSize - 1, outputSize, {lowO, highO}, o)
    end
  end -- if
end

procedure PopulationCount (
  parameter n : cardinal;
  input i : n bits;
  output o : log (n+1) bits
) is
begin
  if n % 2 = 1 then
  print error, "number of bits must be even"
  end; -- if
  loop
    select i then
      if n = 1 then
        o <- i
      | n = 2 then
        o <- (#i[0] + #i[1])  add bits 0 and 1
      else
        local
          constant pairCount = n - (n / 2)
          array pairCount of channel addedPairs : 2 bits
        begin
          for || c in 0..pairCount-1 then
            addedPairs[c] <- (#i[c*2] + #i[(c*2)+1])
          end ||
          AddTree (pairCount, 2, log (n+1), addedPairs, o)
```

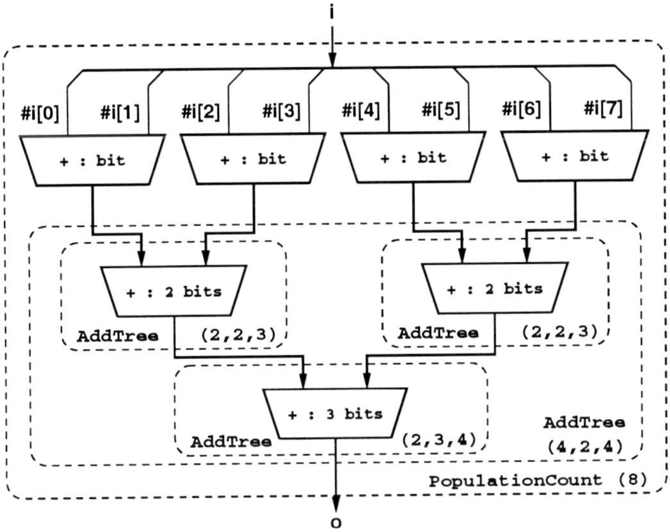

Figure 11.2. Structure of a bit population counter.

```
            end
          end -- if
        end -- select
      end -- loop
    end

    procedure PopCount16 is PopulationCount (16)
```

Commentary on the code

parameterisation: Procedures `AddTree` and `PopulationCount` are parameterised. `PopulationCount` can be used to count the number of bits set in any sized word. `AddTree` is parameterised to allow a recursively defined adder of any number of arbitrary width vectors.

enclosed selection: The semantics of the enclosed handshake of `select` allow the contents of the input `i` to be referred to several times in the body of the `select` block without the need for an internal latch.

avoiding deadlock: Note that the formation of the sum of adjacent bits is specified by a parallel `for` loop.

```
for || c in 0..pairCount-1 then
  addedPairs[c] <- (#i[c*2] + #i[(c*2)+1])
end
```

It might be thought that a serial `for ;` loop could be used at, perhaps, the expense of speed. This is *not* the case: the system will deadlock, illustrating why designing asynchronous circuits requires some real understanding of the methodology. In this case the adder to which the array of `addPairs` is connected requires pairs of inputs to be ready before it can complete the addition and release its inputs. However, if the sum of adjacent bits is computed serially, the next pair will not be computed until the handshake for the previous pair has been completed – which is not possible because `AddTree` is awaiting all pairs to become valid: result deadlock!

11.2.3 A Balsa shifter

General shifters are an essential element of all microprocessors including the Amulet processors. The following description forms the basis of such a shifter. It implements only a rotate right function, but it is easily extensible to other shift functions.

The main work of the shifter is local procedure `rorBody` which recursively creates sub-shifters capable of optionally rotating 1, 2, 4, 8 etc. bits. The structure of the shifter is shown in figure 11.3.

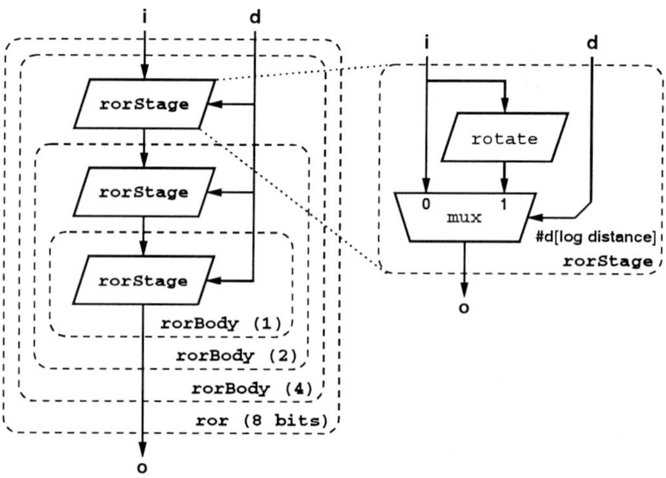

Figure 11.3. Structure of a rotate right shifter.

```
import [balsa.types.basic]

-- ror: rotate right shifter
procedure ror (
  parameter X : type;
  input d : sizeof X bits;
```

```
    input i : X;
    output o : X
  ) is
  begin
    loop
      select d then
        local
          constant typeWidth = sizeof X

          procedure rorBody (
            parameter distance : cardinal;
            input i : X;
            output o : X
          ) is
          local
            procedure rorStage (
              output o : X
            ) is
            begin
              select i then
                if #d[log distance] then
                  o <- (#i[typeWidth-1..distance] @
                        #i[distance-1..0] as X)  {shift}
                else
                  o <- i  {don't shift}
                end -- if
              end -- select
            end
            channel c : X
          begin
            if distance > 1 then
              rorStage (c) ||
              rorBody (distance/2, c, o)
            else
              rorStage (o)
            end -- if
          end
        begin
          rorBody (typeWidth/2, i, o)
        end
      end -- select
    end -- loop
  end

  procedure ror32 is ror (32 bits)
```

Testing the shifter

This next code example builds another test routine in Balsa to exercise the shifter.

```
import [balsa.types.basic]
import [ror]

--test ror32
procedure test_ror32(output o : 32 bits)
is
  variable i : 5 bits
  channel shiftchan : 32 bits
  channel distchan : 5 bits
begin
  begin
    i:= 1;
    while i < 31 then
      shiftchan <- 7 || distchan <- i;
      i:= (i+1 as 5 bits)
    end -- while
  end || ror32(distchan, shiftchan, o)
end
```

11.2.4 An arbiter tree

The final example builds a parameterised arbiter. This circuit forms part of the DMA controller of chapter 12. The architecture of an 8-input arbiter is shown in figure 12.3 on page 212. `ArbFunnel` is a parameterisable tree composed of two elements: `ArbHead` and `ArbTree`. Pairs of incoming sync requests are arbitrated and combined into single bit decisions by `ArbHead` elements. These single bit channels are then arbitrated between by `ArbTree` elements. An `ArbTree` takes a number of decision bits from each of a number of inputs (on the `i` ports) and produces a rank of 2-input arbiters to reduce the problem to half as many inputs each with 1 extra decision bit. Recursive calls to `ArbTree` reduce the number of input channels to one (whose final decision value is returned on port `o`).

```
-- ArbHead: 2 way arbcall with channel no. output
import [balsa.types.basic]
procedure ArbHead (
  sync i0, i1;
  output o : bit
) is
begin
  loop
    arbitrate i0 then o <- 0
    |         i1 then o <- 1
    end -- arbitrate
```

```
    end -- loop
  end -- begin

  -- ArbTree: a tree arbcall which outputs a channel number
  -- prepended onto the input channel's data. (invokes itself
  -- recursively to make the tree)

  procedure ArbTree (
    parameter inputCount : cardinal;
    parameter depth : cardinal;  -- bits to carry from inputs
    array inputCount of input i : depth bits;
    output o : (log inputCount) + depth bits
  ) is
    type BitArray is array 1 of bit
    type BitArray2 is array 2 of bit
    function AddTopBit (hd : bit; tl : depth bits) =
      (#tl @ {hd} as depth + 1 bits)
    function AddTopBit2 (hd : bit; tl : depth + 1 bits) =
      (#tl @ {hd} as depth + 2 bits)
    function AddTop2Bits (hd0 : bit; hd1 : bit; tl : depth bits) =
      (#tl @ {hd0,hd1} as depth + 2 bits)
  begin
    case inputCount of
      0, 1 then print error, "Can't build an ArbTree with fewer than 2 inputs"
    | 2 then loop
        arbitrate i[0] -> i0 then o <- AddTopBit (0, i0)
        |         i[1] -> i1 then o <- AddTopBit (1, i1)
        end -- arbitrate
      end -- loop
    | 3 then local channel lo : 1 + depth bits
      begin
        ArbTree (2, depth, i[0 .. 1], lo) ||
        loop
          arbitrate lo then o <- AddTopBit2 (0, lo)
          | i[2] -> i2 then o <- AddTop2Bits (1, 0, i2)
          end -- arbitrate
        end -- loop
      end
    else local
        constant halfCount = inputCount / 2
        constant halfBits = depth + log halfCount
        channel l, r : halfBits bits
      begin
        ArbTree (halfCount, depth, i[0 .. halfCount-1], l) ||
        ArbTree (inputCount - halfCount, depth,
          i[halfCount .. inputCount-1], r) ||
        ArbTree (2, halfBits, {l,r}, o)
      end -- begin
    end -- case inputCount
  end
```

```
-- ArbFunnel: build a tree arbcall (balanced apart from the las
--  channel which is faster than the rest) which produces a ch
--  number from an array of sync inputs
procedure ArbFunnel (
  parameter inputCount : cardinal;
  array inputCount of sync i;
  output o : log inputCount bits
) is
  constant halfCount = inputCount / 2
  constant oddInputCount = inputCount % 2
begin
  if inputCount < 2 then
    print error, "can't build an ArbFunnel with fewer than 2 inputs"
  | inputCount = 2 then
    ArbHead (i[0], i[1], o)
  | inputCount > 2 then
    local
      array halfCount + 1 of channel li : bit
    begin
      for || j in 0 .. halfCount - 1 then
        ArbHead (i[j*2], i[j*2+1], li[j])
      end ||
      if oddInputCount then
        ArbTree (halfCount + 1, 1, li[0 .. halfCount], o) ||
        loop
          select i[inputCount - 1] then li[halfCount] <- 0
          end -- select
        end -- loop
      else
        ArbTree (halfCount, 1, li[0 .. halfCount-1], o)
      end -- if
    end
  end -- if
end
```

Chapter 12

A SIMPLE DMA CONTROLLER

A simple 4 channel DMA controller is presented as a practical description of a reasonably large-scale Balsa design written entirely in Balsa and so can be compiled for any of the technologies which the Balsa back-end supports. Readers should note that this controller is not the same as the Amulet3i DMA controller referred to in chapter 15. A more detailed description of this controller and the motivation for its design can be found in [8]. A complete listing of the code for the controller can be downloaded from [7].

The simplified controller provides:

- 4 full address range channels each with independent source, destination and count registers.
- 8 client DMA request inputs with matching acknowledgements.
- Peripheral to peripheral, memory to memory and peripheral to memory transfers. Each channel has both source and destination client requests so "true" peripheral to peripheral transfers can be performed by waiting for requests from both parties.

Figure 12.1 shows the programmer's view of the controller's register memory map. The register bank is split into two parts: the channel registers and the global registers.

12.1. Global registers

The global registers contain control state pertaining to the state of currently active channels and of interrupts signalled by the termination of transfer runs. There are 4 global registers:

genCtrl: General control. In this controller, the general control register only contains one bit: the global enable – `gEnable`. The global enable is the only controller bit reset at power-up. All other controller state bits must be initialised before `gEnable` is set. Using a global enable bit in this way allows the initialisation part of the Balsa description to remain small and cheap.

J. Sparsø and S. Furber (eds.), Principles of Asynchronous Circuit Design, 205–218.

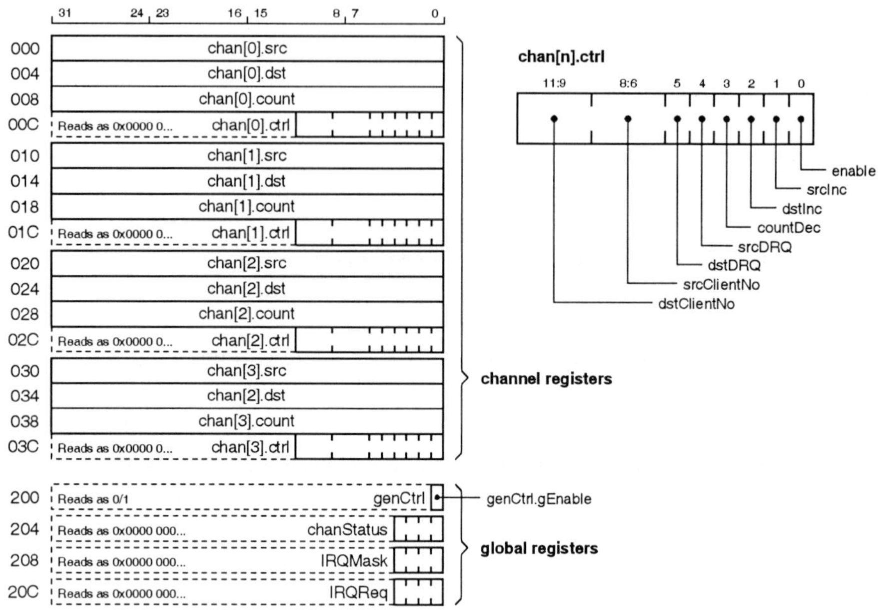

Figure 12.1. DMA controller programmer's model.

chanStatus: Channel end-of-run status. The `chanStatus` register contains 4 bits, one per DMA channel. When set by the DMA controller, a bit in this register indicates that the corresponding channel has come to the end of its run of transfers.

IRQMask, IRQReq: Interrupt mask and status. The `IRQMask` register contains one bit per channel (like `chanStatus`) with set bits specifying that an interrupt should be raised at the end of a transfer run of that channel (when the corresponding `chanStatus` bit becomes set). `IRQReq` contains the current interrupt status for each channel.

The channel status, IRQ mask and IRQ status bits are kept in global registers in order to reduce the number of DMA register reads which must be performed by the CPU after receiving an interrupt in order to determine which channel to service.

12.2. Channel registers

Each channel has 4 registers associated with it in the same way as the Amulet3i DMA controller. The two address registers (`channel[n].src` and `channel[n].dst`) specify the 32-bit source and destination addresses for transfers. The count register (`channel[n].count`) is a 32-bit count of remaining transfers to perform; transfer runs terminate when the count register is decremented to zero. The control register (`channel[n].ctrl`) specifies the updates to be performed on the other three registers and the clients to which this channel is connected. Writing to the control register has the effect of clearing interrupts and end-of-run indication on that channel. The control register contains 8 fields:

enable: Transfer enable. If the enable bit is set, this channel should be considered for transfers when a new DMA request arrives. Channel enables are not cleared on power-up. The `genCtrl.gEnable` bit can be used to prevent transfers from occurring whilst the channel enable bits are cleared during startup.

srcInc, dstInc, countDec: Increment/decrement control. These bits are used to enable source, destination and count register update after a transfer. Source and destination registers are incremented by 4 after transfers (since only word transfers are supported in this version of the controller) if `srcInc` and `dstInc` (respectively) are set. Note that the bottom 2 bits of these addresses are preserved. The count register is decremented by 1 after each transfer if countDec is set. Resetting either `srcInc` or `dstInc` results in the corresponding address remaining unchanged between transfers. This is useful for nominating peripheral (rather than memory) addresses. Resetting `countDec` results in "free-running" transfers.

srcDRQ, dstDRQ: Initial DMA requests. Transfers can take place on a channel when a pair of DMA requests have been received, one for the source client and the other for the destination client (the *requests-pending* registers). The `srcDRQ` and `dstDRQ` bits specify the initial states for those two requests. Setting both of these bits indicates that the source and destination requests should be considered to have already arrived. Resetting one or both of the bits specifies that requests from the corresponding {`src,dst`}`ClientNo` numbered client should trigger a transfer (both client requests are required when both control bits are reset).

srcClientNo, dstClientNo: Client to channel mapping. These fields specify the client numbers from which this channel receives source and destination DMA requests. These fields are only of use when either `srcDRQ` or `dstDRQ` (or both) are reset.

12.3. DMA controller structure

The structure of the simplified DMA controller is shown in Figure 12.2. The simplified DMA controller is composed of 5 units:

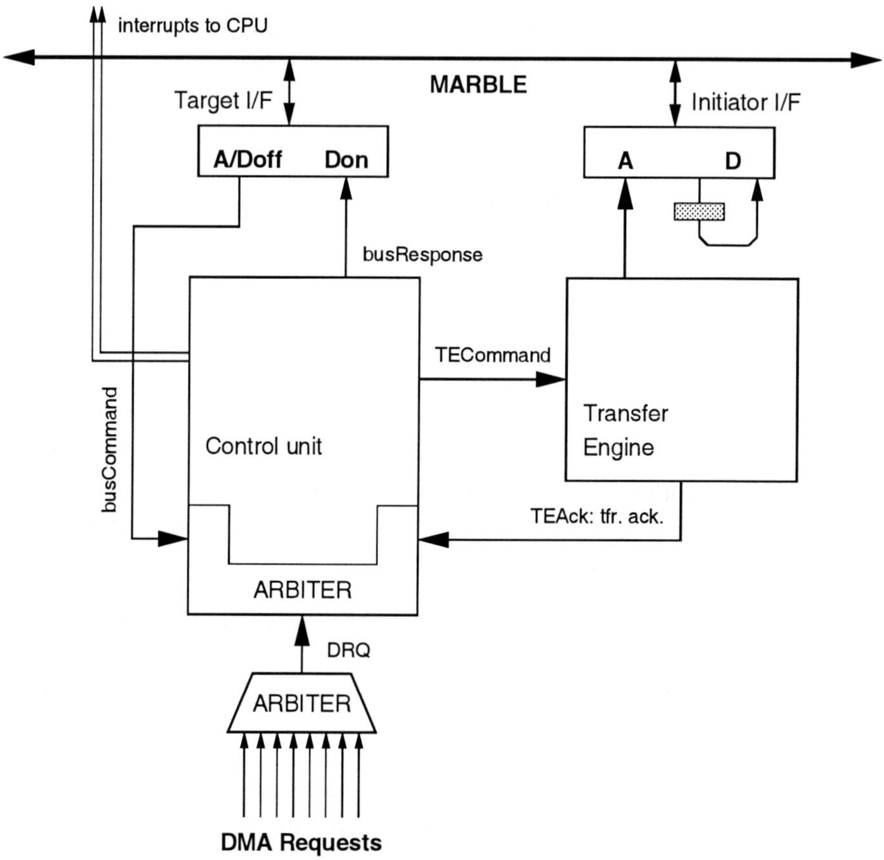

Figure 12.2. DMA controller structure.

MARBLE target interface

The controller is assumed to be attached to the MARBLE asynchronous bus which connects the subsystems in the Amulet3i system-on-chip (see chapter 15). It is relatively easy to provide an interface to other forms of on-chip system connect busses.

The MARBLE target interface provides a connection to MARBLE through which the controller can be programmed. Accesses to the registers from this interface are arbitrated with incoming DMA requests and *transfer acknowledgements* from the transfer engine. This arbitration and the decoupling of transfer engine from control unit allow the DMA controller to avoid potential bus access deadlock situations.

The MARBLE interface used here carries an 8-bit address (8-bit word address, 10-bit byte address) similar to that of the Amulet3i DMA controller. This allows the same address mapping of channel registers and the possibility of extendeding the number of channels to 32 without changing the global register addresses.

MARBLE initiator interface

The initiator interface is used by the DMA controller to perform its transfers. Only the address and control bits to this interface are connected to the Balsa synthesised controller hardware. The data to and from the initiator interface is handled by a latch (shown as the shaded box in Figure 12.2). Only word-wide transfers are supported and so this latch is all that is needed to hold data values between the read and write bus transactions of a transfer. Supporting different transfer widths is relatively easy but has not been implemented in this example in order to simplify the code.

Control unit

Each DMA channel has a pair of register bits, the *requests-pending* bits, which recode the arrival of requests for that channel's source and destination clients. After marking-up an incoming request, the control unit examines the *requests-pending* registers of each channel in turn to find a channel on which to perform a transfer. If a transfer is to be performed, the register contents for that channel are forwarded to the transfer engine and the register contents are updated to reflect the incremented addresses and decremented count. DMA requests are acknowledged straight away when no transfer engine command is issued, or just after the command is issued where a transfer command is issued to the transfer engine. The acknowledgement of DMA requests does not guarantee the timely completion of the related transfer, peripherals must observe bus accesses made to themselves for this purpose. The acknowledgement serves only to confirm the receipt of the DMA transfer request. A request must be removed after an acknowledgement is signalled so that other requests can be received through the request arbitration tree to mark-up potential transfers for other channels.

Transfer engine

The controller's transfer engine takes commands from the control unit when a DMA transfer is due to be performed and performs no DMA request mapping or filtering of its own. The only reason for having the transfer engine is to prevent the potential bus deadlock situation if an access to the register bank is made across MARBLE while the DMA controller is trying to perform a transfer. In this situation, control of the bus belongs to the initiator (usually the CPU) trying to access the DMA controller. This initiator cannot proceed as the DMA controller is engaged in trying to gain the bus for itself. With a transfer engine, and the decoupling of DMA request/CPU access from transfer operations, the control unit is free to fulfil the initiator's register request while the transfer engine is waiting for the bus to become available.

After performing a transfer, the transfer engine will signal to the control unit to provide a new transfer command; it does this by a handshake on the transfer acknowledge channel (marked `TEAck` in Figure 12.2). This channel passes through the control unit's command arbitration hardware and serves to inform the control unit that the transfer engine is free and that the request-pending register can be polled to find the next suitable transfer candidate. The acknowledgement not only provides the self-looping activation required to perform memory to memory transfers but also allows the looping required to service requests for other types of transfer which are received during the period when the transfer engine was busy.

A flag register, `TEBusy`, held in the control unit, is used to record the status of the transfer engine so that commands are not issued to it while a transfer is in progress. This flag is set each time a transfer command is issued to the transfer engine and cleared each time a transfer acknowledgement is received by the control unit. The request-pending registers are not re-examined (and a transfer command issued) if `TEBusy` is set.

Arbiter tree

The DMA controller receives DMA requests on an array of 8 sync channels connected to the input of the ARBITER unit shown in Figure 12.2. This arbiter unit is a tree of 2-way arbiter cells that combines these 8 inputs into a single DMA request number which it provides to the control unit. DMA requests are acknowledged as soon as the control unit has recorded them. Only the successful transfer of data between peripherals should be used as an indication of the actual completion of a DMA operation. When a transfer is begun (i.e. passed from control unit to transfer engine), that transfer's channel registers and requests-pending registers are updated before another arbitrated access to the control unit is accepted. As a consequence, a new request on a channel can arrive (and be correctly observed) as soon as any transfer-related bus activity occurs for that transfer.

12.4. The Balsa description

The Balsa description of the DMA controller is composed of 3 parts: the arbiter tree, the control unit and the transfer engine. The two MARBLE interfaces sit outside the Balsa block and are controlled through the target (`mta` and `mtd`) ports (corresponding to command and response ports) and the initiator address/control (`mia`) port. The top level of the DMA controller is:

```
procedure DMAArb is ArbFunnel (NoOfClients)

procedure dma (
  input mta : MARBLE8bACommand;
  output mtd : MARBLEResponse;
  output mia : MARBLECommandNoData;
  output irq : bit;
  array NoOfClients of sync drq
) is
  channel DRQClientNo : ClientNo
  channel TECommand : array 2 of Word  \textbf{--srcAddr, dstAddr}
  sync TEAck
begin
  DMAArb (drq, DRQClientNo) ||
  DMAControl (mta, mtd, DRQClientNo, TECommand, TEAck, IRQ) ||
  DMATransferEngine (TECommand, TEAck, mia)
end
```

Interrupts are signalled by writing a 0 or 1 to the `irq` port. This interrupt value must then be caught by an external latch to generate a bare interrupt signal.

12.4.1 Arbiter tree

DMA requests from the client peripherals arrive on the sync channels `drq`, these channels connect to the request arbiter `DMAArb`. The procedure declaration for `DMAArb` is given in the top level as a parameterised version of the procedure `ArbFunnel` and was described in chapter 11 on page 202. Figure 12.3 shows the organisation of an 8-input `ArbFunnel`.

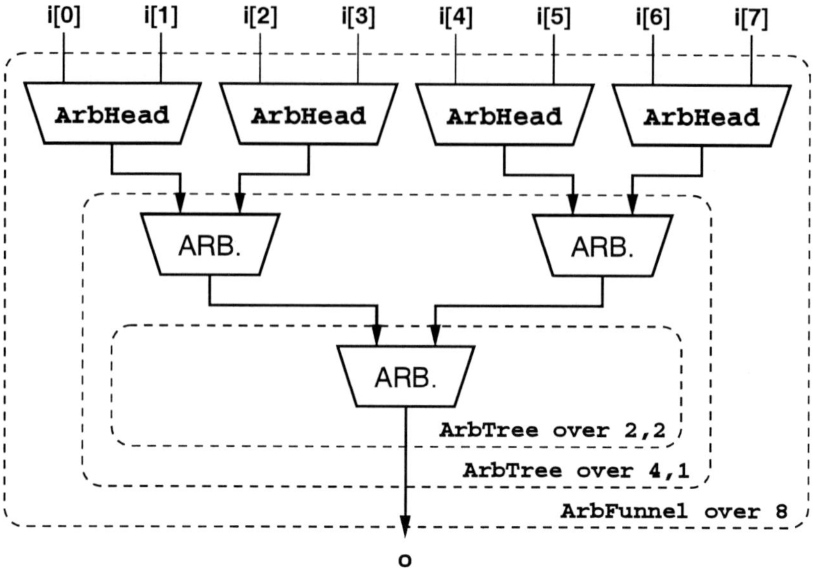

Figure 12.3. 8-input arbiter – ArbFunnel.

12.4.2 Transfer engine

The transfer engine is, like the arbiter unit, quite simple. It exists only as a buffer stage between the control unit and the MARBLE initiator interface. This function is reflected in the sequencing in the Balsa description and the latches used to store the outgoing addresses.

```
procedure DMATransferEngine (
  input command : array 2 of Word;
  sync ack;
  output busCommand : MARBLECommandNoData
) is
  variable commandV : array 2 of Word
begin
  loop
    command -> commandV;
    busCommand <- {commandV[0],read,word};
    busCommand <- {commandV[1],write,word};
    sync ack
  end
end
```

12.4.3 Control unit

The bulk of the controller is contained in the control unit. It contains all the channel register latch bits and register access multiplexers/demultiplexers. The reduced number of channels and single channel type makes this arrangement practical. There are in total 445 bits of programmer accessible state. The ports, local variables and local channels of the control unit's Balsa description are:

```
procedure DMAControl (
  input busCommand : MARBLE8bACommand;
  output busResponse : MARBLEResponse;
  input DRQ : ClientNo;
  output TECommand : array 2 of Word;
  sync TEAck;
  output IRQ : bit
) is
  -- combined channel registers
  variable channelRegisters :
    array NoOfChannels of ChannelRegister
  variable channelR, channelW : ChannelRegister
    array over ChannelRegType of bit
  variable channelNo : ChannelNo
  variable clientNo : ClientNo

  variable TEBusy : bit

  variable gEnable : bit
  variable chanStatus : array NoOfChannels of bit
  variable IRQMask, IRQReq : array NoOfChannels of bit

  variable requestPending :
    array NoOfChannels of RequestPair

  channel commandSourceC : DMACommandSource
  channel busCommandC : MARBLE8bACommand
  channel DRQC : ClientNo
  variable commandSource : DMACommandSource
  . . .
```

The `ChannelRegister` is the combined source, destination, count and control registers for one channel. The variable `channelRegisters` is accessed by reading or writing these 108-bit wide registers (32 + 32 + 32 + 12). The two registers, `channelR` and `channelW`, are used as read and write buffers to the channel registers. This allows the partial writes required for CPU access to individual 32-bit words to fragment only these two registers, not all of the channel registers. The variables `channelNo` and `clientNo` are used to hold channel and client numbers between operations. DMA request arrival and request mark-up can modify clientNo and channel register accesses and ready-to-transfer polling can modify channelNo.

The three channel declarations are used to communicate between a sub-procedure of DMAControl, RequestHandler, which arbitrates requests from the arbiter tree, MARBLE target interface and transfer engine acknowledge for service by the control unit. RequestHandler's description is fairly uninteresting and so will not be discussed.

The body of the control unit, with the less interesting portions removed, is as follows:

```
begin
  Init ();
  -- RequestHandler is an ArbFunnel
  --   with accompanying data
  RequestHandler (busCommand, DRQ, TEAck, commandSourceC,
    busCommandC, DRQC) ||
  loop
    -- find source of service requests
    commandSourceC -> commandSource;
    case commandSource of
      DRQ then DRQC -> clientNo; MarkUpClientRequest ()
    | bus then
      select busCommandC then
        if (busCommandC.a as RegAddrType).globalNchannel
        then . . . -- global R/W from the CPU
        else -- channel regs
          channelNo :=
            (busCommandC.a as ChannelRegAddr).channelNo;
          ReadChannelRegisters ();
          case busCommandC.rNw of
            . . . -- most of CPU reg. access code omitted
            -- CPU ctrl register write
            | ctrl then channelW.ctrl :=
              (busCommandC.d as ControlRegister) ||
              requestsPending[channelNo] := {0,0} ||
              ClearChanStatus ()
            end;
            WriteChannelRegisters ()
          end
        end
      end
    else -- TEAck
      TEBusy := 0;
      if gEnable then AssessInterrupts () end
    end;
    if gEnable and not TEBusy then
      TryToIssueTransfer ()
    end
```

```
    end
  end
```

A number of procedure calls are made by the control unit body, for example, `AssessInterrupts ()`. These procedure calls are to shared procedures whose definitions follow the local variables in `DMAControl`'s description. In Balsa, local procedures which are declared to be "shared" are only instantiated in the containing procedure's handshake circuit in one place. (Normal procedure calls place a new copy of that procedure's body for each call). Calls to shared procedures are combined using a Call component making their use cheaper than normal procedures for whom a new copy of the called procedure's body is placed at each call location.

DMA request handling – MarkUpClientRequest

Incoming DMA requests are marked up in the request pending registers as previously described. The procedure `MarkUpClientRequest` performs this operation by testing all the channels' `srcClientNo` and `dstClientNo` control bits with clientNo (the client ID of the incoming request) in parallel. `MarkUpClientRequest`'s description is:

```
shared MarkUpClientRequest is
 begin
   for || i in 0..NoOfChannels-1 then
     if channelRegisters[i].ctrl.srcClientNo = clientNo
       then requestsPending[i].src := 1
     end ||
     if channelRegisters[i].ctrl.dstClientNo = clientNo
       then requestsPending[i].dst := 1
     end
   end
 end
```

The `for ||` loops in this description performs parallel structural instantiation of `NoOfChannels` copies of the body `if` statements.

Register Access – ReadChannelRegisters, WriteChannelRegisters

The shared procedures used to access the channel registers are very short. They make the only variable-indexed accesses to the channel registers. The two procedures are:

```
shared ReadChannelRegisters is begin
  channelR := channelRegisters[channelNo]
end
```

```
shared WriteChannelRegisters is begin
  channelRegisters[channelNo] := channelW
end
```

Notice that no individual word write enables are present and so, in order to modify a single word, a whole channel register must be read, modified, then written back. The `ReadChannelRegisters` followed by `channelW := channelR` in the description of the CPU's access to the channel registers uses this update method.

Channel status and interrupts – ClearChanStatus, AssessInterrupts

The interrupt output bit is asserted by `AssessInterrupts`. Interrupts are signalled when the `IRQReq` register is non-zero and are reassessed each time an action which could clear an interrupt occurs. `ClearChanStatus` is called during channel control register updates to clear interrupts and channel status (end-of-run) indications. Their descriptions are:

```
shared AssessInterrupts is begin
  IRQ <- (IRQReq as NoOfChannels bits) /= 0
end

shared ClearChanStatus is begin
  chanStatus[channelNo] := 0 ||
  IRQReq[channelNo] := 0;
  AssessInterrupts ()
end
```

Ready-to-transfer polling – TryToIssueTransfer, IssueTransfer

Whenever the DMA controller is stimulated by its command interfaces, it tries to perform a transfer. The request-pending, and `ctrl[n].enable` bits for each channel are examined in turn to determine if that channel is ready to transfer. Incrementing the channel number during this search is performed using a local channel to allow the incremented value to be accessed in parallel from two places. `TryToIssueTransfer`'s Balsa description is:

```
shared TryToIssueTransfer is local
  variable foundChannel : bit
  variable newChannelNo : ChannelNo
begin
  foundChannel := 0 || channelNo := 0;

  while not foundChannel then
    -- source and destination requests arrived?
    if requestsPending[channelNo] = {1,1}
      and channelRegisters[channelNo].ctrl.enable then
```

```
          ReadChannelRegisters ();
          requestsPending[channelNo] :=
            channelR.ctrl.srcDRQ, channelR.ctrl.dstDRQ ||
          foundChannel := 1 ||
          IssueTransfer () ||
          UpdateRegistersAfterTransfer ()
        else
          local
            channel incChanNo : array ChannelNoLen + 1 of bit
          begin
            incChanNo <- (channelNo + 1 as
              array ChannelNoLen + 1 of bit) ||
            select incChanNo then
              foundChannel := incChanNo[ChannelNoLen] ||
              newChannelNo := (incChanNo[0..ChannelNoLen-1]
                as ChannelNo)
            end;
            channelNo := newChannelNo
          end
        end
      end
    end
```

Notice that if a transfer is taken, the `requestPending` bits for that channel are re-initialised from the `ctrl.{srcDRQ,dstDRQ}` control bits for that channel. The procedure `IssueTransfer` actually passes the transfer on to the transfer engine. Its definition is:

```
shared IssueTransfer is begin
  TEBusy := 1 ||
  TECommand <- {channelR.src, channelR.dst}
end
```

The interlock formed by checking `TEBusy` before attempting a transfer, and the setting/resetting of `TEBusy` by transfer initiation/completion ensures that no transfer is presented to the transfer engine (deadlocking the control unit) while it is occupied. The `TEAck` communication back to the control unit also provides stimuli for re-triggering the DMA controller to perform outstanding requests. This re-triggering, combined with the sequential polling of channels, allows outstanding requests (received while the transfer engine was busy) to be serviced correctly. Notice that a static prioritisation on pre-arrived requests is enforced by sequential channel polling.

Register increment/decrement – UpdateRegistersAfterTransfer

After a transfer has been issued, the registers for that transfer's channel must be updated. Procedure `UpdateRegistersAfterTransfer` performs this task:

```
shared UpdateRegistersAfterTransfer is begin
  channelW.ctrl := channelR.ctrl ||
  if channelR.ctrl.srcInc then
    channelW.src := (channelR.src + 1 as Word)
  end ||
  if channelR.ctrl.dstInc then
    channelW.dst := (channelR.dst + 1 as Word)
  end ||
  if channelR.ctrl.countDec then
    channelW.count := (channelR.count - 1 as Word)
  end;
  if channelW.count = 0 then
    chanStatus[channelNo] := 1 ||
    if IRQMask[channelNo] then
      IRQReq[channelNo] := 1
    end ||
    channelW.ctrl.enable := 0
  end;
  WriteChannelRegisters ()
end
```

This procedure uses two incrementers and a decrementer to modify the channel's source address, destination address and count respectively. If the channel's transfer count is decremented to zero, the `chanStatus` bit, interrupt status and channel enable are all updated to indicate an end-of-run.

This concludes the descriptions of the more illustrative aspects of the Balsa DMA controller description. For further details see [8] and, as was noted at the beginning of this chapter, a full code listing is available from [7].

III

LARGE-SCALE ASYNCHRONOUS DESIGNS

Abstract In this final part of the book we describe some large-scale asynchronous VLSI designs to illustrate the capabilities of this technology. The first two of these designs – the contactless smart card chip developed at Philips and the Viterbi decoder developed at the University of Manchester – were designed within EU-funded projects in the low-power design initiative that is the sponsor of this book series. The third chapter describes aspects of the Amulet microprocessor series, again from the University of Manchester, developed in several other EU-funded projects which, although outside the low-power design initiative, never-the-less still had low power as a significant objective.

The chips descibed in this part of the book are some of the largest and most complex asynchronous designs ever developed. Fully detailed descriptions of them are far beyond the scope of this book, but they are included to demonstrate that asynchronous design is fully capable of supporting large-scale designs, and they show what can be done with skilled and experienced design teams. The descriptions presented here have been written to give insight into the thinking processes that a designer of state-of-the-art asynchronous systems might go through in developing such designs.

Keywords: asynchronous circuits, large-scale designs

Chapter 13

DESCALE: *

a Design Experiment for a Smart Card Application consuming Low Energy

Joep Kessels & Ad Peeters

Philips Research, NL-5656AA Eindhoven, The Netherlands

{Joep.Kessels | Ad.Peeters} @philips.com

Torsten Kramer

Kramer-Consulting, D-21079 Hamburg, Germany

Kramer@kramer-consulting.de

Volker Timm

Philips Semiconductors, D-22529 Hamburg, Germany

Volker.Timm@philips.com

Abstract We have designed an asynchronous chip for contactless smart cards. Asynchronous circuits have two power properties that make them very suitable for contactless devices: low average power and small current peaks. The fact that asynchronous circuits operate over a wide range of the supply voltage, while automatically adapting their speed, has been used to obtain a circuit that is very resilient to voltage drops while giving maximum performance for the power being received. The asynchronous circuit has been built, tested and evaluated.

Keywords: low-power asynchronous circuits, smart cards, contactless devices, DES cryptography

*Part of the work described in this paper was funded by the European Commission under Esprit TCS/ESD-LPD contract 25519 (Design Experiment DESCALE).

J. Sparsø and S. Furber (eds.), Principles of Asynchronous Circuit Design, 221–248.

13.1. Introduction

Since their introduction in the eighties, smart cards have been used in a continuously growing number of applications, such as banking, telephony (telephone and SIM cards), access control (Pay-TV), health-care, tickets for public transport, electronic signatures, and identification. Currently, most cards have contacts and, for that reason, need to be inserted into a reader. For applications in which the fast handling of transactions is important, *contactless smart cards* have been introduced requiring only close proximity to a reader (typically several centimeters). The chip on such a card must be extremely power efficient, since it is powered by electromagnetic radiation.

Asynchronous CMOS circuits have the potential for very low power consumption, since they only dissipate when and where active. However, asynchronous circuits are difficult to design at the level of gates and registers. Therefore the high-level design language Tangram was defined [141] and a *silicon compiler* has been implemented that translates Tangram programs into asynchronous circuits.

The Tangram compiler generates a special class of asynchronous circuits called handshake circuits [135, 112]. Handshake circuits are constructed from a set of about forty basic components that use handshake signalling for communication.

Several chips have been designed in Tangram [136, 144] and if we compare these chips to existing clocked implementations, then the asynchronous versions are generally about 20% larger in area and consume about 25% of the power.

In order to find out what advantages asynchronous circuits have to offer in contactless smart cards, we have designed an asynchronous smart card chip. In this chapter, we indicate which properties of asynchronous circuits have been exploited and we present the results. The rest of the chapter is organized as follows. Section 13.2 presents the Tangram method for designing asynchronous circuits. Section 13.3 summarizes the differences in power behaviour between synchronous and asynchronous circuits. Section 13.4 first provides some background to contactless smart cards, then identifies the power characteristics in which contactless devices differ from battery-powered ones, and finally indicates why asynchronous circuits are very suited for contactless devices. Section 13.5 presents the digital circuit, Section 13.6 the results of the silicon, and Section 13.8 the power regulator, which also exploits an asynchronous opportunity. We conclude with a summary of the merits of this asynchronous design.

13.2. VLSI programming of asynchronous circuits

The design flow that is used in the design of the smart card IC reported here is based on the programming language Tangram and its associated compiler and tool set. An important aspect of this design approach is the *transparency* of the silicon compiler [142], which allows the designer to reason about design characteristics such as area, power, performance, and testability, at the programming (Tangram) level.

This section first introduces the Tangram tool set, then briefly describes the underlying handshake technology, and finally illustrates VLSI-programming techniques using the GCD algorithm presented in Chapter 8.

13.2.1 The Tangram toolset

Fig. 13.1 shows the Tangram toolset. The designer describes a design in Tangram, which is a conventional programming language, similar to C and Pascal, extended to include constructs for expressing concurrency and communication in a way similar to those in the language CSP [58]. In addition to this, there are language constructs for expressing hardware-specific issues, like sharing of blocks and waiting for clock-edges.

A compiler translates Tangram programs into so-called *handshake circuits*, which are netlists composed from a library of some 40 handshake components. Each handshake component implements a language construct, such as sequencing, repetition, communication, and sharing.

The handshake circuit simulator and corresponding performance analyzer give feedback to the designer about aspects such as function, area, timing, power, and testability.

The actual mapping onto a conventional (synchronous) standard-cell library is done in two steps. In the first step the component expander uses the component library to generate an abstract netlist consisting of combinational logic, registers, and asynchronous cells, such as Muller C-elements. This step also determines the data encoding and the handshake protocol; generally a four-phase single-rail implementation is generated. In the second step commercial synthesis tools and technology mappers are used to generate the standard-cell netlist. No dedicated (asynchronous) cells are required in this mapping, because all asynchronous cells are decomposed into cells from the standard-cell library at hand.

Similar language-based approaches using handshake circuits as intermediate format are described in [17, 9]. Design approaches in which asynchronous details are not hidden from the designer have also proven successful [80, 21, 83, 30, 64]. A general overview of design methods for asynchronous circuits is given in [69] and in Part I of this book.

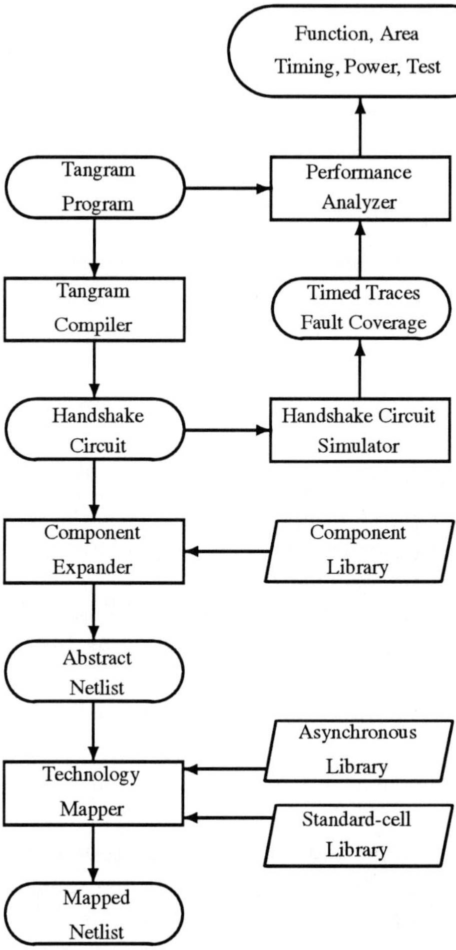

Figure 13.1. The Tangram toolset: boxes denote tools, ovals denote (design) representations.

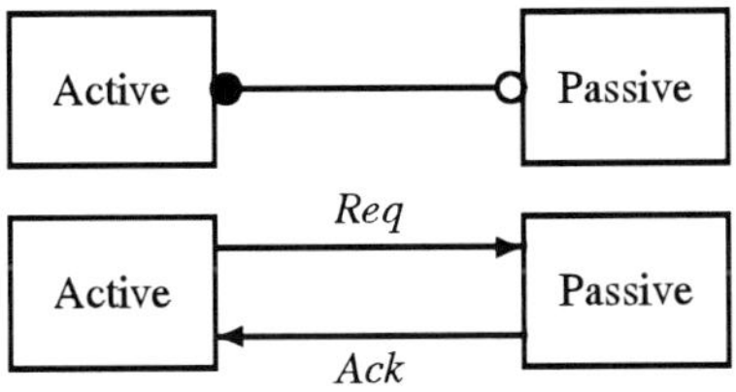

Figure 13.2. Handshake channel: abstract figure (top) and implementation (bottom).

13.2.2 Handshake technology

The design of large-scale asynchronous ICs demands a timing discipline to replace the clock regime that is used in conventional VLSI design. We have chosen handshake signaling [121] as the asynchronous timing discipline, since it supports plug-and-play composition of components into systems, and is also easy to implement in VLSI. An alternative to handshaking would be to compose asynchronous finite-state machines that communicate using fundamental mode or burst-mode assumptions [27, 132].

Fig. 13.2 shows a *handshake channel*, which is a point-to-point connection between an *active* and a *passive* partner. In the abstract figure, the solid circle indicates the channel's active side and the open circle its passive side. The implementation shows that both partners are connected by two wires: a request (*Req*) and an acknowledge (*Ack*) wire. A handshake requires the cooperation of both partners. It is initiated by the active party, which starts by sending a signal via *Req*, and then waits until a signal arrives via *Ack*. The passive side waits until a request arrives, and then sends an acknowledge. Handshake channels can be used not only for synchronization, but also for communication. To that end, data can be encoded in the request, the acknowledge, or in both.

The protocol used in most asynchronous VLSI circuits is a four-phase handshake, in which the channel starts in a state with both *Req* and *Ack* low. The active side starts a handshake by making *Req* high. When this is observed by the passive side, it pulls *Ack* high. After this a return-to-zero cycle follows, during which first *Req* and then *Ack* go low, thus returning to the initial state.

Handshake components interact with their environment using handshake channels. One can build handshake components implementing language constructs. Fig. 13.3 shows two examples: the *sequencer* and the *parallel* component.

The sequencer, when activated via a, performs first a handshake via b and then via c. It is used to control the sequential execution of commands connected to b and c. After receiving a request along a, it sends a request along b, waits for the corresponding acknowledge, then sends a request along c, waits

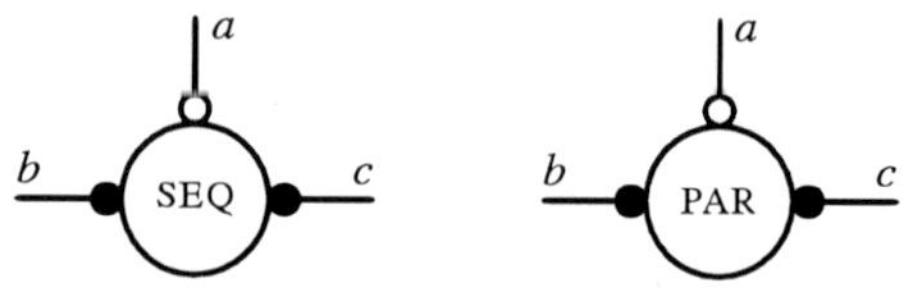

Figure 13.3. Handshake components: sequencer (left) and parallel (right).

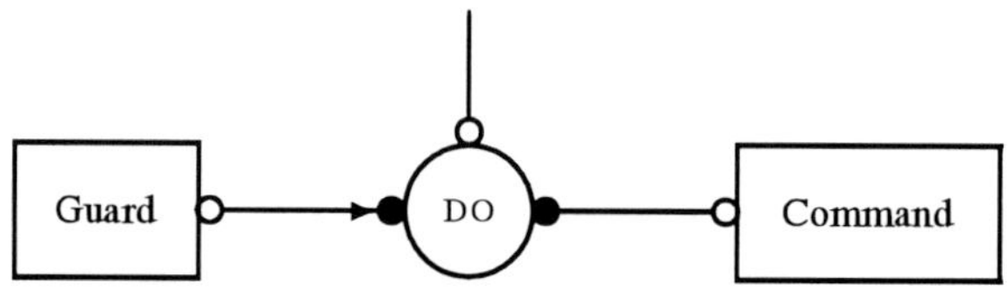

Figure 13.4. Handshake circuit for while loop.

for the acknowledge on c, and finally signals completion of its operation by sending an acknowledge along channel a.

The parallel component, when activated by a request along a, sends requests along channels b and c concurrently, waits until both acknowledges have arrived, and then sends an acknowledge along channel a.

Components for storage of data (variables) and operation on data (such as addition and bit-manipulation) can also be constructed. Tangram programs are compiled into handshake circuits (composition of handshake components) in a syntax-directed way (see also Chapter 8 on page 123). For instance, the compilation of a *while* loop in Tangram, which is written as

```
do  Guard  then  Command  od
```

results in the handshake circuit shown in Fig. 13.4. The do-component, when activated, collects the value of the guard through a handshake on its active data port. When this guard is *false*, it completes the handshake on its passive port, otherwise it activates the command through a handshake on its active control port, and after this has completed, re-evaluates the guard to start a new cycle.

Details about handshake circuits, the compilation from Tangram into this intermediate architecture, the four-phase handshake protocols that are applied, and of the gate-level implementation of handshake components can be found in [141, 135, 112].

13.2.3 GCD algorithm

When designing in Tangram, the emphasis for an initial design typically is on functional correctness. When this is the only point of attention, the result

```
  int = type [0..255]
& gcd_gc : main proc (in?chan <<int,int>> & out!chan int).
  begin x,y:var int ff
  | forever do
      in?<<x,y>>
    ; do x<>y then
        if x<y then y:=y-x
                 else x:=x-y
        fi
      od
    ; out!x
    od
  end
```

Figure 13.5. GCD algorithm in Tangram.

is generally too large, too slow, and too power hungry. The design of a suitable datapath and control architecture, targetting some optimization criteria or cost function, can be approached in a transformational way. One can use the Tangram tool set to evaluate and analyze the design, and based on that, decide which transformations to apply. The transparency of the silicon compiler (‘*What you program is what you get*’) helps in predicting the effect of these transformations.

The GCD algorithm is used as an example to illustrate VLSI programming based on a transparent silicon compiler (see also the discussion in section 8.3.3 on page 127). We start with the algorithm of Fig. 13.5, which is functionally correct but far from optimal when it comes to implementation cost in VLSI. The corresponding handshake circuit is shown in Fig. 13.6.

The cost report for this GCD design, as presented by the Tangram toolset, is the following:

celltype	#occ	area	gate eq. [eqv]	(%)
Delay Matchers	19	2052	38.0	12.5
Memory	16	3744	69.3	22.8
C-elements	12	1242	23.0	7.5
Logic	81	9414	174.3	57.2
Total:	128	16452	304.7	100.0

An important aspect of the design is that it contains four operators in the datapath. We can improve this by changing the Tangram description in such a way that only one subtractor is needed, instead of two. A way to achieve this is to change the timing behaviour of the algorithm, and use a higher number of iterations of the do-loop by either swapping or subtracting the two numbers

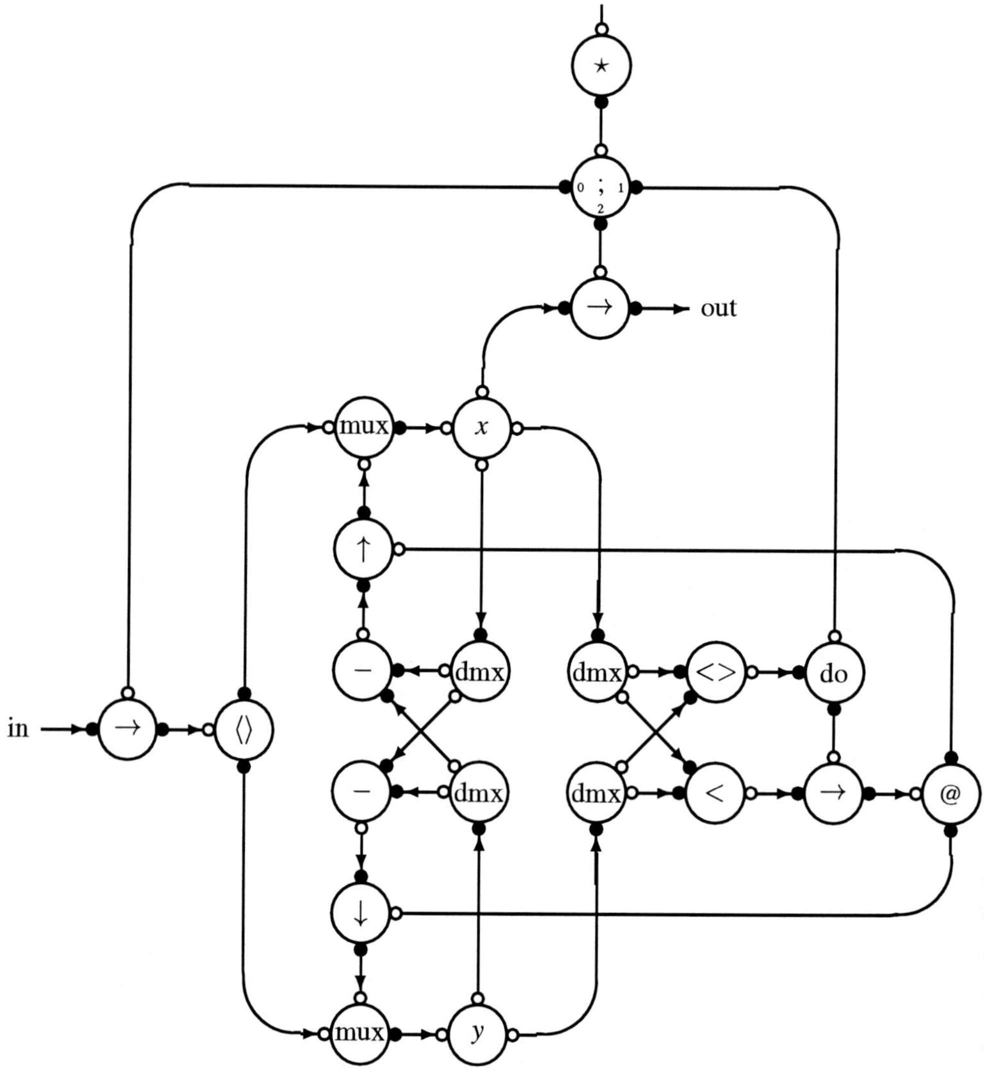

Figure 13.6. Compiled handshake circuit for initial GCD program.

```
  int = type [0..255]
& gcd_gc : main proc (in?chan <<int,int>> & out!chan int).
  begin
    xy : var <<int,int>> ff
  & x = alias xy.0
  & y = alias xy.1
  | forever do
      in?xy
    ; do x<>y then
        if x<y then xy:= <<x,y-x>>
               else xy:= <<y,x>>
        fi
      od
    ; out!x
    od
  end
```

Figure 13.7. GCD algorithm in Tangram with optimized control.

x and y. This requires that x and y are stored in a single flip-flop variable. The new Tangram algorithm thus obtained is shown in Fig. 13.7. Its associated gate-level cost report has improved from 305 to 274 gate equivalents.

celltype	#occ	area	gate eq. [eqv]	(%)
Delay Matchers	14	1512	28.0	10.2
Memory	16	3744	69.3	25.3
C-elements	10	1008	18.7	6.8
Logic	86	8532	158.0	57.7
Total:	126	14796	274.0	100.0

An additional transformation is to compute x<y and y-x using only one operator, and to combine the two assignments to variable xy into one assignment, so as to further simplify the control, at the price of always requiring the worst-case computation time for the conditional expression, even if it involves just a swap of x and y. Furthermore, one can allow one additional step in the computation, so that the termination condition of the loop simplifies from x<>y to y<>0. This final implementation is shown in Fig. 13.8; its handshake circuit is shown in Fig. 13.9, in which the datapath operations have been represented in an abstract way, rather than as separate components.

The handshake circuit of the optimized design has a simpler structure than that of the initial design. The number of logic blocks (and, in the single-rail implementation, the number of delay-matching gate chains) has been reduced from four to two. This improvement is also apparent in the area statistics given below.

```
  int = type [0..255]
& gcd_gc : main proc (in?chan <<int,int>> & out!chan int).
  begin
    xy : var <<int,int>> ff
  & x = alias xy.0
  & y = alias xy.1
  & comp: func(). (y-x) cast <<int,bool>>
  | forever do
      in?xy
    ; do y<>0 then
        xy:= if -comp.1 then <<x,comp.0>> else <<y,x>> fi
      od
    ; out!x
    od
  end
```

Figure 13.8. GCD algorithm in Tangram with optimized datapath.

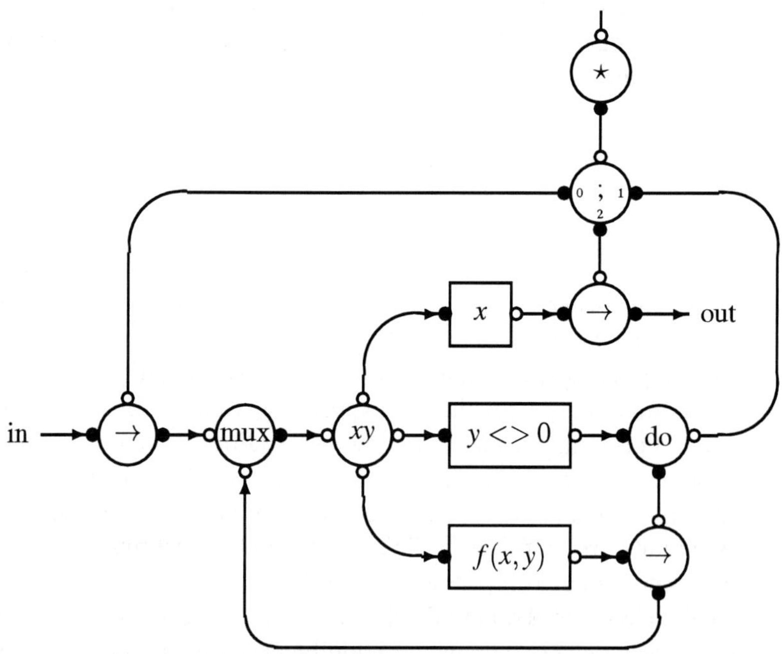

Figure 13.9. Compiled handshake circuit for optimized GCD program.

celltype	#occ	area	gate eq. [eqv]	(%)
Delay Matchers	10	1080	20.0	9.4
Memory	16	3744	69.3	32.7
C-elements	6	576	10.7	5.0
Logic	42	6048	112.0	52.8
Total:	74	11448	212.0	100.0

One may observe that area-wise, the design has improved in the control (the number of C-elements was reduced from 12 to 6), in the logic (the area for combinational logic was reduced from 174 to 112 gate equivalents), and in the total number of delay elements required (which was reduced from 19 to 10).

13.3. Opportunities for asynchronous circuits

When the asynchronous circuits generated by the Tangram compiler are compared to synchronous ones, three differences stand out, leading to four attractive properties of these asynchronous circuits.

1 The subcircuits in a synchronous circuit are clock-driven, whereas they are demand-driven in an asynchronous one. This means that the subcircuits in an asynchronous circuit are only active when and where needed. Asynchronous circuits will therefore generally dissipate less power than synchronous ones.

2 The operations in a synchronous circuit are synchronized by a central clock, whereas they are synchronized by distributed handshakes in an asynchronous circuit. Therefore

   a) a synchronous circuit shows large current peaks at the clock edges, whereas the power consumption of an asynchronous circuit is more uniformly distributed over time;

   b) the strict periodicity of the current peaks in a synchronous circuit leads to higher clock harmonics in the emission spectrum, which are absent in the spectrum of an asynchronous design.

3 Synchronous circuits use an external time reference, whereas asynchronous circuits are self-timed. This means that asynchronous circuits operate over a wide range of the supply voltage (for instance, from 1 up to 3.3 V) while automatically adapting their speed. This property, called *automatic performance adaptation*, implies that asynchronous circuits are resilient to supply voltage variations. It can also be used to reduce the power consumption by adaptive voltage scaling, which means adapting the supply voltage to the performance required [100]. Adaptive volt-

age scaling techniques are also applied in synchronous circuits, but then special measures must be taken to adapt the clock frequency.

Asynchronous circuits also have drawbacks. The most important one is that these circuits are unconventional: designers and mainstream tools and libraries are all oriented towards synchronous design methods. Additional drawbacks of asynchronous circuits come from the fact that they use gates to control registers (latches and flip-flops), instead of the relatively straightforward clock-distribution network in synchronous circuits. Although this enables the low power consumption it also leads to circuits that are typically larger, slower, and harder to test. Testability (for fabrication defects) is probably the most fundamental issue of these. For a discussion on testing asynchronous circuits we refer to [61, 120].

Property 2.b was the main reason for Philips Semiconductors to design a family of asynchronous pager chips [114]. However, as is addressed in the next section, it is the other three properties that can be used most advantageously in contactless smart card chips.

13.4. Contactless smartcards

Contactless smart cards have a number of advantages when compared to contacted ones: they are convenient and fast to use, insensitive to dirt and grease and, since their readers have no slots, they are less amenable to vandalism.

The communication between a contactless smart card and reader is established through an electromagnetic field, emitted by the reader. The card has a coil through which it can retrieve energy from this field. The amount of energy that can be retrieved depends on the distance to the reader, the number of turns in the coil, and the orientation of the card in the field.

Fig. 13.10 shows the functional parts of a contactless smartcard consisting of a VLSI circuit (in the dotted box) and an external coil. The tuned circuit formed by the coil and capacitor C_0 is used for three purposes:

- receiving power;
- receiving the clock frequency (equal to the carrier frequency); and
- supporting the communication.

The complete circuit consists of a power supply unit and a digital circuit with a buffer capacitor (C_1) for storing energy.

Several standards for contactless smart cards currently exist:

- ISO/IEC 10536, which specifies *close coupling* cards, operating at a distance of 1cm from the reader.

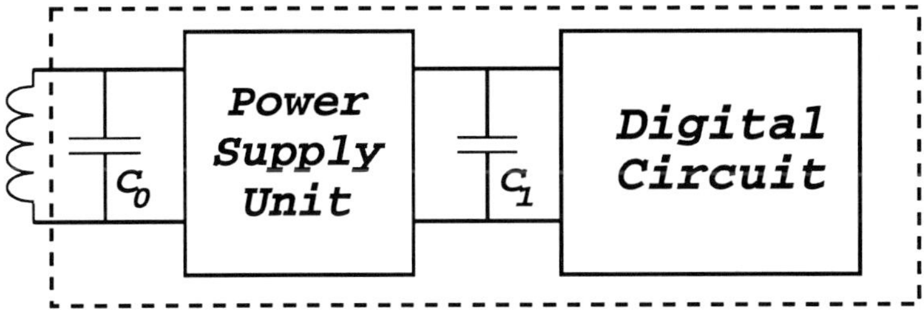

Figure 13.10. Contactless smart card.

Table 13.1. Main characteristics of ISO/IEC 14443 standard.

ISO 14443 standard	A (Mifare)	B
Carrier frequency	13.56 MHz	13.56 MHz
Throughput (up and down)	106kbit/sec	106kbit/sec
Down link (reader to card)	ASK 100%	ASK 10%
encoding	Miller	NRZ
Up link (card to reader)	ASK	BPSK
frequency	847.5 kHz	847.5 kHz
modulation	Manchester	NRZ

- ISO/IEC 14443, for so-called *proximity integrated circuit cards* (PICCs), operating at a distance of up to 10cm from the reader, typically using 5 turns in the on-card coil. This standard defines two types, A and B, the main characteristics of which are given in Table 13.1.

- ISO/IEC 15693 specifies *vicinity integrated circuit cards* (VICCs), operating at some 50cm from the reader, and typically requiring a coil with a few hundreds of turns.

The Mifare [63] standard (ISO/IEC 14443 type A) has hundreds of millions of cards in operation today. Fig. 13.11 shows a Mifare card with both the chip and the coil visible. Mifare is a proximity card (it can be used up to 10 cm from the reader) supporting two-way communication. Performance is important, since the transaction time must be less than 200 msec. One of the first companies to deploy Mifare technology *en masse* was the Seoul Bus Association, which has millions of such bus cards in use, generating hundreds of millions of transactions per month.

This chapter reports an asynchronous Mifare smart card IC that was reported earlier in [73]. Both synchronous [116] and asynchronous [2] circuits for smart cards of the ISO/IEC 14443 type B standard have also been reported. Due to the 100% ASK modulation scheme in the type A standard, a Mifare IC is exposed to periods during which no power comes in at all, in contrast to the type B standard, which is based on 10% ASK modulation.

Since on average (over time) contactless smart card chips receive only a few milliwatts of power, power efficiency is very important. Although low power is also important in battery-powered devices, there are two crucial differences between these two types of device.

1 To maximize the battery life-time in battery-powered devices one should minimize the *average* power consumption. In contactless devices, however, one should in addition minimize the *peak* power, since the peaks must be kept below a certain level, which depends on the incoming power and the buffer capacitor.

2 The supply voltage is nearly constant in battery-powered devices whereas in contactless ones it may vary over time during a transaction due to fluctuations in both the incoming and the consumed power.

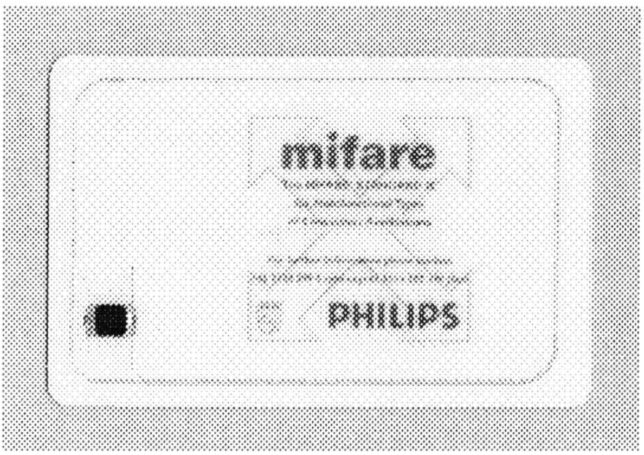

Figure 13.11. Mifare card, showing IC (bottom left) and coil.

In the bullets below, we give some facts about conventional synchronous chips for contactless smart cards, which, as we will see later, offer opportunities for improvement by using asynchronous circuits.

- A synchronous circuit runs at a fixed speed dictated by the clock, despite the fact that both the incoming and the effectively consumed power vary over time. Synchronous circuits must, therefore, be designed so as to allow the most power-hungry operations to be performed when minimum power is coming in. Consequently, if too much power is being

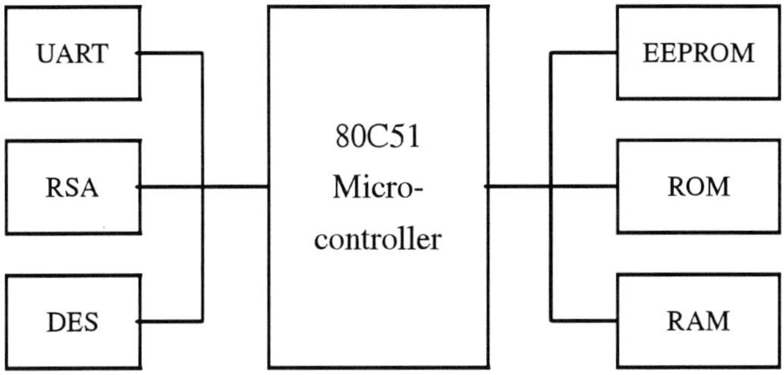

Figure 13.12. **Global design of the smart-card circuit.**

received, that superfluous power is thrown away. If, on the other hand, too little power is being received, the supply voltage drops making the circuit slower and, as soon as the circuit has become too slow to meet the time requirements set by the clock, the transaction must be canceled. For this reason contactless smart card chips contain subcircuits that detect when the voltage drops below a certain threshold and then abort the transaction.

- Currently, the performance of the microcontroller in a contactless smart card chip is usually not limited by the speed of the circuit, but by the RF-power being received.
- A synchronous circuit requires a buffer capacitor of several nanofarads and the area needed for such a capacitor is of the same order of magnitude as the area needed for the microcontroller.
- The communication from the smart card to the reader is based on modulating the load, which implies that normal functional load fluctuations may interfere with the communication.

13.5. The digital circuit

We have built the digital circuit shown in Fig. 13.12 that consists of:

- an 80C51 microcontroller;
- three kinds of low-power memory, the sizes and access times of which are given in Table 13.2 (64 bytes can be written simultaneously in one write access to the EEPROM);
- two encryption coprocessors:
  - an RSA converter [119] for public key conversions and

- a triple DES converter [96] for private key conversions;

- a UART for the external communication.

The EEPROM contains program parts as well as data such as encryption keys and e-money. Both the ROM and the RAM are equipped with matching delay lines and for the EEPROM we designed a similar function based on a counter. These delay lines have been used to provide all three memories with a handshake interface, which made it extremely easy to deal with the differences in access time as well as variations in both temperature and supply voltage. An additional advantage is that the controller automatically runs faster when executing code from ROM than when executing code from EEPROM.

The circuit is meant to be used in a *dual-interface* card, which is a card with both a contacted and a contactless interface. Apart from the RSA converter, which will not be used in contactless operation, all circuits are asynchronous. In contactless operation, the average supply voltage will be about 2 V. The simulations, however, are done at 3.3 V, which is the voltage at which the library has been characterized.

Table 13.2. Memory sizes and access times.

Memory type	Size [kbyte]	Access time [ns] read	write
RAM	2	10	10
ROM	38	30	
EEPROM	32	180	4,000

13.5.1 The 80C51 microcontroller

The 80C51 microcontroller is a modified version of the one described in [144, 143]. The four most important modifications are described below.

To deal with the slow memories a *prefetch unit* has been included in the 80C51 architecture. At 3.3 V the average instruction execution time in free-running mode is about 100 ns provided it takes no time to fetch the code from memory. If, however, code is fetched from the EEPROM and the microcontroller has to wait during the read accesses, the performance would be drastically reduced, since most instructions are one or two bytes long, taking 180 or 360 ns to fetch. To avoid this performance degradation a form of *instruction prefetching* has been introduced in which a process running concurrently to the 80C51 core is fetching code bytes as long as a two byte FIFO is not full.

The prefetch unit gives an increase in performance of about 30%. A simplified version of the prefetch unit is described in Section 13.5.2.

We also introduced *early write completion*, which means that the microcontroller continues execution as soon as it has issued a write access. This has been introduced to prevent the microcontroller from waiting during the 4 msec it takes to do a write access to the EEPROM (for instance to change the e-money), but also to speed up the write accesses to the RAM. To exploit this feature when doing a write access to the EEPROM, the corresponding code must be in the ROM.

The controller has been provided with an *immediate halt* input signal by which its execution can be halted within a short time. This provision is necessary to deal with the fact that the information, which the reader sends to the card, is coded by suppressing the carrier during periods of 3 μsec. Since the card does not receive any power during these periods, the controller has to be halted immediately (only some basic functions continue to operate). In the synchronous design this halting function came naturally, since the clock would stop during these periods.

We have introduced a *quasi synchronous* mode, in which the microcontroller is, at instruction level, fully timing compatible with its synchronous counterpart. In this mode, the asynchronous microcontroller waits after each instruction until the number of clock ticks required by a synchronous version have elapsed. This mode is necessary when time-dependent functions are designed in software. Since this mode is under software control, the microcontroller can easily switch modes depending on the function it is executing. This feature was also of great help to demonstrate the *guaranteed performance*, which is the maximum clock rate at which each instruction terminates within the given number of clock ticks. For most programs, the *free-running performance* is about twice as high as the guaranteed performance.

We have compared the asynchronous circuit as compiled from Tangram with a synchronous circuit that is functionally equivalent, and which has been compiled from synthesizable VHDL to the same CMOS technology. These ICs have a comparable performance. The asynchronous microcontroller nicely demonstrates the three properties of asynchronous circuits that we want to exploit in the design of the smart-card chip:

- The average power consumption of the asynchronous 80C51 is about three times lower than the power consumption of its synchronous counterpart when delivering the same performance at the same supply voltage.

- Fig. 13.13 shows the current peaks of both the synchronous and the asynchronous 80C51 at 3.3 V, where the asynchronous version is running in quasi synchronous mode, giving a performance that is 2.5 times higher

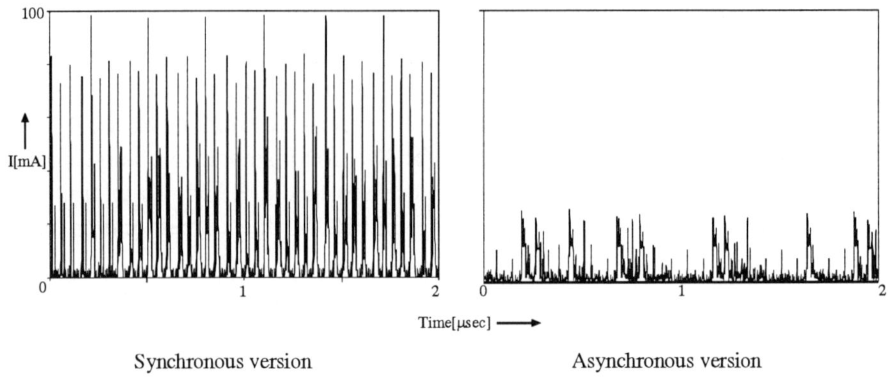

Figure 13.13. **Current peaks of 80C51 microcontroller.**

than the synchronous design (the synchronous version runs at 10 MHz and the asynchronous version at 25 MHz). Despite the fact that the figure does not give a fair impression of the average power being consumed, it clearly shows that the current peaks of the asynchronous 80C51 are about five times smaller than those of the synchronous equivalent.

- The performance adaptation property of asynchronous circuits is demonstrated in Fig. 13.14, which shows the free-running performance of the microcontroller, when executing code from ROM, as a function of the supply voltage. As is expected, the performance depends linearly on the supply voltage. When the supply voltage goes up from 1.5 to 3.3 V, the performance increases from 3 to 8.7 MIPS (about a factor 3). Since the ROM containing the program does not function properly when the supply voltage is below 1.5 V, we could not measure the performance for lower values. We observed, however, that the DES coprocessor, which does not need a memory, still functions correctly at a supply voltage level as low as 0.5 V.

  The figure also shows the supply current as a function of the supply voltage. Note that the current increases in this range from 0.7 to 6 mA (about a factor 9). Since in CMOS circuits the current is the product of the transition rate (performance) and the charge being switched per transition (both of which depend linearly on the supply voltage), the current increases with the square of the voltage. From this it follows that the power, being the product of the current and the voltage, goes up with the cube of the voltage.

From this data one can compute the third curve showing the energy needed to execute an instruction, which increases with the square of the supply voltage from 0.35 to 2.25 nJ.

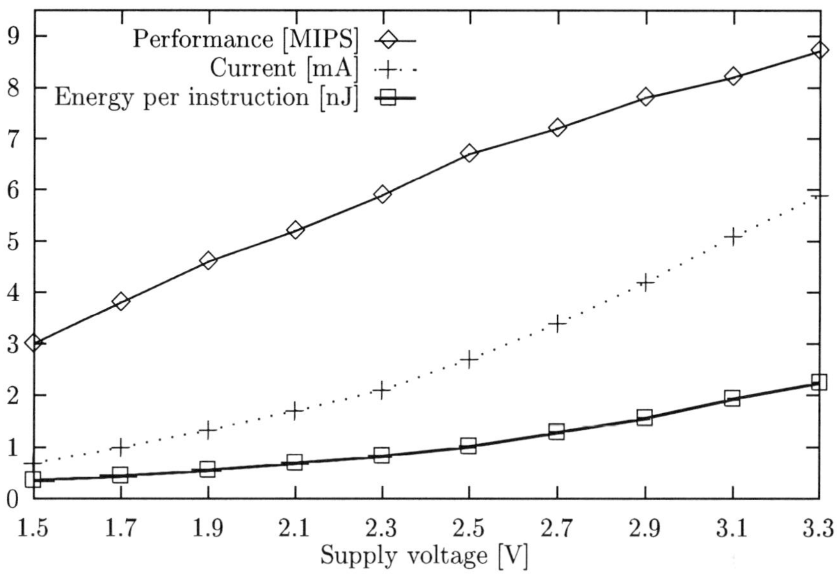

Figure 13.14. Measured performance of the asynchronous 80C51 for various supply voltages.

13.5.2 The prefetch unit

Fig. 13.15 gives the Tangram code of a simplified version of the prefetch unit. The prefetch unit communicates with the 80C51 core through two channels: it receives the address from which to start fetching code bytes via channel `StartAddress` and then it sends these bytes through channel `CodeByte`. Since the prefetch unit plays the passive role in both communications, it can probe each channel to see whether the core has started a communication through that channel. The state of the prefetch unit consists of program counter, `pc`, and a two-place buffer, which is implemented by means of an array `Buffer`, an integer `count`, and two one-bit pointers `getptr` and `putptr`.

```
forever do
   sel probe(StartAddress)
       then (  StartAddress?pc
            || putptr := getptr
            || count := 0
            || AbortMemAcc()
            )
    or probe(CodeByte) and (count>0)
       then CodeByte!Buffer[getptr]
          ; (  getptr := next(getptr)
            || count := count-1
            )
    or MemAck
       then Buffer[putptr] := MemData
          ; (  putptr := next(putptr)
            || count := count+1
            || pc := pc+1
            || CompleteMemAcc()
            )
   les
 ; if (count<2) and -MemReq
   then MemReq := true
   fi
od
```

Figure 13.15. Tangram code of a simplified version of the prefetch unit.

The prefetch unit executes an infinite loop and in each step it first executes a selection command (denoted by `sel` ... `les`), in which it can select among three *guarded commands*, which are separated by the keyword `or`. Each guarded command is of the form

```
Guard  then  Command,
```

in which `Guard` is a Boolean condition, and `Command` a command that may be executed only if the guard holds. A command is said to be *enabled* if the corresponding guard holds. Executing a selection command implies waiting until at least one of the commands is enabled, then selecting such a command —in an arbitrated choice— and executing it.

In the first guarded command, channel `StartAddress` is probed to find out whether the core is sending a new start address. In that case, program counter `pc` is set to the address received, the buffer is flushed, and a possible outstanding memory access is aborted (by resetting both `MemReq` and the delay counter). All four subcommands in this guarded command are executed in parallel ('`A || B`' means execute commands `A` and `B` concurrently, whereas '`A ; B`' means execute `A` and `B` sequentially).

The second guarded command takes care of sending the next program byte via channel `CodeByte` to the core if the core is ready to receive that byte and the buffer is not empty. The third guarded command gets enabled if `MemAck` goes high indicating that the data signals in a read access are valid. In that case the value read from memory is put in the buffer after which the memory handshake is completed.

After each event, if the buffer is not full and no memory access is being performed a next memory access is started. Since `(count<2) V -MemReq` is an invariant of the loop, the last (conditional) command in the loop can be simplified to unconditional assignment `MemReq:= (count<2)`.

Note that the value `(pc-count)` is equal to the program counter in the core, since it is set to the destination address in the case of a jump, increased by 1 if a code byte is transferred to the core, and kept invariant if a code byte is read from memory. Therefore the core does not need to hold the program counter and instead, when the information is needed for a relative branch, it can retrieve the counter value from the prefetch unit. Clearly, this feature requires an extension of the Tangram code shown in Fig. 13.15.

13.5.3 The DES coprocessor

A transaction may need up to ten single DES conversions, where each conversion takes about 10 ms if it is executed in software. Therefore a hardware solution is needed, since these conversions would otherwise consume about half of the transaction time.

Fig. 13.16 shows the datapath of the DES coprocessor. The processor supports both single- and triple-DES conversions and, for the latter type of conversion, contains two keys: a foreground and a background key. Single-DES conversions use the foreground key, whereas triple-DES conversions use the foreground key for the first and third conversion and the background key for the second conversion. The foreground key is stored in register `CD0` consisting of 56 flipflops (the DES key size is 56 bits), whereas the background key resides in variable `CD1` consisting of 56 latches. The text value resides in variable `LR` consisting of 64 latches (DES words contain 64 bits).

A single-DES conversion consists of 16 steps and, in each step, the key is permuted and a new text value is computed from the old text value and the key. In order to have the key return to its original value at the end of a conversion, the key makes two basic permutations in 12 of the 16 steps and only one in the remaining 4, where 28 basic permutations are needed for a complete cycle. The permutations are performed in flipflop register `CD0`.

Most of the area is taken by the combinational circuit called *DES*. Since this circuit is also dominant in power dissipation, one should minimize the number of transitions at its inputs. For this purpose, we have introduced two latch

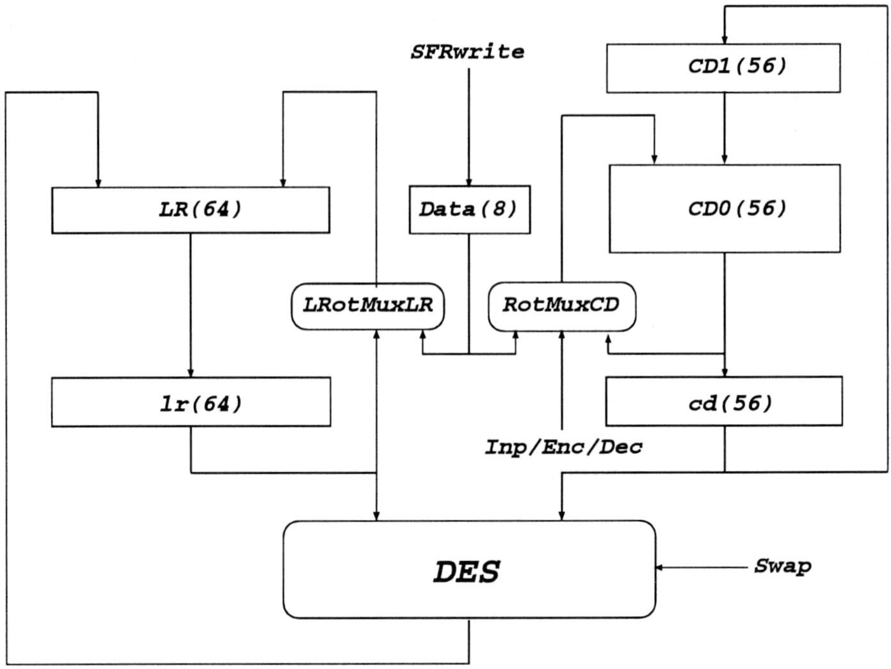

Figure 13.16. DES coprocessor architecture.

registers: cd for the key and lr for the text. If two basic permutations are done in one step, cd hides the effect of the first one from combinational circuit *DES*. In addition, all inputs of combinational circuit *DES* change only once in each step by loading the two registers lr and cd simultaneously and then storing the result in register LR as described by the following piece of Tangram text.

```
( lr:= LR || cd:= CD0 ) ; LR:= DES(lr,cd)
```

Therefore, latch register lr also serves as a kind of slave register. Latch register cd also serves a functional purpose, since the two keys are swapped by executing the following three transfers.

```
cd:= CD0 ; CD0:= CD1 ; CD1:= cd
```

The size of the DES coprocessor is 3,250 gate equivalents, of which 57% is taken by the combinational logic and 35% by latches and flip-flops. Consequently, the overhead in area due to the asynchronous design style (delay matching and C-elements) is marginal at 8%. At 3.3 V, a single-DES conversion takes 1.25 μs and 12 nJ.

Fig. 13.17 shows the simulated current of the DES coprocessor at 3.3 V (the microcontroller is active before and after the DES computation). The real current peaks will be much smaller due to a lower supply voltage (the DES processor functions properly at a supply voltage as low as 0.5 V) as well as the buffer capacitor (the resolution in the simulation is 1 ns).

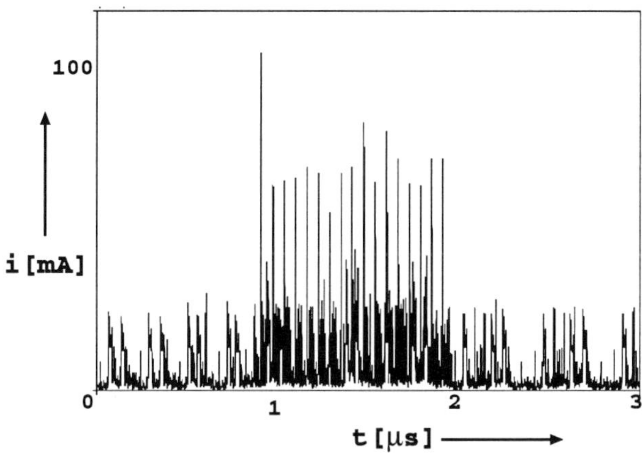

Figure 13.17. DES coprocessor current at 3.3 V.

The conversion time, of a few microseconds, is so small that we used the handshaking mechanism to obtain synchronization between the microcontroller and the coprocessor. After starting the coprocessor, the microcontroller can continue executing instructions, and only when reading the result will it be held up in a handshake until the result is available. Note that a synchronous design would require a form of busy-waiting.

13.6. Results

Fig. 13.18 shows the layout of the chip, which is in a five-layer metal, 0.35 μm technology and has a size of $4.52 \times 4.16 \approx 18$ mm^2, including the bond pads. Many bond pads are only included for measurement and evaluation purposes. A production chip only needs about 10 bond pads.

The two horizontal blocks on top form the buffer capacitor (in a production chip, the capacitor would only require about one quarter of the area). The memories are on the next row, from left to right: two RAMs, one ROM and the EEPROM, which is the large block to the right. The asynchronous circuit is located in the lower left quadrant, near the centre.

Table 13.3a gives the area of the blocks constituting the *contactless digital circuit*, which is the asynchronous circuit together with the memories.

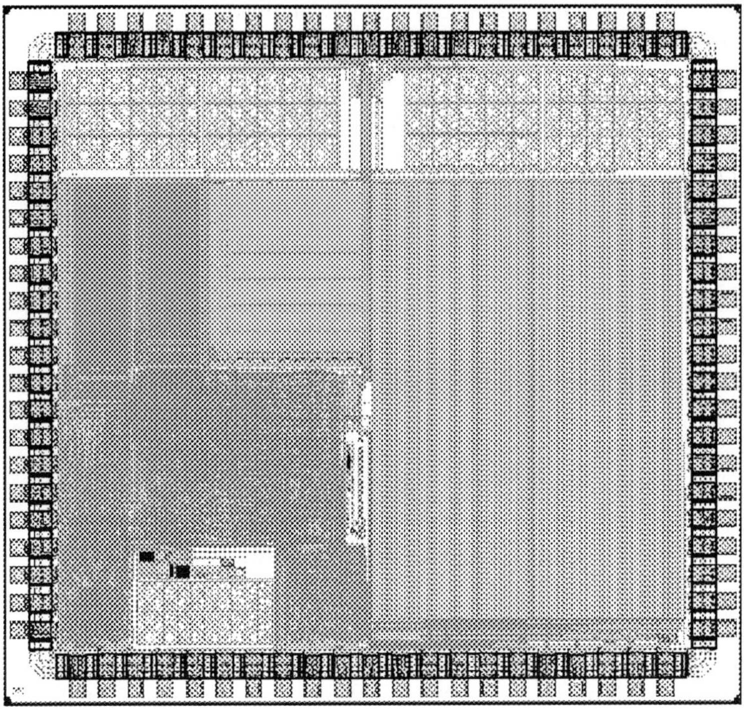

Figure 13.18. Layout of smart card chip.

The other modules are either synchronous or analog circuits, where the synchronous modules are not used in contactless operation. From this table it follows that the asynchronous logic takes only 12% of the total contactless digital circuit.

Table 13.3a. Areas of the contactless digital circuit blocks.

Block	Area [mm^2]
RAM	1.2
ROM	1.0
EEPROM	5.6
Async. circ.	1.1
Total	8.9

Table 13.3b. Areas of the asynchronous modules.

Module	Area [GE]
CPU	7,800
Pref. Unit	700
DES	3,250
UART	2,040
Interfaces	3,680
Timer	1,080
Total	18,550

The sizes of the different asynchronous modules are given in Table 13.3b. In the standard cell library used, a gate equivalent (GE) is 54 μm^2 with a typical layout density of 17,500 gates per mm^2.

Table 13.3c. Power of the contactless digital circuit.

Block	Power
Core	56%
ROM	27%
RAM	17%

Table 13.3d. Effect of asynchronous design on power and area at different levels.

Level	Power	Area
Async. circ.	−70%	+18%
Async. + Mem.	−60%	+2%

Table 13.3c shows the power dissipation of the digital circuit blocks when the controller is executing code from ROM (being the normal situation).

Table 13.3d shows the effects on power and area of an asynchronous design at two different levels. The asynchronous circuit gives a reduction in power dissipation of about 70% for 18% additional area. At the level of the contactless digital circuit, however, we obtain a power reduction of 60% for only 2% additional area. Note that this analysis does not include the synchronous RSA converter and the analog circuits needed in a production chip, such as for instance the buffer capacitor and the power supply unit. Therefore at chip level the relative reduction in power dissipation will be about the same, whereas the overhead in area will be reduced even further.

13.7. Test

The testing of asynchronous circuits for manufacturing defects is known to be a difficult problem [61, 120]. The main problem is that asynchronous circuits have many feedback loops that cannot be controlled by an external signal. This makes the introduction of scan testing expensive, and forces the designer to become involved in the development of a functional test, either in producing the patterns directly or in implementing the design-for-test measures.

A functional test approach was chosen for the chip described in this chapter. During test, the microcontroller is connected to an external ROM that contains a test program. This program computes a signature, and a circuit is said to be defective if the signature is not correct. In addition, current measurements are performed to increase the test coverage.

The functional tests were developed using the test for the 80C51 microcontroller [144] as a starting point. Both this test and its extension were developed using the test-evaluation tool described in [152]. This tool evaluates the *structural* test coverage (controllability and observability) of *functional* traces.

For the datapath logic, the tool can be used to achieve a 100% coverage for the stuck-at-input fault model, even though actually achieving this may be a real challenge to the test engineer. For the 80C51 microcontroller, however, with the inherent controllability and observability of its registers and buses, this appears to be feasible.

In the absence of full-scan, it is known that a 100% coverage for the stuck-at-input fault model can only be achieved for the asynchronous control logic by using a combination of functional patterns and current measurements in a circuit that has been modified so as to be pausable in the middle of carefully selected handshakes [120]. Since this modification has not been implemented in the smartcard circuit reported here, the fault coverage can never be 100%.

The exact fault coverage of the traces that were used here is not known, as this would demand unrealistic levels of compute power using a verification tool, but is estimated to be around 90% for the asynchronous control and datapath subsystem.

13.8. The power supply unit

Fig. 13.19 shows the power supply unit consisting of a rectifier and a power regulator, which are both completely analog circuits. The design of the rectifier is conventional, and of the regulator we discuss only those aspects of the behaviour that are relevant to the design of the digital circuit without going into the details of its design.

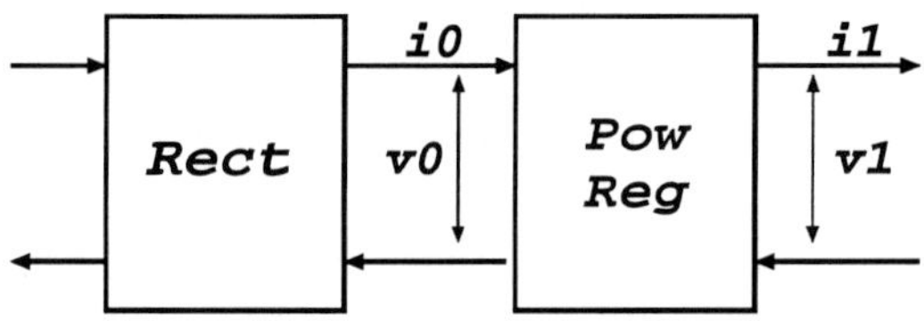

Figure 13.19. Power supply unit.

To avoid interference with the communication, a power regulator has been designed that shows an almost constant load at its input. Fig. 13.20 shows Spice-level simulation results of such a power regulator when the input voltage V_0 is fixed at 5V. On the horizontal axis we have the activity (number of transitions per second) of the digital circuit. The input load is almost constant, since input current i_0 is almost constant over the whole range.

When the activity is low, output voltage V_1 is constant at about 3V. In this range, too much power is coming in and the regulator functions as a voltage source with output current i_1 increasing when the activity increases. The superfluous power is shunted to ground. However, i_1 reaches a *saturation point* when it reaches i_0. From this point on, no more power is shunted to ground and the regulator starts to function as a current source with output voltage V_1 decreasing when the activity increases. The regulator delivers maximum power in the middle of the range where both the outgoing voltage and the outgoing current are high. Note that these simulation results assume constant incoming RF power. The variations in the incoming RF power during a transaction, however, are an additional source of fluctuations in V_1, since these variations result in shifts of the saturation point.

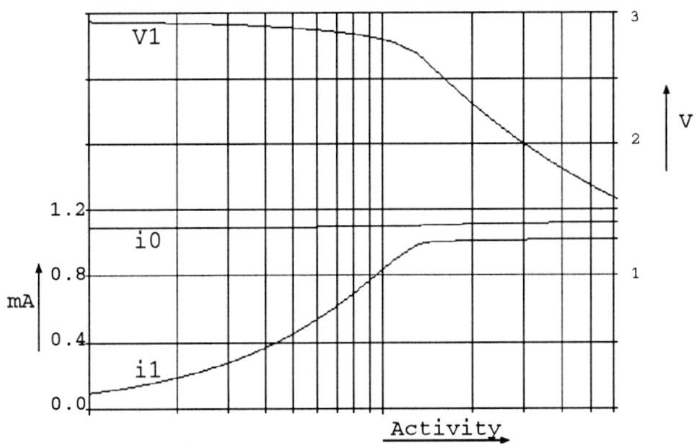

Figure 13.20. Power regulator behaviour.

A power source with these characteristics burdens the designer of a synchronous circuit with the problem of trading off between performance and robustness. Going for maximum performance means assuming a supply voltage of 3 V in which case a transaction must be aborted if the voltage drops below 2.5 V, for example. On the other hand, if the designer opts for a more robust design by choosing 2 V as the operating voltage, performance is lost when the regulator delivers 3 V. Such trade-offs are not needed for an asynchronous circuit, since it always automatically gives the maximum performance for the power received.

13.9. Conclusions

We have designed, built and evaluated an asynchronous chip for contactless smart cards in which we have exploited the fact that asynchronous circuits:

- use little average power,
- show small current peaks, and
- operate over a wide range of the supply voltage.

Measurements and simulations showed the following advantages of this design when compared to a conventional synchronous one:

- The asynchronous circuit gives the maximum performance for the power received. This comes mainly from the fact that the asynchronous design needs less of what is the main limiting factor for the performance, namely power. Compared to a synchronous design, the asynchronous circuit needs about 40% of the power for less than 2% additional area. In addition, the automatic speed adaptation property of asynchronous circuits saves the designer from trading off between performance and robustness. Due to this property the asynchronous circuit will give free-running instead of guaranteed performance, where the difference between the two is about a factor two.
- The asynchronous design is more resilient to voltage drops, since it still operates correctly for voltages down to 1.5 V.
- The current peaks of an asynchronous circuit are less pronounced, making the requirements with respect to the buffer capacitor more modest.
- The combination of the power regulator with the asynchronous circuit gives little communication interference. In this case, the smaller current peaks and the self-adaptation property are of importance.

Acknowledgements

We gratefully acknowledge the other members of the Tangram team: Kees van Berkel, Marc Verra and Erwin Woutersen, and we thank Klaus Ully for helping us to get the DES converter right.

Chapter 14

AN ASYNCHRONOUS VITERBI DECODER *

Linda E. M. Brackenbury
Department of Computer Science, The University of Manchester
lbrackenbury@cs.man.ac.uk

Abstract Viterbi decoders are used for decoding data encoded using convolutional forward error correcting codes. Such codes are used in a large proportion of digital transmission and digital recording systems because, even when the transmitted signal is subjected to significant noise, the decoder is still able efficiently to determine the most likely transmitted data.

This chapter descibes a novel Viterbi decoder aimed at being power efficient through adopting an asynchronous approach. The new design is based upon serial unary arithmetic for the computation and storage of the metrics required; this arithmetic replaces the add-compare-select parallel arithmetic performed by conventional synchronous systems. Like all Viterbi decoders, a history of computational results is built up over many data bits to determine the data most likely to have been transmitted at an earlier time. The identification of a starting point to this tracing operation allows the storage requirement to be greatly reduced compared with that in conventional decoders where the starting point is random. Furthermore, asynchronous operation in the system described enables multiple, independent, concurrent tracing operations to be performed which are decoupled from the placing of new data in the history memory.

Keywords: low-power asynchronous circuits, Viterbi, convolution decoder

14.1. Introduction

The PREST (Power REduction for Systems Technology) [1] project was a collaborative project where each partner designed a low power alternative to

*The Viterbi design work was supported by the EPSRC/MoD PowerPack project GR/L27930 and the EU PREST project EP25242, and this support is gratefully acknowledged.

J. Sparsø and S. Furber (eds.), Principles of Asynchronous Circuit Design, 249–272.

an industry standard Viterbi decoder. The Manchester University team's aim was to obtain a power-efficient design through the use of asynchronous timing.

The Viterbi decoder function [148] was chosen as it is a key function in digital TV and mobile communications. It detects and corrects transmission errors, outputting the data stream which, according to its calculations, is the most likely to have been transmitted. Viterbi coding is popular because it can deal with a continual data stream and requires no framing information such as a block start or finish. Furthermore, the way in which the output stream is constructed means that even if output data is incorrectly indicated, this situation will correct itself in time.

14.2. The Viterbi decoder

14.2.1 Convolution encoding

In order to understand the functions performed by the decoder, the way the data is encoded needs to be described. This is illustrated in figure 14.1. The input data stream enters a shift register which is initially set to zero. The 2-bit register stores the previous two input bits and these are combined with the current bit in two modulo-2 adders to provide the two binary digits which form the encoded output for the current input bit.

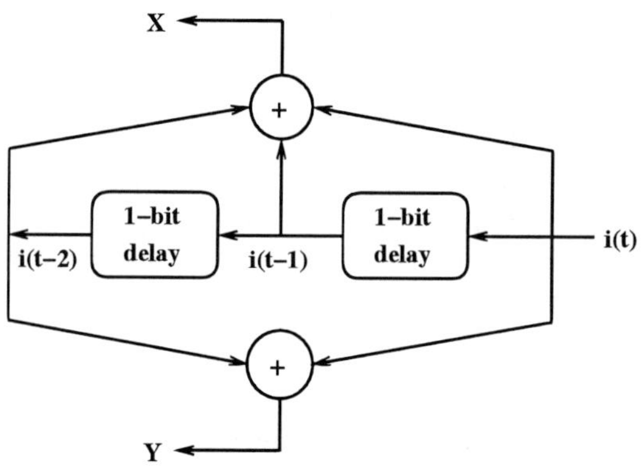

Figure 14.1. Four-state encoder.

For example, suppose that the input stream comprises 011 with the current input bit ('1') on the right and the oldest bit ('0') on the left, then the encoded X-output, which adds the bits in all three bits, is 0 and the encoded Y-output, which adds the current and oldest bit, is 1; so '01' would be transmitted in this case. As each input bit causes two encoded bits to be transmitted, the code is referred to as $1/2$ rate.

The use of three input bits (called the constraint length k) means that the encoder has $2^{(k-1)} = 4$ possible states, ranging from state $S0$ when both previous bits are zero to $S3$ when both these bits are ones. So, a current state n will become state $2n$ modulo 4 if the current input bit is a zero and $(2n+1)$ modulo 4 if it is a one; for example, if the current state is 2, the next state will be 0 or 1. That is, each state leads to two known states. This state transition information versus time is normally drawn in the form of a trellis diagram as shown in figure 14.2 for a four-state system. States are also referred to as nodes, and the node-to-node connection network is referred to as a butterfly connection due to its predictability, regularity and shape.

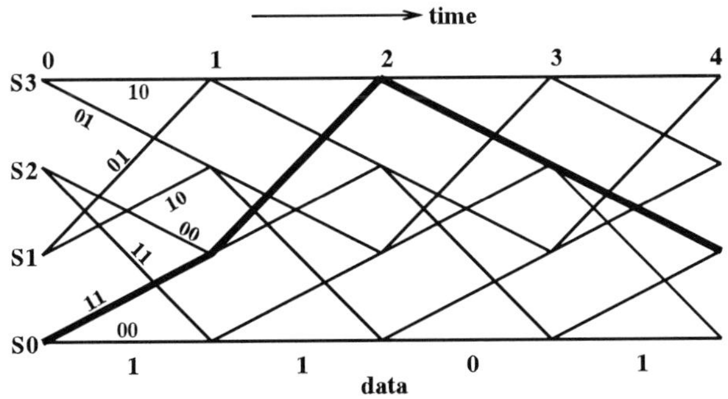

Figure 14.2. Four-node trellis.

By knowing the starting state of the trellis at any time and the subsequent input stream, the route taken by the encoder and the state it reaches can be traced. For example, starting at state 0 at time 0, an input pattern of 1 then 1 then 0 followed by 1 will cause the trellis path to travel from state $S0 \rightarrow S1 \rightarrow S3 \rightarrow S2 \rightarrow S1$; this route is indicated by the thicker black line on figure 14.2.

This figure also shows the encoder outputs associated with each path or branch on the trellis. For example, moving from state $S3$ to $S2$, corresponding to a current input of '0', the input stream must be 110 (with the oldest bit leftmost), so the encoded X and Y outputs are 0 and 1; this path is labelled with 01 to indicate the encoder output in moving from $S3$ to $S2$. The other paths are calculated in a similar way and labelled appropriately.

14.2.2 Decoder principle

The decoder attempts to reconstruct the most likely path taken by the encoder through the trellis. It does this by constructing a trellis and attaching weights to the nodes and each possible path at each timeslot; these indicate how likely it is that the node and path were the route taken by the encoder. Consider again the four-node example above. The encoded output for transmission resulting from the same input sequence of 1 then 1 then 0 then 1 would be 11 followed by 01, 01 and finally 00 (from an initial encoder state of all-zeros). Suppose that instead of receiving this sequence, the decoder receives corrupted data of 11 followed by 00, 01 and 00. Also assume that node 0 has an initialised weight of zero and that the other nodes have a higher initialised weight, say 2, at time 0; this corresponds to the encoder starting in state $S0$.

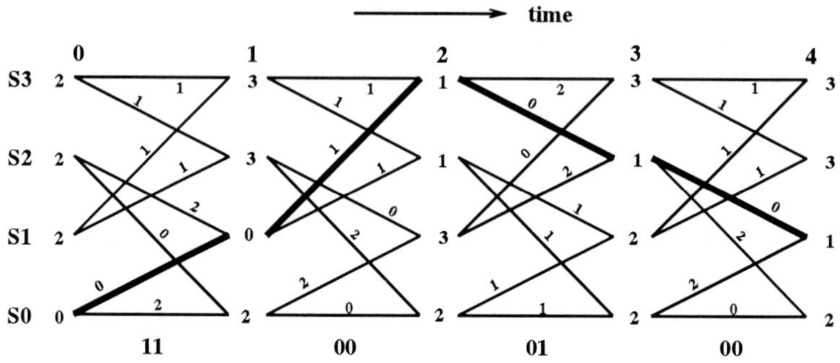

Figure 14.3. Decoding trellis.

For each branch, the distance between each received bit and the bit expected by the branch gives the weight of the branch. Where X and Y are encoded as single bits, referred to as hard coding, the number of bits difference between the received bits and the ideal encoded state for the branch gives the branch weight. The weights allocated in each timeslot for the received data are shown in figure 14.3. The received bits in the first timeslot are 11. Branches representing an encoded output of 11 are given a weight 0 while the branches representing an encoded output of 00 are given a weight of 2, and branches with ideal encoded outputs of 01 or 10 are given a weight of 1; these branch weights represent the distance between the branch and the received input. The branch weights are then added to the node weights to give an overall weight. Thus for example, state $S2$ has a node weight of 2 at time 0 and its branches have weights of 0 and 2 giving an overall weight of 2 and 4 going into the receiving node at the end of the timeslot. The overall weight indicates how likely it is that the route through the encoder corresponded to this sequence; the lower the overall weight the more likely this is.

From the trellis diagram, it can be seen that each state can be reached from two other states and that each state will therefore have two incoming overall weights. The weight that is chosen is the lower of these two since this represents the most likely branch that the encoder would traverse in reaching a particular state. So for example, state $S1$ can be entered from either state $S0$ or $S2$ and at time 1, the overall node plus branch weight from $S0$ is $0+0=0$ while from $S2$ it is $2+2=4$. The weight arising from $S0$ is chosen as this is the lower and therefore more likely route into $S1$; the weight arising from the $S2$ branch is discarded and the new weight for node $S1$ at the end of this first timeslot is that from $S0$, i.e. 0.

This process continues in a similar way for each timeslot with weights given to each branch according to the difference between the encoded branch pattern and the received bits. This difference is then added to the node weight to give an overall weight with the next weight for a node equal to the lower incoming weight. This results in the weights given in figure 14.3.

To form the output from the decoder, a path is traced through its trellis using the accumulated history of node weight information. The weights at time 4 indicate that the encoder is most likely to be in state $S1$ as this has the lowest node weight. Furthermore, this state was most likely to have been reached from $S2$. Continuing tracing backwards from $S2$ taking the lower overall node count, the most likely path taken from $S2$ is $S3$, $S1$ and $S0$ (initialisation). Note that despite having received corrupted data, this is exactly the sequence taken by the encoder in sending this data!

The output data can be 'rescued' from the decoder by noting that to reach states $S0$ and $S2$ the current data input to the encoder is a '0' while to reach states $S1$ and $S3$, the current encoder input is a '1'; that is, the least significant state bit indicates the state of the current data. Thus since the optimum states the decoder reaches at time 1, 2, 3 and 4 is $S1$, $S3$, $S2$ and $S1$ respectively, the decoder would output a data stream of 1101 as being the most likely input to the encoder.

14.3. System parameters

In practice the decoder designed is larger and more complicated than indicated by the simple example given. The encoder uses the current bit and the six previous bits in various combinations to provide the two encoded output streams; this spreads each input bit out over a longer transmission period, increasing the probability of error-free reception in the presence of noise. If the current bit is bit 0 with the oldest bit in the shift register being bit 6, then the encoded X-output is obtained by adding bits 0, 1, 2, 3 and 6 and the encoded Y-output from adding bits 0, 2, 3, 5 and 6. Since the constraint length is 7, the system has 64 nodes and therefore 128 paths or branches. Thus state n, where n is an integer from 0 to 63, leads to states $2n$ modulo 64 and $(2n+1)$ modulo 64.

Furthermore, the received bits are not hard coded (just straight ones and zeros) but soft coded. Three bits are used to represent values, with 100 used for a strong zero value and 011 for a strong one value. Noise in transmission means that a received character can be indicated by any 3-bit value from 0 to 7 and in interpreting the received value, it is helpful to regard the number as signed. Thus 011 indicates a strong one while 000 denotes received data that weakly indicates a one. Similarly, 100 implies a strong zero while 111 denotes a probable but weak zero. To make the text easier to follow hereafter, unsigned values will be used with the code for a 3-bit perfect zero taken as 000 (i.e. value 0) while a 3-bit perfect one is taken as 111 (i.e. value 7).

The interface to the asynchrous decoder is synchronous with the validity of the encoded characters on a particular positive clock edge indicated by a Block-Valid signal; encoded data is only present when this signal is high. Code rates other than the 1/2, primarily described in this chapter, are also possible. These are achieved by using the Block-Valid with a Puncture and a Puncture-X-nY signal; if Block-Valid is active (high) then a high Puncture signal indicates that only one of the X or Y symbols is valid and Puncture-X-nY indicates which. Both encoded characters are present if Block-Valid is high and Puncture is low. All the code rates originate from the 1/2 rate encoder with data for the other code rates then obtained by omitting to send some of these encoded characters. In this way, in addition to the 1/2 code rate, the system receives and decodes rates of 2/3 (two input bits result in the transmission of three encoded characters), 3/4, 5/6, 6/7 and 7/8 (7 input bits defined by 8 encoded characters with the remaining 6 not transmitted). As the code rate progressively increases, less redundancy is included in the transmitted data resulting in an increased error rate from the decoder; a rate of 1/2 will yield the most error-free output.

The system is expected to be operated from a 90 MHz clock. Since this clock is used by all code rates, the code rate not only defines the ratio of the number of input bits to the number of transmitted encoded characters but also specifies the ratio of the number of clocks containing encoded information (of 1 or 2 valid characters) to the number of clock cycles. For example, a 3/4 rate signifies that three input bits result in four transmitted characters and also that three clocks out of four contain some encoded information. Thus in a 3/4 code, every fourth clock contains no encoded information (Block-Valid is low). Any clock for which Block-Valid is high is said to contain a symbol. Thus with a 90 MHz clock rate, a 1/2 rate code which has valid data on alternate clock cycles yields a data rate of 45 MSymbols/sec and a 7/8 code rate is equivalent to 90x7/8 Msymbols/sec = 78.75 Msymbols/sec.

14.4. System overview

To perform the decoder computation previously outlined, the Viterbi decoder comprises three units as shown in figure 14.4. The Branch Metric Unit (BMU) receives the transmitted data and computes the distance between the ideal branch pattern symbols of (0,0), (0,7), (7,0) or (7,7) and the received data; these weights are then passed to the Path Metric Unit (PMU). The PMU holds the node weights and performs the node plus branch weight calculations,

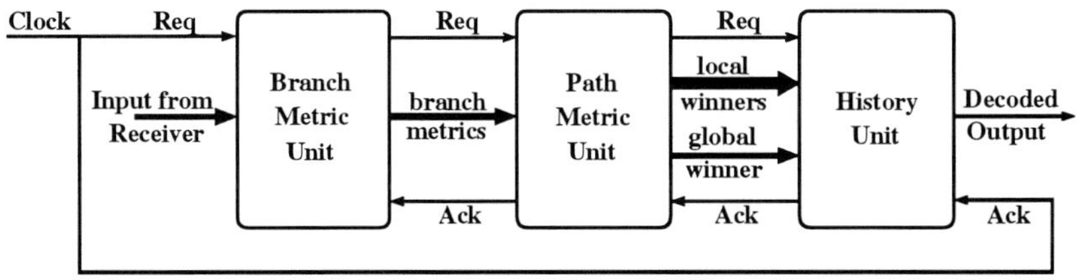

Figure 14.4. Decoder units.

selecting the lower overall weight as the next weight for a node in a particular timeslot. The computed node weights are then fed back (within the PMU) and become the node weights for the next timeslot.

As well as computing the next slot node weights, the PMU remembers whether the winner at a node came from the upper or lower branch leading in to it; this is only a single bit per node. On each timeslot, this local winner information is passed to the History Unit (HU), called the Survivor Memory Unit in synchronous systems. This information enables the HU to keep a history of the routes taken through the trellis by each node. In addition to this local node information, in our asynchronous design the state with the lowest node weight is identified in the PMU and its identity passed to the HU. Locating this global winner gives a known starting point in searching back through the trellis history to find the best data to output.

Conventional synchronous designs do not undertake an overall node winner identification and consequently start the search at a random point. They need to store a relatively large number of timeslots in order to ensure that there is sufficient history to make it likely that the correct route through the trellis is eventually found. In the asynchronous design, the identification of the overall node winner in the PMU was relatively easy to perform and it seemed the natural way to proceed. It has had the desirable effect of enabling the amount of timeslot history kept in the HU to be reduced and it also reduces the activity in the HU, saving power.

The HU uses both the overall winning node information (the global winner) and the local node winners in order to reconstruct the trellis and trace back the path from the current overall winner to find the node indicated by the oldest timeslot stored; the bit for this node is then output. The HU can be visualised as a circular buffer. Once data is output from the oldest slot, this information is overwritten with the current (newest) winner data so that the next oldest data becomes the oldest data in the next timeslot.

Figure 14.4 shows the bundled-data interface used between units; four-phase signalling is used for the Request and Acknowledge handshake signals. The Clock signal is required because the external system to the decoder is synchronous. Input data is supplied and output data removed on the positive clock edge provided that the bits on that clock edge are valid as indicated by Block-Valid. In practice, all units contain buffering at their interfaces in order to decouple the operation of the units, to cater for local variations in operating times, and to satisfy latency requirements imposed by the external system.

14.5. The Path Metric Unit (PMU)

14.5.1 Node pair design in the PMU

The PMU performs the core of the computation in the decoder and is the starting point for the design. Conventionally, the computation performed is that of Add (node to branch weight), Compare (upper and lower weights to a node) and Select (lower weight as next weight for a node). Because of the butterfly connection, the branch weights associated with nodes j and $j+32$ and their connections lead to nodes $2j$ and $2j+1$, as shown in figure 14.5 where BMa and BMb represent the branch weights; it should be noted that since a branch represents ideal convolved characters of (0,0), (0,7), (7,0) or (7,7), it is only necessary to compute a total of four weights in any system representing their distance from the received soft-coded characters.

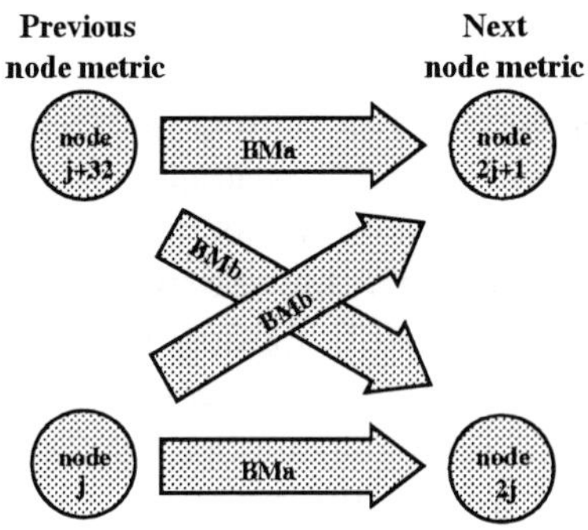

Figure 14.5. Node pair computation.

As the logic for this pair of nodes is self-contained and all the logic can be similarly partitioned into self-contained pairs of nodes, the basic unit of logic design in the PMU is the node pair; this is then replicated the required number of times (32 in this system). Furthermore, since 8-bit parallel arithmetic is normally used, in a 64-node system this leads to 512 data signals in the butterfly connection and 1024 interconnections within the PMU.

In an effort to simplify this routing problem and to achieve an associated power reduction from this simplification, serial arithmetic was proposed for the asynchronous design; in principle, this would reduce the butterfly to just

Table 14.1. Serial unary arithmetic.

number		transition representation
zero	=	000000
one	=	111111
two	=	000001
three	=	111101
four	=	000101
five	=	110101
six	=	010101

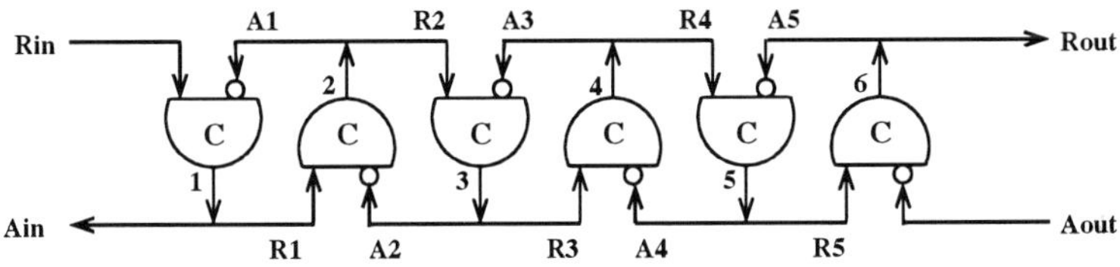

Figure 14.6. Six-bit 2-phase dataless FIFO.

64 wires. The adoption of serial arithmetic significantly impinges upon the node pair design and the way weights are stored. A conventional binary representation of values where serial arithmetic was performed on them was not a practical option as it would lead to a system throughput appreciably below that specified. This led to the idea of indicating values by the number of entries occupied in a FIFO buffer, so for example a count of five would require that the bottom five stages of the buffer showed that they were full; note that the buffer stages don't need to store any data but merely require a full/empty indication.

The speed and simplicity of this full/empty dataless FIFO scheme is further enhanced by adopting serial unary arithmetic for representing the data in the buffer (rather than a '1', say, to represent full and a '0' for empty). This is essentially a 2-phase representation for values, so that the number of transitions considering the full/empty bits as a whole represents the count. This is illustrated in table 14.1 for a 6-stage FIFO where the input enters on the left hand side (and exits from the right hand side).

The FIFOs used to hold the path and state metrics are Muller pipelines as shown in figure 14.6 (see also figure 2.8 on page 17). The encoding of a metric, M, is simply the state of an initially empty Muller pipeline after it has been exposed to M 2-phase handshakes on its input. Since a Muller C-gate in the technology used has a propagation delay of around 1 nsec the FIFO can transfer data in and out at a rate of around 500 MHz.

Using serial unary arithmetic, the major design component in a node pair for adding the node and branch weights and transferring the smaller to be the new node weight is an increment-decrement unit. The basic scheme for a node pair is illustrated in figure 14.7.

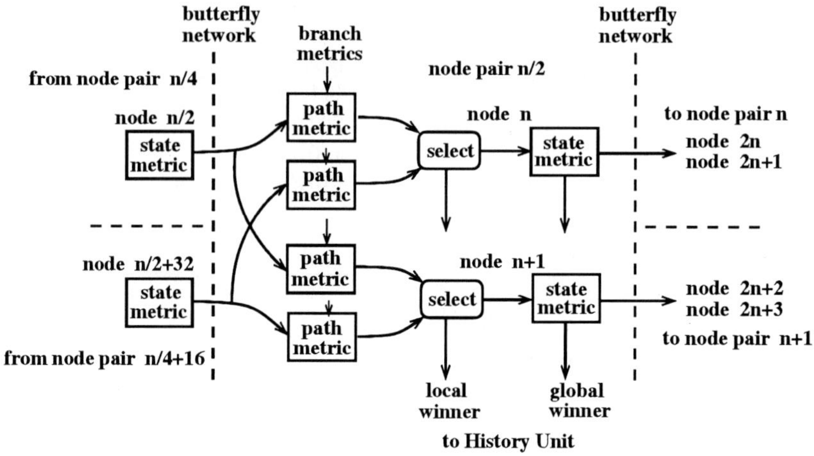

Figure 14.7. Node pair logic.

The new weights for each state are stored in the State Metric FIFOs on the right hand side of figure 14.7. When the global and local winners of these have been sent to the HU and acknowledged, the next timeslot commences with the parallel loading of the branch weights into the Path Metric FIFOs on the left in figure 14.7; these overwrite any existing content in these FIFOs. Parallel loading here, rather than serial entry, was selected on the grounds of speed and the need to clear the Path Metric FIFOs of any existing count.

The branch weights loaded are those computed by the BMU. The BMU first computes the conventional branch weights based on the difference between the two ideal 3-bit characters expected on the trellis branch and the two received values. The BMU then translates this to a transition pattern. This is made more complicated by the fact that the external environment to the Path Metric FIFOs is sometimes in the $< 1 >$ state necessitating a 2-phase inversion of the computed pattern.

Once the branch weights are loaded in, the node weights are then added to them. The node weights are transferred serially from the State Metric FIFOs across the butterfly connection into the Path Metric FIFOs. For each event transferred, two Path Metric FIFOs are incremented by one and the State Metric FIFO decremented by one. The transfer is complete when the feeding State Metric FIFO is empty. The Path Metric FIFOs for the node pair can commence the comparison and selection of the lower count as the node weight for the receiving State Metric FIFO once the receiving State Metric FIFO is empty.

The simplest way of performing this comparison and selection of the lower Path Metric FIFO count is to pair transitions (events) in the upper and lower Path Metric FIFOs connected to a receiving State Metric FIFO. Each paired event decrements each Path Metric FIFO by one and produces a transition which is used to increment the State Metric FIFO. The observant reader will note that the pairing of events to produce an output transition is exactly the action performed by a two-input Muller C-gate and this in principle is all that is required for the Select element shown in figure 14.7.

The pairing action ceases when the lower count in the two Path Metric FIFOs has decremented to zero. At this point, this Path Metric FIFO is the local path winner and the new node weight in the receiving State Metric FIFO is complete. The identity of the winning Path Metric FIFO (upper or lower) is needed to reconstruct the trellis; this information is buffered in the PMU and sent to the HU when all local winners are known and the overall winning node has been identified. This completes the actions required in the current timeslot and the PMU is then free to commence the next timeslot.

14.5.2 Branch metrics

The proposed scheme is only viable if the numbers that need to be transferred between the FIFOs can be kept small. A simulator was written in order to establish the minimum values that were consistent with meeting the bit error rates specified for the industrial standard device.

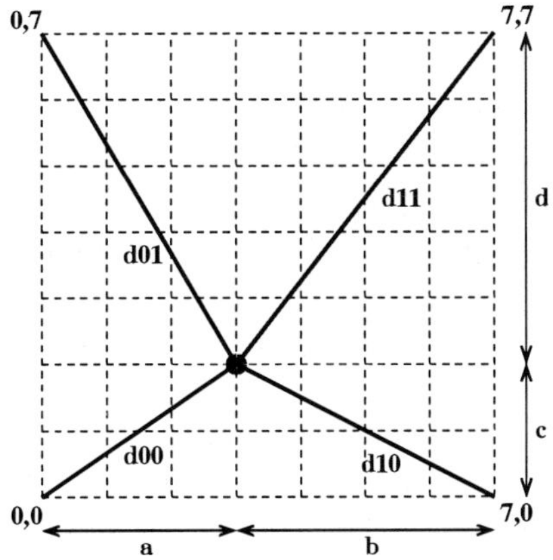

Figure 14.8. Computing the branch metric.

Table 14.2. Branch metric weight generation.

	received 3-bit character							
	0	1	2	3	4	5	6	7
Weight referenced to 0:	0	0	0	0	1	3	5	7
Weight referenced to 7:	7	5	3	1	0	0	0	0

In the BMU, the distance of the incoming data from the ideal branch representations of (0,0), (0,7), (7,0) and (7,7) needs to be computed. This calculation is depicted in figure 14.8. The incoming data is assumed to have a value of (a,c) which does not correspond to any of the ideal points. The squares of the distances $d00$, $d01$, $d10$ and $d11$ are a^2+c^2, a^2+d^2, b^2+c^2 and b^2+d^2 respectively. Only the relative values of these quantities are of interest. Substituting $(7-a)$ for b and $(7-c)$ for d in the quadratic expressions gives squared distances of a^2+c^2, $a^2+(7-c)^2$, $(7-a)^2+c^2$ and $(7-a)^2+(7-c)^2$. Expanding out and subtracting a^2+c^2 gives distance values of 0, $49-14c$, $49-14a$ and $98-14a-14c$. Dividing by 7, adding $a+c$ and then substituting back b for $(7-a)$ and d for $(7-c)$ yields the linear linear metrics $a+c$, $a+d$, $b+c$ and $b+d$.

Thus in this particular system, the Euclidian distance squared is equivalent to the Manhattan distance, which is a somewhat surprising and unexpected result. It indicates that to use squared distances offers no advantage over using the much simpler linear distances; indeed, using squared distances followed by scaling to reduce the number size (which is adopted in some systems) introduces unnecessary circuit complexity and some inaccuracy.

The linear weights are further minimised by subtracting the x and y distance to the nearest ideal point from the branch weights, so the smallest linear metric is always made zero. For example, if the incoming soft codes are exactly 7,7 in figure 14.8 then linear metrics of 14, 7, 7 and 0 are generated. However, if the incoming bits are at co-ordinate 5,6 (due to noise) then the metrics become 11, 6, 8 and 3 which by subtracting 3 from all values (2 for x value and 1 for y value) become branch metrics of 8, 3, 5 and 0. Using decrementing of the smallest distance to the nearest point, which then becomes zero, the values generated for weights for *each* 3-bit character are reduced as shown in table 14.2.

The maximum metric is now 14, and this will always arise when the incoming data coincides with one of the ideal input combinations. This is still too large for a system which needs to operate serially at close to 90 MHz. Therefore the metric is further scaled non-linearly and, to preserve the relative value of the weights, this is performed separately on each of the two incoming 3-bit soft values.

Referring to table 14.2, weights of zero remain zero while weights of 1, 3, 5 and 7 are scaled to 1, 2, 3 and 4. Clearly, the weights for both 3-bit soft codes need to be added to obtain the overall branch metric weight. Thus for example, the incoming soft-codes of 7,7 generate weights of 4+4, 4+0, 0+4 and 0 = 8, 4, 4 and 0, while incoming soft-codes of 5,6 generates weights of 2+3, 2+0, 0+3 and 0 = 5, 2, 3 and 0. Although the scaling here is non-linear and will therefore introduce some inaccuracy, simulation showed that this was not significant in relation to the results obtained with and without the scaling.

Furthermore, simulation also reveals that, unsurprisingly, paths with the largest weights are rarely involved in the most likely path to be chosen during the path reconstruction to find the output. Therefore, weights generated can be limited or capped and a limit of 6 is used in the BMU. Thus the weights actually generated and loaded by the BMU for a 7,7 soft code input are 6, 4, 4 and 0; however, the weights for a 5,6 input code remain as above.

Six-bit FIFOs are used throughout the PMU and again numbers in the PMU are capped at 6. To deal with cases where the serial addition of the node metric to a branch weight would exceed this number, logic referred to as the Overflow Unit is placed at the input of each Path Metric FIFO. This receives the incoming request handshake but does not pass it to the FIFO, instead it returns it to the sending State Metric FIFO as an acknowledge signal.

14.5.3 Slot timing

The overall or global winner from the PMU in a particular timeslot is the node having the lowest state metric count. In the same way that the BMU values can be adjusted so that the lowest weight is zero, the state metric values can also be decremented so that their minimum value is zero. As a result, numbers in the BMU and PMU are guaranteed to range only between zero and six. Furthermore if the soft bits contain no noise, which is the situation most of the time, then one (and only one) State Metric FIFO will contain a zero count indicating the best path through the trellis. This means that in the majority of timeslots, there is no need to perform any subtraction on the state metrics.

Detecting that a count is zero is in itself easy since it is indicated by an all-zeros state in the FIFO. This, as well as establishing the local winner, is done locally within a node and is timed from the control signals applied to the node; each node has a control section which generates the timing signals required by the node, and the timing within a node is independent of the timing of all other nodes.

A slot commences with the loading of the BMU branch weights. The node timing then passes to the next stage where the state-to-path metric transfer is performed. Following this, detecting that the sending State Metric FIFOs and the State Metric FIFO to receive the lower path metric count are empty causes the generation of a state-to-path metric done signal. The node timing then moves on to the next phase which generates a path-to-state metric enable. If at the time this signal is activated one of the Path Metric FIFOs for the node is empty, then a flip-flop is set indicating that this node is a global winner candidate; in this case no transfer to the State Metric FIFO is required and the path-to-state metric done signal is generated; this signal is used to clock the upper/lower branch (local) winner into a flip-flop and also to set a 'local winner found' flip-flop.

If neither Path Metric FIFO is empty then the path-to-state enable signal allows the transfer to its State Metric FIFO until one of the Path Metric FIFOs becomes empty; at this point, the path-to-state metric done signal is generated, setting the 'local winner' and the 'local winner found' flip-flops only.

The 'local winner found' and 'global winner found' signals now progress up the PMU logic hierarchy to the top level because all information needed to be passed to the HU has to be present before the request out signal to the HU is generated by the PMU. Furthermore, when the local and global winner data have been assembled for the HU, all nodes need to be informed that the slot has ended and the timing can be shifted to the start of the next slot. It should therefore be noted that while the timing within the nodes is local, the communication of the winner information to the HU and the subsequent release of the nodes has to be global.

14.5.4 Global winner identification

The formation of the 'all local winners found' and the global winner identification is partitioned across the various levels of the PMU logic hierarchy. At the level of a pair of node pairs, the four global winner candidate signals are input to a priority function which produces a 2-bit node address and a 'global winner found' signal if one of the inputs was a candidate. This logic is repeated with four such pairs of node pairs so that using the 'global winner found' signal generated by each pair of node pairs, the two bit address obtained indicates which pair of node pairs contains the global winner; these are then combined with the two-bit address generated by that pair of node pairs to form a 4-bit node address. Finally, this logic is repeated at the top level where there are four sets of four pairs of node pairs. Again the 'global winner found' signal from each set is used in a priority logic function to produce the two-bit address of the winning set and this is combined with the four address bits identifying the winning node within the winning set; this is the 6-bit node identification that is sent to the HU as the global winner.

The 'local winner found' signals only require combining in NAND or NOR gates and this is done at the node pair, four pairs of node pairs and top levels. At the top level, all the 'local winner found' signals must be true in order to generate the request out signal to the HU. Since the global winner identification is generated from the node whose local winner is the first to be indicated, the global winner is guaranteed to be identified prior to the last local winner being found. The acknowledge signal from the HU, in response to the PMU request out, causes a reset signal to all nodes which resets the global candidate and the 'local winner found' flip-flops and moves the timing on.

For the cases where no State Metric FIFO contains zero, this is detected; it indicates that noisy data has been received. Here, the global winner can be identified by performing a subtraction such that one or more State Metric FIFOs with the smallest count become zero. This could be time consuming and a decision was taken not to perform the decrement in the current timeslot. Instead the local winners are sent to the HU in the normal way but a Not-Valid signal accompanies the request handshake indicating that while the local winner information is genuine the global winner identification should be ignored.

The decrementing of all the state metric weights is performed in the next cycle by all the Overflow Units (which precede the Path Metric FIFOs). A signal to this unit indicates that decrementing is required. This results in the first incoming request from a State Metric FIFO to transfer its count to the Path Metric FIFO being ignored. The Overflow Unit sends an acknowledge back to its sending State Metric FIFO but does not pass the request on to its Path Metric FIFO. This effectively decrements the State Metric FIFOs by one by discarding the first item sent by them to the Path Metric FIFOs.

Only a count of one is decremented in this way on any timeslot. This may still leave all state metrics with a non-zero count in them but simulation revealed that this was highly unlikely. Furthermore, if the Overflow Units were used to decrement the state metrics by the smallest count then either considerable logic to determine the size of this count would be needed, or time consuming logic which decremented all state metrics by one and then tested to see if all these metrics were still non-zero (repeating these steps if necessary) would be required. Instead the much simpler approach of detecting a zero-valued state metric and identifying when all state metric counts are non-zero is used.

In retrospect, it would have been better to have decremented the State Metric down to zero in the current timeslot. The decrementing has to occur at some point and postponing to the next timeslot merely shifts when the operation is performed. More importantly, the failure to identify the global winner in the case of all State Metrics FIFOs holding a non-zero count means that information which is in the PMU is not passed to the HU and therefore the HU has less information on which to base its decisions as to the data output.

14.6. The History Unit (HU)

The global and local winner information from the PMU to the HU is accompanied by a Request handshake signal in the normal way. Having specified the interface to the PMU, the design of the asynchronous HU can be decoupled from the design of the rest of the system. As previously mentioned, the identification of a global winner means that the number of timeslots of local and global winner history that need be kept by the HU can be reduced compared with systems that need to start the tracing back through the trellis information from a random point. A rule of thumb for the minimum number of timeslots that need to be stored for determining the correct output is around 5 times the constraint length. With a length of seven for the system described, a minimum history of 35 timeslots is required and this was confirmed by simulation. On this basis, a 65-slot HU was developed.

14.6.1 Principle of operation

Figure 14.9 illustrates the principle of operation of the HU which, for simplicity, is shown as having only four states and storing only five time slots indicated by the rectangular outline. At each time step, indicated by T1, T2, etc., the PMU supplies the HU with the local winner information (an arrow points back from each state to the upper/lower winner state in the previous time step) and a global winner indicated by a sold circle.

Consider the situation when the latest data to have been supplied is at T6. The global winner at T6 is S3, and following the arrows back, the global winner at T2 is S0. The next data output bit is therefore 0 (the least significant bit of S0's state number), and this is output as the buffer window slides forward to time step T7. At T7 the received data has been corrupted by noise, and the global winner is (erroneously) S3. Following the local winners back, the backtrace modifies the global winners in time steps T6 and T5, but the path converges with the old path at T4. The next data output (from the global winner at T3) is 1 and is still correct. Moving on to T8, the global winner is S0 and, tracing back, the global winners at T5, T6 and T7 are changed, with T5 and T6 resuming their previous correct values. The noise that corrupted the path at T7 has no lasting effect and the output data stream is error-free.

14.6.2 History Unit backtrace

The data structure used in the HU is illustrated in table 14.3. Each of the 65 slots contains 64 bits of local winner information and a 6-bit global winner identifier. There is also a 'global winner valid' flag which indicates whether or not the global winner has been computed. The 65 slots form a circular buffer with the start (and end) of the buffer stepping around the slots in sequence.

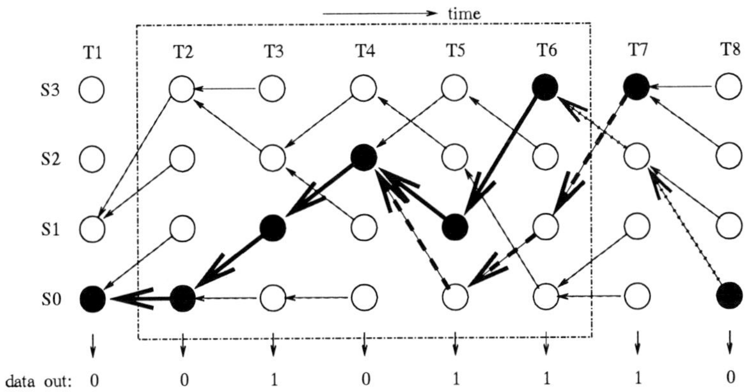

Figure 14.9. History Unit backtrace example.

Table 14.3. History Unit data structure.

slot number	local winner (64 bits)	global winner (6 bits)	valid (1 bit)	
0	L00[63..0]	G00[5..0]	V00	
1	L01[63..0]	G01[5..0]	V01	
...				
18	L18[63..0]	G18[5..0]	V18	
19	L19[63..0]	G19[5..0]	V19	
20	L20[63..0]	G20[5..0]	V20	← *head*
21	L21[63..0]	G21[5..0]	V21	
22	L22[63..0]	G22[5..0]	V22	
...				
64	L64[63..0]	G64[5..0]	V64	

At each step the next output bit is issued from the least significant bit of the current head-slot global winner identifier. Then the new local and global winner information is written into the head slot and the head pointer moves to the next slot. The new local and global winner information is used to initiate a backtrace, which updates the current optimum path held in the global winner memories.

The trellis arrangement produces a simple arithmetic relationship between one state and the next state so that, given a global winner identity in one slot, the previous global winner identity is readily computed. The parent identity can be derived from the child identity by shifting the child state one place to the right and inserting the relevant local winner bit into the most significant bit position. For example, if the global winner is node 23 in a slot, then the global winner in the previous slot will be node 11 (if the current slot local winner for node 23 is 0) or node 11+32 = node 43 (if the local winner for node 23 is 1).

Where the current global winner is the child of the previous global winner, the current winner continues the good path already stored in the global winner memories. This makes it unnecessary to search back through the local winner information in order to reconstruct the trellis and hence saves power. Therefore, when data is received from the PMU, if the incoming global winner is the child of the last winner, then it is only necessary to output data from the oldest global winner entry and then to overwrite this memory line with the incoming local and global winner information.

However, if sufficient noise is present (or noise has been present and the data now switches back to a good stream), then there may be a discontinuity between the incoming and previous global winner; this is recognised by the current global winner not being the child of the previous winner. In this case, the global winner memories do not hold a good path and this path is reconstructed using the local winner information. Here, the output data is read out and the winner information is written in as before. In addition, starting from the current global winner, this node identification is used to select its upper/lower branch winner from the current local winner information. The parent identity is then constructed as described above. This computed parent identity is compared with the global winner identity for the previous slot. If they are the same then the backtrace has now converged onto the good path kept in the global winner memories and no further action is required. If, however, they are not the same then the computed parent identity needs to overwrite the global winner in the previous timeslot. The backtrace process now repeats in order to construct a good path to the next previous timeslot, and this process continues until the computed parent identity does match the stored global winner.

Backtracing slot by slot thus proceeds until the computed path converges with the stored path. The algorithm is shown in a Balsa-like pseudo-code in figure 14.10. (Note, however, that Balsa does not have the ‘<<’ and ‘>>’ shift operators; they are borrowed here from C to improve clarity.) In practice, simulation shows that path convergence usually occurs within eight or fewer slots. So, although the most recent items may be over-written, the oldest items tend to be static and the output data from the oldest slots does not change. Overwriting the entire path is a rare occurrence and, in this circumstance, the data output from the system is almost certainly erroneous.

No backtrace is commenced in any slot where the global winner is invalid; the global winner entry is marked as invalid but the local winner information is written in the normal way. Any subsequent backtrace that encounters an invalid global winner will force a not equivalence with the incoming computed global winner at that slot, so that the computed global winner replaces the invalid stored value and the entry is then marked as valid.

```
loop
  c := head;                          -- child starts at head
  data_out <- Gc[0];                  -- output lsb of Gc
  Lc := In.local_winners;             -- update head local winners,
  Gc := In.global_winner;             --   global winner, and
  Vc := In.global_winner_valid;       --   global winner valid bit
  head := head + 1;                   -- step head pointer to next slot
  if Vc then                          -- backtrace only from valid head
    p := (c-1) % 65;                  -- parent slot number
    while (c /= head                  -- detect buffer wrap-around
           and (not Vp                -- over-write invalid parent
           or  Gp /= (Lc[Gc] << 5) + (Gc >> 1)))) -- not converged
    then
      Gp := (Lc[Gc] << 5) + (Gc >> 1));          -- update parent
      Vp := TRUE;                     -- mark as valid
      c  := p;                        -- next slot
      p  := (c-1) % 65                -- next parent slot number
    end -- while
  end -- if
end -- loop
```

Figure 14.10. History Unit backtrace sequential algorithm.

14.6.3 History Unit implementation

The type of memory used in the HU is the dominant factor in determining its implementation. Initially, RAM elements were considered for this storage as single and dual-port read elements were present in the available cell library. However, their use makes it difficult to keep track of incomplete backtraces when a new backtrace needs to be started. In addition, the global and local winner memories need to be separate entities but this introduces some inefficiency in the address decoding. Furthermore, there are difficulties in providing the many specific timed signals required to drive the memory. The RAM timings are equivalent to many simple gate delays. Such gates would be used to form the reference timing signals, and it is not clear that the gate propagation delays due to supply voltage changes vary in the same way as the RAM delays.

For these reasons, the memory was implemented with flip-flop storage and the system is shown in figure 14.11. It comprises 65 lines made up of 64 slots of replicated storage and one further slot which, on reset, becomes the slot holding the head token; the head slot receives the new local and global winner information from the PMU. The control block holds the global winner identification plus the token handling and backtrace logic.

The concurrent asynchronous algorithm is illustrated in Balsa-like pseudo-code in figure 14.12, which represents a single stage in the History Unit. The complete HU comprises 65 such stages, one of which is initialised to be the

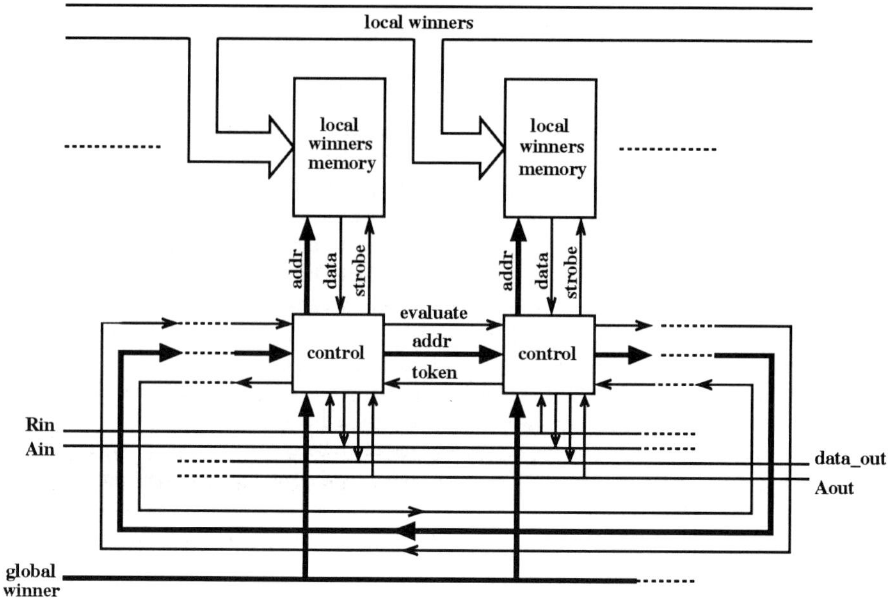

Figure 14.11. History Unit implementation.

head of the circular buffer. This can be compared with the sequential algorithm shown earlier in figure 14.10. The transformation from the sequential to the concurrent algorithm is illustrative of high-level asynchronous design methodologies even when the design is being carried out manually (as was this Viterbi decoder) and not in a high-level language such as Balsa.

The head slot contains the oldest winner information and determines the 1-bit data output from the system. Remembering that odd states signify a '1' input and even states a '0' input, the head slot outputs the least significant global winner bit on the *data-out* line. This data enters a buffer and the acknowledge *Aout* signifies its acceptance. The head is then free to write the current winner information to its memory. The *Token* signal then passes (leftwards) to the adjacent slot which now becomes the new head.

The parent node of the current global winner is computed as described and this is passed (rightwards) to the adjacent slot with an *Evaluate* signal. The computed parent is compared with the stored winner in the previous stage. Equivalence results in no further backtracing and the backtrace is said to be retired. Not equivalence causes overwriting of the global winner and this winner accompanied by *Strobe* is used to address (*Addr*) the local memory. The data bit returned on *Data* is used to compute the parent of this winner which is then passed rightwards to the preceding timeslot with an *Evaluate* signal. This process repeats until the backtrace converges with the existing global winners

```
loop
  arbitrate In then
    if head then                  -- head...
      data_out <- G[0];           --    output next data bit
      L := In.local_winners ||    --    update local values
      G := In.global_winner ||
      V := In.global_winner_valid;
      if V then                   --    if global winner is valid
        addrOut <- (L[G] << 5) + (G >> 1) -- start backtrace
      end; -- if V
      sync tokenOut ||            --    pass on head gtoken
      head := false               --    and clear head Boolean
    end -- if head
  | addrIn then                   --    backtrace input?
    if not head then
      if addrIn /= G or not V then  -- path converged? If not...
        G := addrIn ||            --    update global winner
        V := true;
        addrOut <- (L[G] << 5) + (G >> 1) -- propagate backtrace
      end -- if ..
    end -- if not head
  | tokenIn then                  --    head token arrived
      head := true                --    set head Boolean
  end -- arbitrate
end -- loop
```

Figure 14.12. History Unit backtrace stage.

and can be retired. A backtrace has to be forcibly retired if it is in danger of running into the head slot; an arbiter is used to test for this in the control and it is the only place where arbiters need to be used in the system. Fortunately, the meeting of the head and backtrace processing is a rare occurrence.

It should be noted that, unlike a conventional system, path reconstruction is only undertaken if necessary and then for only as long as required; both strategies save power. Furthermore, the use of asynchronous techniques within the HU enables the writing of winner information from the PMU to be independent of and run concurrently with any path reconstruction activity. The use of flip-flop storage rather than RAM has resulted in a simpler and more flexible design. It also has the distinct advantage of enabling multiple backtraces, whose frontiers are all at different slots, to be run concurrently.

14.7. Results and design evaluation

The asynchronous Viterbi decoder system was implemented as described using the industrial partner's cell library which was designed to operate from 3.3V. Non-standard elements such as the Muller C-gate were constructed from the cell library; the only full-custom element which had to be designed was an arbiter.

Following simulation, the decoder was fabricated on a 0.35 micron CMOS process. Results for a 1/2 code rate, random input bit stream with no errors show an overall power dissipation of 1333 mW at 45 Msymbols/sec. Of this, the dissipation in the PMU dominates at 1233 mW while the HU takes only 37 mW. The difference (about 60 mW) between these figure and those for the overall consumption are accounted for by the dissipation in the BMU and in the small amount of 'glue' logic prior to the BMU.

Errors in the input data to the decoder result in only small variations in the dissipation with the overall dissipation falling slightly with an increasing input error rate; internally, this decrease comprises a small reduction in the PMU dissipation and a smaller rise in the HU dissipation. The results for other code rates are a scaled version of those obtained for the 1/2 code rate as might be expected. For example, a 3/4 code rate which receives 3 symbols for every 4 clocks exhibits 1.5 times the dissipation of the 1/2 rate code with its 2 symbols every 4 clocks.

The asynchronous PMU performs approximately the same amount of work regardless of the number of errors in the data stream. This results from capping numbers at 6. This means that for a good data stream, all nodes have a weight of six except the one node on the good (correct) path. Thus the PMU is almost permanently 'saturated' and practically all work performed relates to paths which will never be selected! Errors cause some spread in the node weights with the higher weights (4, 5 and 6) predominating and the slightly smaller counts on some nodes accounts for the slight drop in dissipation in the PMU under error conditions.

The asynchronous PMU dissipation is very high and does not compare well with synchronous PMUs using conventional parallel arithmetic to perform the add, compare and select operation [15]. In order to understand why this occurs, it is necessary to examine the operation and logic used in the asynchronous PMU in more detail. With good (i.e. no error) data, 63 nodes have weights of 6 and one node has a weight of 0. This translates to PMU activity on each timeslot where all State Metric FIFOs but one contain counts of 6 which are then transferred to Path Metric FIFOs, whose counts of 6 are in turn removed from the Path Metric FIFOs, paired and transferred to the receiving State Metric FIFOs. Entering or removing a count of 6 from a FIFO involves 21 changes of state in the stages. Furthermore, the number of transitions actually involved is higher due to (around 5) internal transitions on the elements making up the C-gates forming each FIFO stage. Thus, each of 63 nodes experiences around 650 transitions just on the data path per timeslot. The control and other overheads on the data path can be expected to form (say) an additional 30% of logic. This indicates a node activity of around 850 transitions/timeslot and overall the PMU can be expected to make a maximum of 54,400 transitions on each timeslot.

Unfortunately, the design of the FIFOs, particularly the Path Metric FIFOs, has led to high capacitive loading on the true and inverse C-gate outputs at each stage. A dissipation of 1233 mW with 54,400 transitions per slot and an energy cost of $5.45 \times C$ joules per transition, where C is the average track plus gate loading capacitance (in Farads), indicates an average loading of 92 fF and this is confirmed by measurements.

By contrast, the HU power efficiency is excellent and is the real success story in this design. Its dissipation is low and is far smaller than that in any other system the designers are aware of. The HU dissipation demonstrates that by keeping a good path, very little computing is required to output the data stream when there are no errors. Furthermore, when lots of noise is present, so that the backtrace process is active with many good paths in the process of being reconstructed concurrently, the dissipation in the HU only rises slightly; this indicates that accessing the local winner memory in flip-flops and overwriting the global winner information is not a power-costly operation.

The HU dissipation also compares favourably with a (synchronous) system built along similar principles to the HU described here but using RAM elements from the library instead of flip-flops. Due to the limitations of the RAM devices previously mentioned, these introduce additional complexity because only one backtrace is performed at any time; it is therefore necessary to keep track of the depth reached by incomplete backtraces which are abandoned for a new backtrace leaving a partially complete global winner path reconstruction. The difference in dissipation between the asynchronous HU using flip-flops and the other using RAMs reflects the power cost of accessing the local winner RAM and the associated significant additional computation involved in the backtrace. This points to the power efficiency of storing the HU information in a manner best suited to the task.

14.8. Conclusions

As in many asynchronous designs, the system design has had to be approached from first principles and has caused a complete rethink about how to implement the Viterbi algorithm. This has resulted in a number of novel features being incorporated in the PMU and HU units. In the PMU, the decision to use serial unary arithmetic has enabled the conventional parallel add, compare and select logic to be dispensed with and replaced by dataless FIFOs which perform the arithmetic serially.

While the PMU is an interesting and different design from that conventionally used, its power consumption is not good. Its design illustrates that power efficiency at the algorithmic and architecture levels needs to be combined with efficient design at the logic, circuit and layout levels to realise the true potential of a system. This is demonstrated by a synchronous PMU constructed along similar architectural principles to those described but implemented using a low-power logic family and full custom layout which dissipates only 70 mW at 45 Msymbols/sec [15]. It is clear that while a full custom design of the asynchronous PMU datapath would reduce the current power levels significantly, a major revision of the PMU logic for the datapath, paying particular attention to loading, is required for a design which has better power efficiency than other systems.

The identification of a global winner is probably the most important advance in the PMU design. This has meant that both a good path and a local winner history can be kept by the HU, leading to a greatly reduced amount of overall storage required to deduce the output data. The use of flip-flop storage has also greatly contributed to the power efficiency of this unit and it does demonstrate the power advantages of optimising design at all levels in the design hierarchy down to and including the logic.

The HU also illustrates the advantages of asynchronous design in that the placing of current information is decoupled from any backtracing operations of which there may be many running concurrently. Furthermore, the speed of the backtracing is only dependent on the logic required to perform this operation and not on any other internal or external system timing. Such a decoupled, multiple backtracing activity would clearly be more difficult to organise in the context of a synchronous timing environment.

14.8.1 Acknowledgement

As in any large project, a number of people have been engaged in the design and implementation of the Viterbi decoder described in this chapter. It is therefore a pleasure to acknowledge the other colleagues in the Amulet group in the Computer Science Department at Manchester University involved at all stages in this project, namely Mike Cumpstey, Steve Furber and Peter Riocreux. I am also grateful to them for comments on the draft of this chapter.

14.8.2 Further reading

Further information on the Viterbi algorithm may be found in [148], [71] and [70]. Futher information on the PREST project is in [1].

Chapter 15

PROCESSORS *

Jim D. Garside

Department of Computer Science, The University of Manchester

jgarside@cs.man.ac.uk

Abstract Computer design becomes ever more complex. Small asynchronous systems may be intriguing and even elegant but unless asynchronous logic can not only be competitive with 'conventional' logic but can show some significant advantages it cannot be taken seriously in the commercial world.

There can be no better way to demonstrate the feasibility of something than by doing it. To this end several research groups around have the world have been putting together real, large, asynchronous systems. These have taken several forms, but many groups have chosen to start with microprocessors; a processor is a good demonstrator because it is well defined, self-contained and forces a designer to solve problems which are already well understood. If an asynchronous implementation of a microprocessor can compare favourably with a synchronous device performing an identical function then the case is proven.

This chapter describes a number of processors that have been fabricated and discusses in some detail some of the solutions employed. The primary source of the material is the Amulet series of ARM implementations – because these are the most familiar to the author – but other devices are included as appropriate. The later parts of the chapter widen the descriptions to include memory systems, cacheing and on-chip interconnect, illustrating how a complete asynchronous System on Chip (SoC) can be produced.

Keywords: low-power asynchronous circuits, processor architecture

*The majority of the work described in this chapter has been made possible by grants from the European Union Open Microprocessor systems Initiative (OMI). The primary sources of funding have been OMI-MAP (Amulet1), OMI/DE-ARM (Amulet2e) and OMI/ATOM (Amulet3). Without this funding none of these devices would have been made and this support is gratefully acknowledged.

J. Sparsø and S. Furber (eds.), Principles of Asynchronous Circuit Design, 273–315.

15.1. An introduction to the Amulet processors

Most of the examples in this chapter are based around the Amulet series of microprocessors, developed at the University of Manchester. All of these have been asynchronous implementations of the ARM architecture [65] and, as such, allow direct comparisons with their synchronous contemporaries. It should be noted that the primary intention, as in other ARM designs, was to produce power-efficient rather than high-performance processors.

Brief descriptions of the three fabricated Amulet processors and some other notable examples are given below.

15.1.1 Amulet1 (1994)

Figure 15.1. Amulet1.

Amulet1 [158] (figure 15.1) was a feasibility study in asynchronous design, using techniques based extensively on Sutherland's Micropipelines [128]. Although two-phase signalling was used for communications standard, transparent latches were used internally rather than Sutherland's capture-pass latch (see figure 2.11 on page 20). The external two-phase interface proved difficult to interface with external commodity parts.

Amulet1 comprised 60,000 transistors in a 1.0μm, 2-layer metal process and ran the ARM6 instruction set (with the exception of the multiply-accumulate operation). It achieved about half the instruction throughput of an ARM6 manufactured on the same process with roughly the same energy efficiency (MIPS/W).

Figure 15.2. Amulet2e.

15.1.2 Amulet2e (1996)

Amulet2e [44] (figure 15.2) was an ARM7 compatible device with complete instruction set compliance. In addition to the CPU it included an asynchronous 4 KByte cache memory and a flexible external interface making it much easier to integrate with commodity parts. A few other optimisations such as (limited) result forwarding and branch prediction were added.

Internally this device used four-phase rather than two-phase handshake protocols. It occupied 450,000 transistors (mostly in the cache memory) in a $0.5\mu m$ 3-layer metal process; although about three times faster than Amulet1 it was still only half the performance of a contemporary synchronous chip.

15.1.3 Amulet3i (2000)

Amulet3i [48] was intended as a macrocell for supporting System on Chip (SoC) applications rather than a stand-alone device. It is an ARM9 compatible device comprising around 800 000 transistors in a $0.35\mu m$ 3-layer metal process. It comprises an Amulet3 CPU, 8 KBytes of pseudo-dual port RAM, 16 KBytes of ROM, a powerful DMA controller and an external memory/test interface, all based around a MARBLE [4] asynchronous on-chip bus.

Figure 15.3. DRACO.

Amulet3i achieves roughly the same performance as a contemporary, synchronous ARM with an equal or marginally better energy efficiency. It was integrated with a number of synchronous peripheral devices, designed by Hagenuk GmbH, to form the DRACO (DECT Radio Communications Controller) device (figure 15.3).

15.2. Some other asynchronous microprocessors

Several other groups around the world have also been developing asynchronous microprocessors over the past decade or so. In this section a (non-exhaustive) selection of these are briefly described.

Caltech has produced two asynchronous processors: the 'Caltech Asynchronous Microprocessor' (1989) [86] was a locally-designed 16-bit RISC which was the first single chip asynchronous processor; the 'MiniMIPS' [88] was an implementation of the R3000 architecture [72]. Both of these processors were custom designed using delay-insensitive coding rather than the 'bundled data' used in the Amulets. This, together with a design philosophy aimed at speed rather than low-power consumption results in high-performance, high-power devices.

Another MIPS-style microprocessor is the University of Tokyo's 'TITAC-2' (1997) [130] (figure 15.4) which is an R2000. This was developed in another

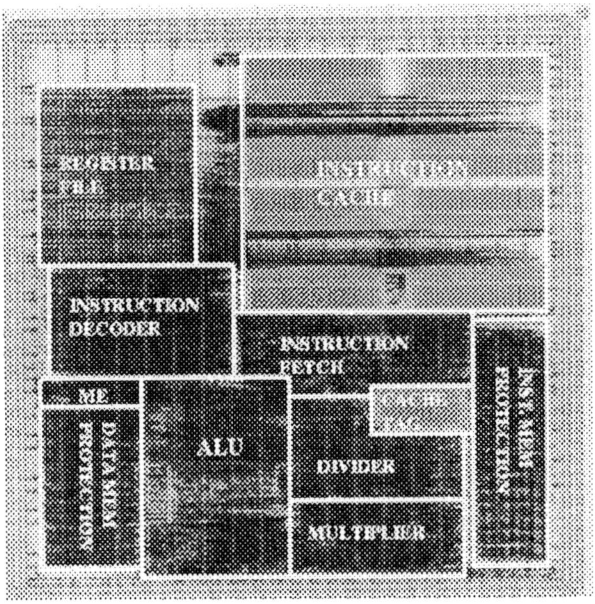

Figure 15.4. TITAC-2. (Picture courtesy of the University of Tokyo.)

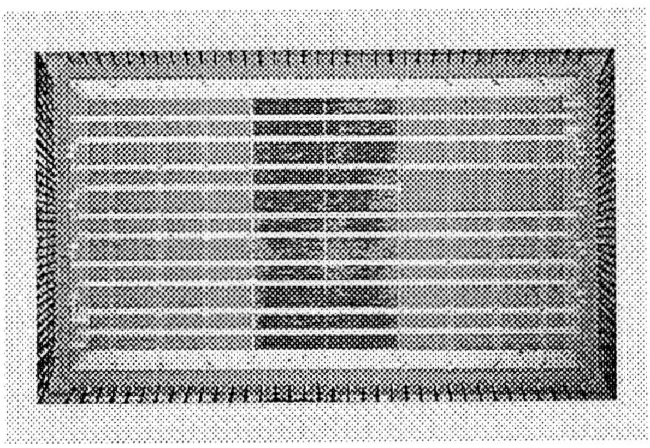

Figure 15.5. ASPRO-216. (Picture courtesy of the TIMA Laboratory,CIS Group, IMAG (Grenoble).)

different design style (quasi-delay insensitive). As may be apparent from the figure TITAC-2 employed considerable manual layout.

'ASPRO-216' (1998) [117] from IMAG in Grenoble is slightly different in that it is a 16-bit signal processor rather than a general-purpose microprocessor. More significantly its design was largely automated and synthesised from a CHP(Communicating Hardware Processes) [84, 118] description. This shows in the more 'amorphous' appearance of the processor, although the chip is dominated by its memories (figure 15.5).

All the devices mentioned above have been research prototypes. Commercial take up of asynchronous processor technology has been slower; nevertheless there are some significant examples.

Philips Research Laboratories in Eindhoven have been developing the 'Tangram' [135] circuit synthesis system, primarily aimed at low-performance, very low power systems. This has been used to produce an asynchronous implementation of the 80C51 (1998) [144] which has been deployed in commercial pager devices where its low power and low EMI properties are particularly attractive. It is also intended for use in smartcard applications (see chapter 13 on page 221).

Although not strictly a microprocessor the Sharp DDMP (Data-Driven Media Processor) (1997) [131] merits inclusion here. Intended for multimedia applications this provides a number of parallel processing elements which are employed or left idle according to the demand at any time. Asynchronous technology was attractive here because of the ease of power management.

Finally the DRACO device (figure 15.3) was designed specifically for commercial use although not (yet) marketed due to company reorganisation. As a processor in a radio 'base station' the low EMI properties of asynchronous logic were the reasons for adoption of this technology.

15.3. Processors as design examples

Why build an asynchronous microprocessor? Part of the answer must be the various advantages of low power, low EMI etc. claimed for any asynchronous design and demonstrated in the commercial devices mentioned above, but why is a processor a good demonstration of these features?

In many ways it is not. A better demonstrator of asynchronous advantages may well be an application with a regular structure which is amenable to very fine grain pipelining: some signal processing or network switching applications have these characteristics. However there is a great deal of appeal in constructing a microprocessor. Firstly, it is a well-defined and self-contained problem; it is easy to define what a microprocessor should do and to demonstrate that it fulfils that specification. Secondly, it forces an asynchronous designer to confront and solve a number of implementation problems which might not occur if a 'tailor made' demonstration was chosen. Lastly, it is often possible to compare the result with contemporary, synchronous devices in order to quantify and assess the results of the work. Of course, microprocessors are also an intensely fast-moving and competitive market in which it is hard to compete, even in a familiar technology!

15.4. Processor implementation techniques

15.4.1 Pipelining processors

Pipelining [56] the operation of a device such as a microprocessor is an efficient way to improve performance. At its simplest, pipelining can subdivide a time-consuming operation into a number of faster operations whose execution can be overlapped. If done 'perfectly' a five-stage pipeline can speed up an operation by (almost) five times with only a small hardware cost in pipeline latches.

A typical synchronous pipeline should divide the logic into equally timed slices. If there is an imbalance in the partitioning the slowest pipeline stage sets the clock period for the whole system; faster logic is slowed down by the clock resulting in some time wastage (figure 15.6). This is more emphasised if the timing of a particular pipeline stage is, in some way, variable or 'data dependent'. Data dependencies can be quite common in a microprocessor: a simple example is an ALU operation, where a 'move' operation is faster than an addition because the former operations require no carry propagation. A more 'exaggerated' example is a memory access where a cache hit is much faster than a cache miss.

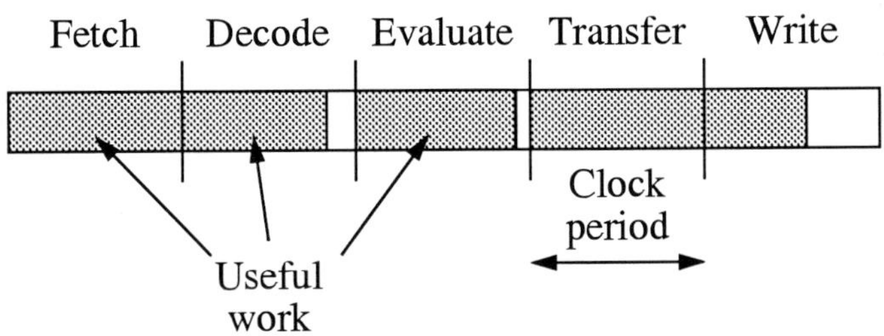

Figure 15.6. Synchronous pipeline usage.

Normally in most of such cases the clock must allow for the slowest possible cycle. This either slows down the clock or causes the designer to invest considerable hardware in speeding up the worst-case operations, for example adding fast carry propagation mechanisms. Whilst the latter is a good trade if a high proportion of the operations are slow this is poor economics if the worst-care operations are rare.

Another possible solution open to the synchronous designer is to allow certain slow operations to occupy more than one clock cycle; this is clearly expedient when a cache miss occurs and the processor must idle until an off-chip memory can be read. However multi-cycle operations introduce the need for system-wide clock control to stall other parts of the system; even then the timing resolution is still limited to whole clock cycles.

Asynchronous pipelining is conceptually much easier. Not only is all control localised, but it is implicitly adaptable to pipeline stages with variable delays. This means that rare, worst-case operations can be accommodated without either excessive hardware investment or a significant impact in the overall processing time by altering the delay of a pipeline stage dynamically. This, combined with the fact that operations can flow at their own rate rather than being stacked one to a stage, gives the pipeline considerable 'elasticity'. In such a pipeline a cache miss still occupies a single cycle, although that cycle will be a particularly slow one!

Another clear example of this is visible in the ARM instruction set [65] where data processing operations and address calculations may specify an operand shift before the ALU operation. In early ARM implementations this was contained in a single pipeline stage and hidden by the slower memory access time. To avoid a performance penalty more modern synchronous designs have the options of:

- an additional shifter pipeline stage (increases latency);
- stalling for an extra clock when a shift is required (increases complexity).

An asynchronous pipeline can simply stretch the execution cycle when required. As the additional time is unlikely to be as long as the ALU stage delay this is more flexible than either of the synchronous options, and the overall impact is small because shift/operate operations are quite rare.

'Average case' performance. The elasticity of an asynchronous pipeline has led to the myth that an asynchronous pipeline can perform its processing in an 'average' rather than worst case time for each pipeline stage. This is true only if the unit under consideration is kept constantly busy. This will not be true in the general case: on some occasions the unit will be ready before its operands and at other times it will have completed before subsequent stages can accept its output; in each case an interlude of idleness is enforced. This effect can be reduced by providing fast buffers between processing elements to allow some 'play' in the timing, but true average case behaviour can only be achieved with buffers of infinite size. Any buffering increases the pipeline latency and should therefore be used with circumspection.

In practice an asynchronous pipeline should be balanced in a similar fashion to a synchronous pipeline. The difference is that occasional, time consuming operations can be accommodated without either pipeline disruption or significant extra hardware. An added bonus is that the pipeline latency, especially when filling an 'empty' pipeline, can be reduced because the first instruction is not retarded by the clock at each stage (figure 15.6). The problem then devolves into partitioning the system and solving the resulting problem of internal operand dependences.

15.4.2 Asynchronous pipeline architectures

Once an asynchronous pipeline has been developed it is very easy to add pipeline stages to a design. However, pipelining indiscriminately can be a Bad Thing, at least in a general-purpose processor. The major reason for this is the need to resolve dependencies [56], where one operation must gather its operands from the results of previous instructions; if many operations are in the pipeline simultaneously it is quite likely that any new instructions will be forced to wait for some of these to complete. Resolving dependencies in an asynchronous environment can be a relatively complex task and is discussed in a later section.

A less obvious consequence is the increase in latency in the pipeline, not only due to added latch delays but because, in some circumstances, the pipeline needs to drain and refill. Consider a processor pipeline with a fast FIFO buffer acting as a prefetch buffer; this initially seems like a good idea as it can help reduce stalls due to (for example) occasional slow cycles in the prefetch (e.g. cache miss) and execution (e.g. multiplication) stages. However, at least in a general purpose processor, this benefit is masked because a single stall can fill up the prefetch buffer and, typically shortly thereafter, a branch requires it to be drained. This was an architectural error evidenced in Amulet1, which suffered a noticeable performance penalty due to its four-stage prefetch buffer. Of course in other applications, where pipeline flushes are rare, the ease of adding buffering can provide significant gains; however experience has shown that for a general purpose CPU a more conventional approach producing a reasonably balanced pipeline based around a known cycle time (such as the cache read-hit cycle time) is the 'best' approach.

One definite advantage of an asynchronous pipeline is that the pipeline flow can be controlled locally. Consider that bane of the RISC architecture the multi-cycle instruction; in a synchronous environment such an operation must be able to suspend large parts or all of the processing pipeline, necessitating a widespread control network. In an asynchronous environment the system need not be aware of operations in other parts of the system; instead a multi-cycle operation simply appears as a longer delay, possibly causing a stall if other processing elements require interaction with the busy unit(s).

In the ARM architecture there is one case where this ability is very useful; the multiple register load and store operations (LDM/STM) can transfer an arbitrary subset of the sixteen current registers to or from memory. Amulet3 implements this by generating several local 'instruction' packets for the execution stages for a single input handshake. At this point it is likely that the prefetch will fill up the intervening latches and stall, but this is a natural consequence of the pipeline's operation.

There are other examples of this behaviour in Amulet3 (figure 15.7), notably the Thumb decoder which ingests 32-bit packets which can contain either one ARM instruction or two of the 'compressed', 16-bit Thumb instructions. In the latter case two output handshakes are generated for each input. This provides an advantage over earlier (synchronous) Thumb implementations, which fetch instructions sixteen bits at a time, because it reduces the number of instruction fetch memory cycles and, with a slow memory, uses the full available bandwidth. The power consumption in the memory system is also reduced commensurately.

Local control is also possible in instructions such as 'CMP' (compare) which do not need to traverse the entire pipeline length; it is just as easy to remove a packet from the pipeline by generating no output handshakes as it is to generate extra packets. In Amulet3 comparison operations only affect the processor's flags which reside in the 'execute' stage and therefore cause no activity further down the pipeline.

A final benefit of the localised control is that the pipeline operation can be regulated by any active stage. Both Amulet2 and Amulet3 have retrofitted a 'halt' (until interrupted) instruction into the ARM instruction set (implemented transparently from an instruction which branches to itself). This can be detected and 'executed' anywhere within the pipeline with the same effect of causing the processor to stall. Indeed Amulet2 instigates halts in its execution unit whereas Amulet3 halts by suspending instruction prefetch, but the overall effect is the same. Halting an asynchronous processor (or part thereof) is equivalent to stopping the clock in a synchronous processor and, in a CMOS process, can drop the power consumption to near zero. This facility therefore makes power management particularly easy and – in many cases – near automatic.

15.4.3 Determinism and non-determinism

Before examining specific architectural techniques which can be employed in an asynchronous processor it is worth considering something of the design philosophies employed. The most significant is probably that of non-

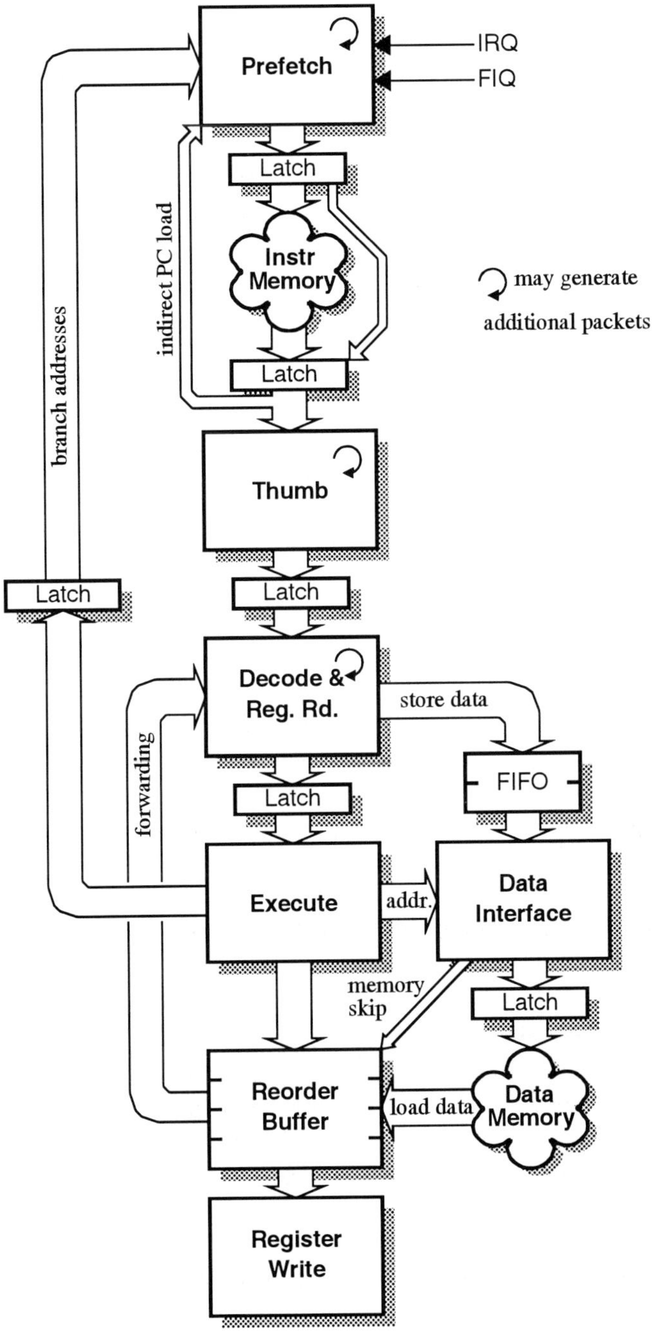

Figure 15.7. Amulet3 core organisation.

determinism because, unlike a synchronous processor, an asynchronous processor can behave non-deterministically and yet still function correctly.

An advantage in the analysis and design of a synchronous system is that the state in the next cycle can be determined entirely from the current state. This *may* also be true in an asynchronous system, but the timing freedom means that this is not the only choice of action. Within a small asynchronous state machine it is possible to achieve the same behaviour with internal transitions ordered differently (e.g. the inputs to a Muller C-element can change in any order) and this is also true on a macroscopic level. The first example used here is a processor's prefetch behaviour, chosen because different philosophies have been chosen in different projects.

All the Amulet processors have had a non-deterministic prefetch depth. This is achieved by allowing the prefetch unit to run freely, normally only constrained by the rate at which the instruction memory is able to accept instructions. In order to branch the prefetch process is 'interrupted' and a new programme counter value inserted; this is an asynchronous process which requires arbitration and therefore happens at a non-deterministic point.

An alternative approach, for example used in the ASPRO-216 processor [117], is to prefetch a fixed number of instructions. This can be done by prompting the prefetch unit to begin a new fetch for each instruction which completes. If a branch is required this can be signalled, if not this too is signalled. In effect the processing pipeline becomes a ring in which a number of tokens are circulated and reused. (See also section 3.8.2 on page 39.)

Having a deterministic prefetch simplifies certain tasks, notably dealing with speculative prefetches and branch delay slots. As it is possible to say exactly how many instructions will be prefetched following a branch these can be processed or discarded with a simple counting process. However keeping tokens flowing around a ring places an extra constraint on the elasticity of the pipeline which – in some circumstances – may sacrifice performance.

With a non-deterministic prefetch depth it is possible to have fetched zero or more instructions speculatively, although there will be an upper bound set when the pipeline 'backs up'. In an architecture without delay slots (such as ARM) the lower limit is not a problem, but some means other than instruction counting must be provided to indicate that the prefetch stream has changed. The Amulet processors do this by 'colouring' the prefetch streams. To illustrate: imagine the processor prefetches a stream of 'red' instructions. Eventually, as these are executed, a branch is encountered which causes the execution unit to request prefetches starting at a new address and in a different colour ('green', say). Subsequent red instructions must then be discarded, the first green instruction being the next operation completed. A subsequent green branch will cause another colour change; because all the former red operations

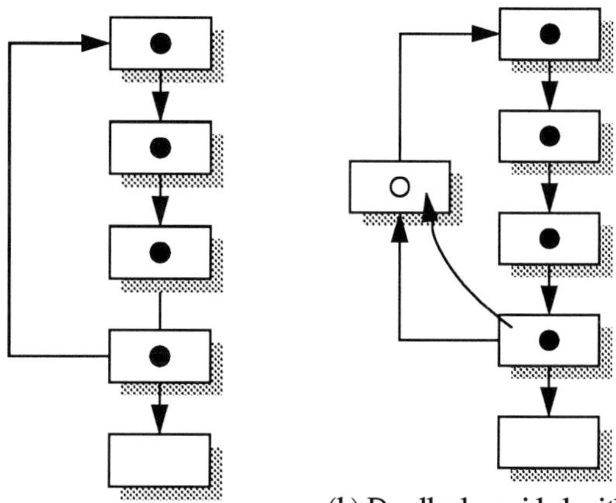

Figure 15.8. Branch deadlock and its prevention.

must have been flushed at this point it is possible to switch back to red, thus only two colours (i.e. a single colour bit) is required.

Before leaving this issue there is one other, less obvious, problem with a non-deterministic prefetch depth which can cause deadlock if not considered in the design. In this architecture the act of branching uses an arbiter to insert a token into the processor's pipeline. If the pipeline is already full – and there is nothing to limit this – then the arbiter cannot acknowledge the operation and thus the pipeline deadlocks (figure 15.8(a)). Because a branch could occur when the pipeline is not full the deadlock is not inevitable, but it could happen each time a branch is attempted.

This problem is easily solved; an extra latch which is normally empty can decouple the branch operation from the main pipeline flow (figure 15.8(b)), leaving 'normal' operation to continue. As a second, valid branch cannot be taken until after the first has been accepted and its target fetched and decoded, a single latch will always prevent deadlocks here.

This class of problem is generic. A manifestation of a similar problem became evident early in the design of Amulet1 where the instruction fetch competed non-deterministically for the bus with data loads and stores. In a situation where the processing pipeline is full it is possible that an instruction fetch can occupy the memory bus but be unable to complete because there is no latch ready for the result. If a data transfer is pending then it is blocked, resulting in the pipeline remaining full (figure 15.9(a)). This can be rectified if the instruction fetch is throttled so that it cannot gain the bus until it is known that it can relinquish it again (figure 15.9(b)). The converse of the problem does

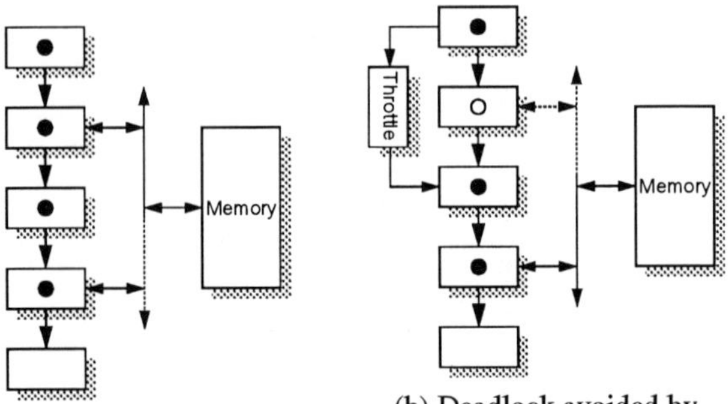

(a) Deadlock

(b) Deadlock avoided by throttling prefetch

Figure 15.9. Bus contention deadlock and its prevention.

not occur because the data transfer, once started, cannot be stalled, so only instruction prefetch requires throttling.

Amulet3, with its separate instruction and data buses, does not exhibit this problem within the processor, although the memory system can still suffer an analogous deadlock. This is described further in the section on memory.

If a deterministic, asynchronous solution is preferred this could be implemented by adding a data transfer phase to every instruction. There is still an asynchronous advantage here in that an 'unwanted' data access would be very fast but there is a price in limiting the adaptability of the processor's pipeline.

Counterflow Pipeline Processor. Although most asynchronous microprocessor designs have a 'conventional' architecture (other than lacking a clock!) it may be practicable to implement radically different processor structures, and several have been studied. One interesting – and highly non-deterministic – idea is the Counterflow Pipeline Processor (CFPP) [126] in which instructions flow freely along a pipeline containing processing units towards the register bank while operands flow equally freely towards them. This is intended to reduce stalls in waiting for operand dependences between instructions; as well as evaluating and carrying its result to completion an instruction can cast the result backwards to be picked up as required by subsequent operations.

Whilst expensive in the number of buses required, the CFPP allows considerable flexibility in the number and arrangement of functional units (figure 15.10). Functional units may be ALUs of full or limited function, memory access units, multipliers etc. The only rule is that the operation must be attempted by one of the available units, so stalls are only needed if an uncompleted operand reaches the last appropriate functional unit still without all its operands.

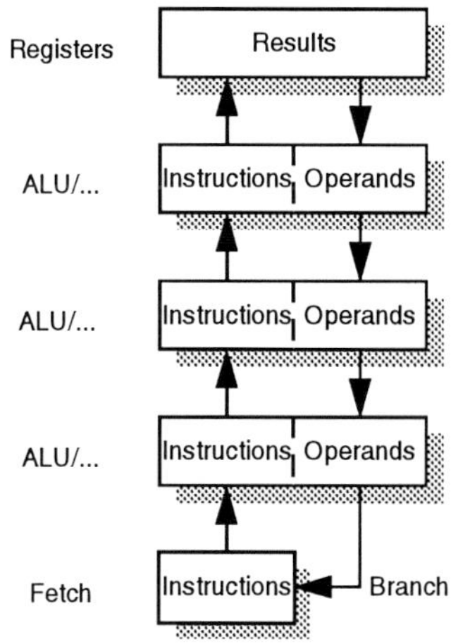

Figure 15.10. Principle of the Counterflow Pipeline Processor.

To ensure that an instruction does not miss any of its operands, passing in the other direction, a degree of synchronisation is needed at each pipeline stage. Because there is no 'favoured' direction in the pipeline an arbiter is required at each stage to ensure that two packets do not 'cross over' without checking each other's contents. Because every movement of both instructions and data requires an arbitration, fast, efficient arbiters are essential for such a performance architecture.

Of course deepening the pipeline to accommodate many functional units increases the penalty due to branches – which need to propagate 'backwards' to the fetch stage – so good branch prediction is important.

Arbitration and deadlock. It is possible to build an asynchronous processor which is as deterministic in its operation sequences as its synchronous counterpart (i.e. deterministic except for such events as interrupts). Alternatively it is possible to make a highly non-deterministic processor.

Each scheme has both advantages and disadvantages. Enforcing synchronisation could lead to a reduction in performance; for example the memory interface may not begin a prefetch until told that an instruction does not require the memory for a data transfer, with a consequent reduction in available bandwidth. On the other hand the predictability of the system's behaviour can make testing significantly easier by reducing the reachable state space of the system. The decision as to what (if anything) should be allowed to be non-deterministic is a decision for the designer which must be reviewed in the particular circumstances. However it must be remembered that every non-determinism has an associated arbiter (which, theoretically, can require an infinite resolution time) and is likely to introduce a potential deadlock which must be identified and prevented.

In the general case a good rule for avoiding deadlock is to examine carefully any instances of arbitration for a 'shared' resource (such as the bus) and ensure that no unit can be granted access until it is certain that it can relinquish it again regardless of the behaviour of other parts of the system. Each arbiter increases the number of states reachable by the processor and makes the design problem harder, but it increases the system's flexibility. Non-determinism can be beneficial if used with caution.

15.4.4 Dependencies

When a processor is pipelined beyond a certain depth it is necessary to insert interlocks to ensure that dependences between instructions are satisfied and programmes execute correctly. Even devices such as the MIPS R3000 [72] – which was a 'Microprocessor without Interlocked Pipeline Stages' is interlocked in the sense that the programmer/compiler could use a clock cycle count to ensure correct operation; an expedient which is disqualified in an asynchronous environment. Similar constraints are applied in the ARM architecture.

PC pipeline. ARM does not use the clock explicitly in the way MIPS does, but there is one aspect of the architecture which is similar. The Programme Counter (PC) is available to the programmer in the general-purpose register set and when it is read the value obtained is the address of the current instruction plus two instructions. This is a historical consequence of the early ARM implementations where there were two clock cycles between generating the instruction's address and executing the operation. Compatibility with this must be maintained, even in an asynchronous processor where the prefetch and execution are autonomous.

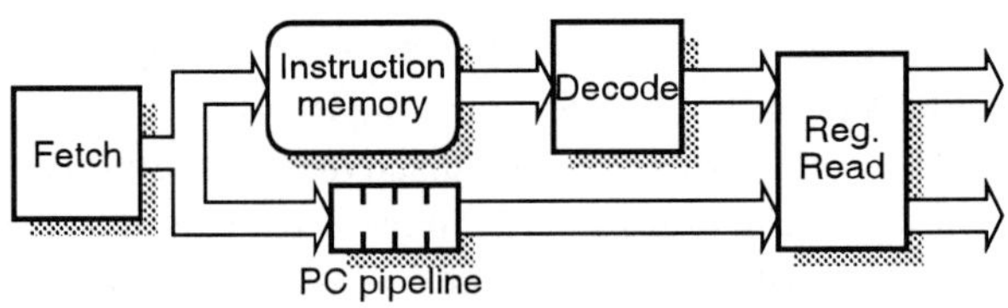

Figure 15.11. PC pipeline.

Because the generation and subsequent, possible use of the PC are unsynchronised in an Amulet processor a method of transmitting the value must be found. To do this all Amulet processors have maintained a copy of the PC with each fetched instruction (figure 15.11). These flow down the pipeline with the instruction and can be read whenever the PC may be needed. The PC may be required explicitly as an operand or address, implicitly in branch instructions, or to find a return address in the case of exceptions such as interrupts or memory aborts. Different Amulet cores have varied the exact value (e.g. PC+8) held with this 'PC pipeline' in attempts to minimise the later calculation overheads, but in Amulet3 the PC is held without any premodification which allows any of the required values PC+2, PC+4, PC+8 to be calculated with a simple incrementer.

It is worth noting that the PC values need not be bundled with the instruction directly. The PC pipeline can be a separate, asynchronous pipeline from the instruction fetch which can have a different depth, providing that a 'one in, one out' correspondence is maintained. This is a feature which is exploited in Amulet3 to throttle the prefetch unit and prevent instruction fetches causing a deadlock; this mechanism is described more fully in section 15.5.2.

Register dependencies. The greatest register dependency problems are read-after-write dependencies. One of these occurs in the case of a fragment of code such as:

```
LDR   R1, [R0]       ; Load ...
ADD   R2, R1, R3     ; ... and read
```

In this example it is essential that the value is (at least apparently) loaded into R1 before the subsequent instruction uses it. As soon as the execution path is pipelined there is a risk that this will not be assured and this uncertainty is increased in an asynchronous device where the load could take an arbitrary time.

Three solutions for ensuring register bank dependencies are satisfied are given below.

- Don't pipeline.
- Lock.
- Forward.

The first solution was the approach taken in the earliest, synchronous ARM implementations. This involves reading register operands, performing an operation and writing the result back in a single cycle so that a subsequent operation always has a 'clean' register view. This is simple but makes the evaluation cycle very long and is unacceptable in a high-performance processor.

A locking approach allows selective pipelining of the instruction execution by retarding instructions whose operand registers are not yet valid. This involves setting a 'lock' flag when an instruction is decoded which wishes to modify a register and clearing the flag again when the value finally arrives. A subsequent instruction can test the locks on its operands and be delayed until they are all clear. This mechanism is eminently suited to an asynchronous implementation because a stalled instruction is simply caused to wait, which can be done without recourse to arbitration.

In practice it is convenient to allow more than one result to be outstanding for a single register. Partly this is a consequence of the ARM's extensive use of conditional instructions, such as:

```
CMP      R1, R2    ; Set flags
MOVNE    R0, #-1   ; If R1 ? R2
MOVEQ    R0, #1    ; If R1 = R2
```

In such a case a single lock flag (on R0 in this instance) is inadequate and some form of semaphore is needed. It turns out that the operation of such a semaphore is fairly simple to implement in an asynchronous environment provided that testing and incrementing are mutually exclusive. At issue time an instruction therefore:

- attempts to read its operands and waits until they are all unlocked, then
- locks its register destination(s) by incrementing the semaphore.

The semaphore is decremented again when the result is returned; this action may take the semaphore to zero and, possibly, free up a waiting instruction. This can happen at any time.

The example above illustrates another potential problem in an asynchronous system: the two 'MOV' operations are mutually exclusive and so only one will be executed. As this is not known at issue time both have incremented the semaphore and so both must decrement it, otherwise R0 will be permanently locked. In general if a speculative operation is begun it must complete – the 'write' operation therefore always takes place, although sometimes the register contents are not changed and only the unlocking is performed.

The principle of semaphores was designed and implemented as the 'lock FIFO' in Amulet1 and Amulet2. In these processors the semaphore also held the destination register addresses to avoid carrying them with the instruction. As the instructions flowed (largely) in order, results and destinations could be paired at write time.

The lock FIFO was implemented as an asynchronous pipeline as shown in figure 15.12. Because the cells in each latch (horizontal) are transparent latches the entries are copied from one to the next, thus ensuring the outputs of the OR gates are glitch free. These can then be used to stall any read operations on the particular register until the write is complete. The only hazard would be if a register was actively locked whilst a read was being attempted; this is prevented by a sequencing of read-then-lock by the instruction decoder.

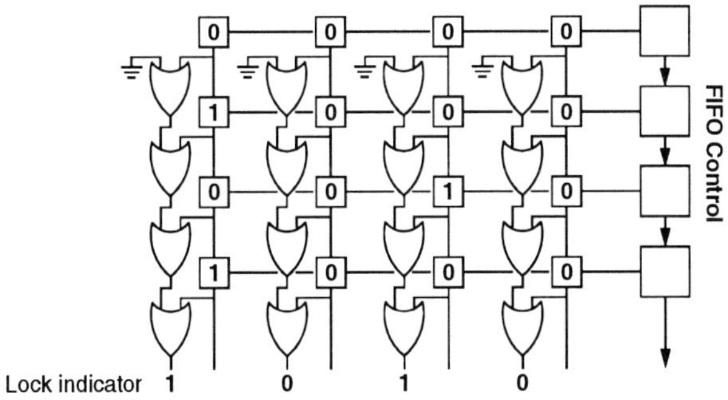

Figure 15.12. Lock FIFO.

Unlocking can happen safely at any time, both relative to the reading or locking of other registers. The destination address, in its decoded form, is already available at the bottom of the data FIFO.

Reorder buffer. Although the lock FIFO works successfully it can introduce inefficiency in that it enforces in-order completion on the instructions and stalls each instruction until its operands are available. Therefore it is an effective, cheap method to guarantee functionality but is less than ideal for high-performance architectures.

In Amulet3 register dependencies are resolved using an asynchronous reorder buffer [66, 51]. Whilst the major incentive for this was to facilitate a less intrusive page fault mechanism (see below) this allows instructions to complete in an arbitrary order and results can be forwarded at any time. It is therefore a significant step towards a complete out-of-order asynchronous processor.

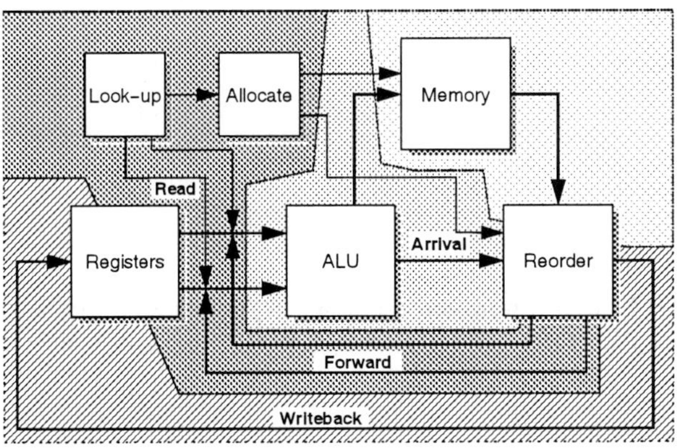

Figure 15.13. Reorder buffer position in Amulet3.

Table 15.1. Reorder buffer allocation.

Instruction		Slot0	Slot1	Slot2	Slot3
LDR	R0, [R2]	*R0*	?	?	?
MOV	R4, #17	R0	*R4*	?	?
LDR	R7, [R0+4]!	R0	R4	R0	*R7*
ADD	R7, R7, R4	*R7*	R4	R0	R7
CMP	R3, R4	*R7*	R4	R0	R7
ADDNE	R7, R7, R6	R7	*R7*	R0	R7
SUB	R1, R7, R0	R7	R7	*R1*	R7

The reorder buffer is positioned between the various data processing units and the register bank (figure 15.13) and crosses several timing domains. An instruction first encounters the reorder buffer control at decode time when its operands may be discovered and forwarded; any results are allocated space at this time. Subsequently an instruction may be subdivided (and, in principle, issued at any time) and results can arrive separately and independently. Operands can be forwarded any number of times between when they arrive and when they are overwritten. Finally an ordered writeback occurs where the results are retired and the reorder buffer slots freed for reallocation.

In the decode phase the operand register numbers are compared with a small content addressable memory (CAM) to determine which reorder buffer slots may contain appropriate values for forwarding. This list always terminates with the register itself, so a value is always available from somewhere. Once this is map is known the reorder buffer slots can safely be reassigned.

The assignment process cyclically associates each reorder buffer entry with a particular register address and the instruction packet carries forward just the reorder buffer 'slot' number. Consider the following code fragment:

```
LDR    R0, [R2]       ;
MOV    R4, #17        ;
LDR    R7, [R0+4]!    ;
ADD    R7, R7, R4     ;
CMP    R3, R4         ;
ADDNE  R7, R7, R6     ;
SUB    R1, R7, R0     ;
```

Assuming that the reorder buffer has four entries and the next free entry is (arbitrarily) 0, the reorder buffer assignment will proceed as shown in table 15.1. In each case the italicized entry is the latest one to have been assigned.

Note:

- the second load (LDR) operation uses the ARM's base writeback mode and therefore requires two destinations;
- the comparison has no register destinations;
- the same register address can appear multiple times in the reorder buffer;
- a slot is assigned even if the instruction is conditional (ADDNE) and may not produce a valid result.

The instruction decoder still retains the reorder buffer map prior to the start of the instruction. In parallel with the reassignment it proceeds as follows, using the final instruction as an example:

- The locations of the appropriate registers are examined; these may be in the process of being reassigned but cannot be overwritten yet because the instruction execution has not begun.

  For R7 try: slot 1, slot 0, slot 3, register bank.

  For R0 try: slot 2, register bank.
- Try to read from each location in the list until a value is obtained.

Note that a list of possibilities is required because the assigned slots need not contain a valid value. The most obvious cause of invalidation in an ARM is an instruction which fails its condition code check (e.g. the value of R7 in slot1), but other conditions – such as a preceding branch – can also result in an instruction being abandoned.

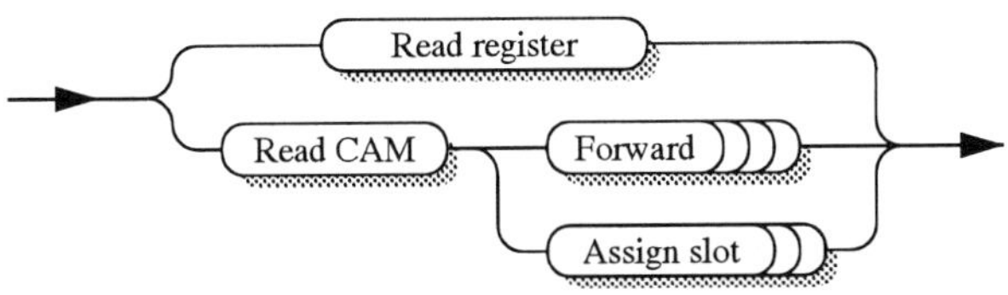

Figure 15.14. Register processes in the Amulet3 decode unit.

The flow of control through the decode phase is summarised in figure 15.14. Note that the forwarding time can vary depending on external factors while the slot assignment time depends on the number of slots which are required (from zero to two) and (occasionally) may have to wait for a slot to be available. Reorder buffer slots are assigned serially, even within a single instruction which simplifies the asynchronous implementation. There is a small performance impact on more 'complex' instructions, but these are relatively rare.

Subsequent to assignment the instruction packets proceed to further stages carrying their reorder buffer slot number. Although Amulet3 issues only single instructions in order the ARM instruction set effectively allows two semi-independent operations via the internal execution unit and via the external data interface (figure 15.15). Each of these may produce a result at any time so each has its own port into the reorder buffer. Whilst these inputs are asynchronous they are guaranteed to target different slots and can therefore be independent.

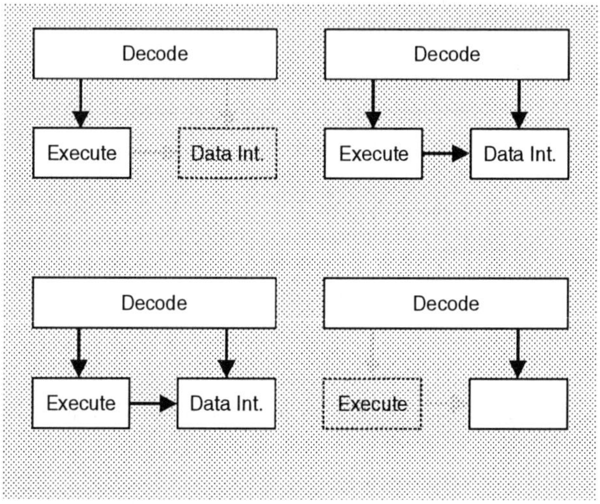

Figure 15.15. **Sub-instruction routing.**

The writeback process simply copies results back to the register bank. On its arrival each result 'fills' a reorder buffer slot. The 'writeback' process therefore waits until a particular slot is ready, copies it out, and moves to the next one. The fact that the slots may become ready in an arbitrary order is not a concern.

Superimposed on this process is a forwarding mechanism which waits until a result is ready and copies it back into the decode stage. This process can happen before, during or after the writeback process; in fact the processes are asynchronous and concurrent. The key is that both processes use non-destructive copying and therefore leave the data available for the other as required. A result in the reorder buffer remains until it is overwritten.

Either of the above processes may be required to wait if they need a result which has not yet arrived. In order to control this in a non-interacting way there are two separate 'flags' which indicate the presence of data in a slot. One flag is raised to indicate that the slot is 'full'; this is cleared when the writeback process has copied the result to the register bank. This is at the heart of the control circuit shown in figure 15.16 (which will be described shortly). The

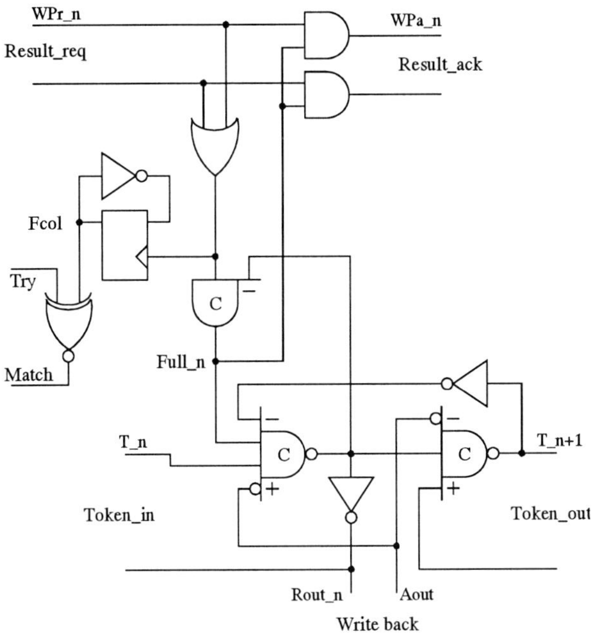

Figure 15.16. **Reorder buffer copy-back circuit.**

Table 15.2. 'Fcol' state as results arrive in a reorder buffer.

Result Number	0	1	2	3
0	$0 \rightarrow 1$	0	0	0
1	1	$0 \rightarrow 1$	0	0
2	1	1	$0 \rightarrow 1$	0
3	1	1	1	$0 \rightarrow 1$
4	$1 \rightarrow 0$	1	1	1
5	0	$1 \rightarrow 0$	1	1
6	0	0	$1 \rightarrow 0$	1
7	0	0	0	$1 \rightarrow 0$
8	$0 \rightarrow 1$	0	0	0

second flag ('Fcol') is merely *changed* to indicate the arrival of a result. As the state of this flag alternates on each pass through the cyclic reorder buffer it is possible for a forwarding request to test to see if a result has arrived without changing the flag's state (table 15.2). This is essential because a value may be forwarded zero or more times, the number not being known at issue time.

As a final embellishment a result returned to the reorder buffer may be invalid (due, for example, to a condition code failure). Each slot is therefore accompanied by a flag which can prevent the data being written to the register bank (the 'full' flag is still cleared) or being forwarded. In the latter case further forwarding may be attempted from a less recent result, culminating in the use of the default register value.

The writeback circuit (figure 15.16) is a good example of the working of such asynchronous control circuits. It operates as follows.

- The arrival of a result (top left) causes the 'Full' bit to be set and acknowledges itself (top right). The request comes from one of a number of mutually exclusive sources.

  This event also toggles the forwarding colour ('Fcol') which allows any instructions issued subsequent to this one to use the result.

  Note that the input can be on any of the mutually exclusive input channels; two are shown here but more are possible.

- The input request can be removed at any time leaving the slot marked as 'Full'.

- When it becomes this slot's turn to copy back a token is received (bottom left) which initiates an output request (bottom centre). If the token is received first the circuit waits for the result to arrive. The token is acknowledged.

  This process also allows the 'Full' bit to be reset, waiting for the input request to fall if it has not already done so.

- When the copy back is complete the (broadcast) acknowledge is picked up by the circuit to complete the token input handshake and pass the token to the next, similar circuit to the right. The next slot cannot attempt to output until the four-phase output acknowledge is complete.

These slots are connected in a ring and reset so that there is a single token present for the first stage to output after which it runs freely, emptying each slot in turn when it contains a result.

The experimental (and often unsystematic!) way in which Amulet3 was designed meant that the original state machine was defined and refined to a version close to that shown in figure 15.16 as a schematic and only subsequently subjected to a more formal analysis. The analysis was carried out using Petrify (which was introduced in section 6.7 on page 102).

At first this caused problems due to the choices within the system. The output channel 'calls' the register bank and it was expedient to broadcast the output acknowledge to all the copy back circuits and use the (unique) active request to sensitize the relevant location. This proved hard to model in a single subcircuit. However on a suggestion from one of the Petrify developers (Alex Yakovlev) it proved easier to model the whole system than a single part. The resultant signal transition graph (STG) (see figure 15.17) clearly shows the four implemented subcircuits which pass control around cyclically. The rest of the processor is abstracted in the four small rings which, when an output handshake has been completed, can reset the circuit to 'Full' again at an arbitrary time.

The analysis with Petrify both verified the circuit's operation and removed a redundant transistor in each subcircuit.

There are two hazards in the asynchronous processes as described here. The first is that there is no local way of preventing a slot being overwritten before it has been emptied. In Amulet3 this is guaranteed elsewhere by ensuring that a slot is never assigned until it has been freed by the copy back process. In effect a count of free slots is maintained which is decremented when a slot is assigned and incremented when it is released again. Because assignment and release are both cyclic and in-order it is not necessary to pass individual slot identification, the presence of a token is adequate. This 'throttle' is implemented as a simple, dataless FIFO which also acts as an asynchronous buffer between the unsynchronized processes. This is shown in figure 15.18 for a system with four reorder buffer slots; in the state shown one result has been produced but not yet committed to the register bank, two more are being generated and the decoder could issue another instruction which generated a single result.

The other hazard is because the forwarding and writeback processes are not synchronised, therefore the register value which is read (as a default) could be changed during the read process, resulting in 'random' data being read. However this can only happen if there is a valid value for that register in the reorder buffer and therefore it is certain that this value will be forwarded in preference to the register contents. Providing that the implementor can ensure that the 'random' data does not cause problems within the circuit, the mechanism is secure against this asynchronous interaction.

Studies of ARM code indicated that for the Amulet3 architecture a reorder buffer with five or more entries was unlikely ever to fill [50]. Amulet3 therefore implements a four entry buffer; however the mechanism described is extensible to any chosen size.

15.4.5 Exceptions

An 'exception' is an unexpected event which occurs in the course of running a programme. The Amulet processors are compatible with the ARM instruc-

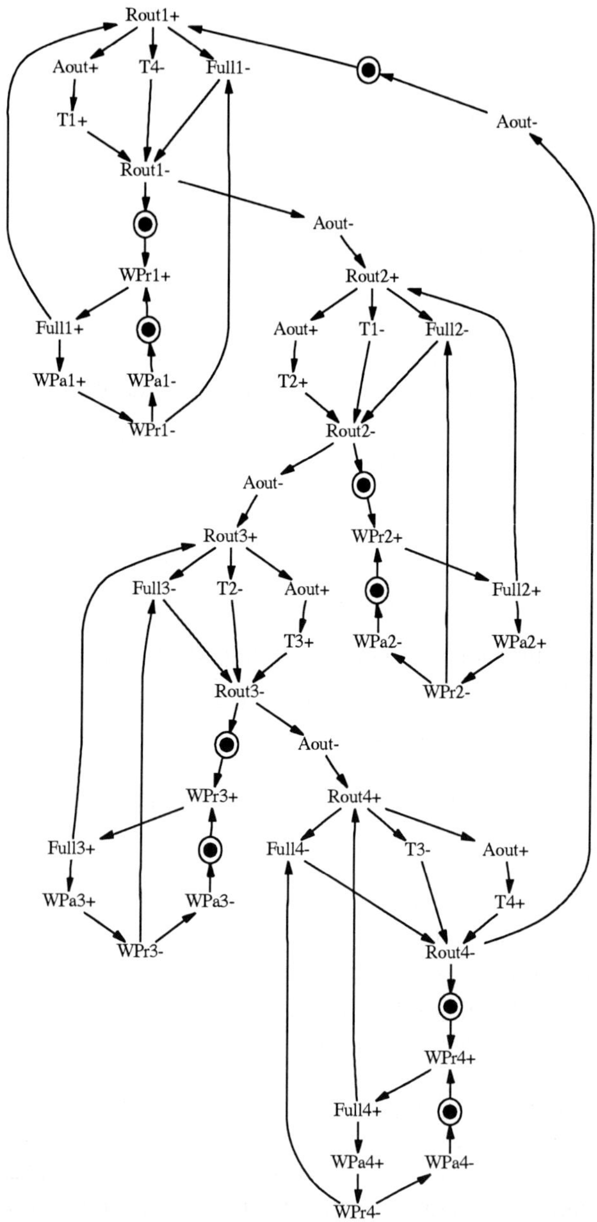

Figure 15.17. **An STG for all four reorder buffer token-passing circuits.**

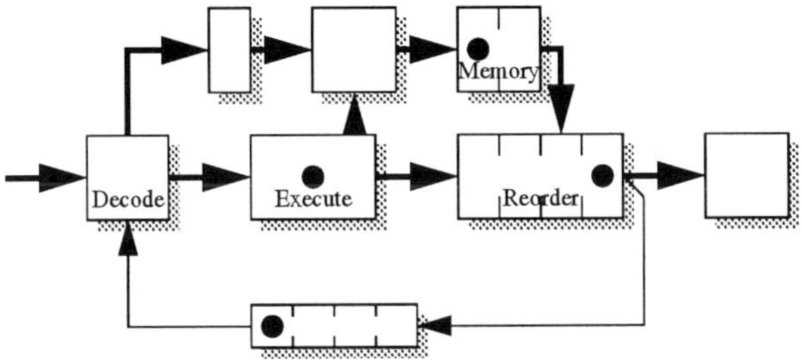

Figure 15.18. **Token passing throttle on reorder buffer.**

tion set and, therefore, the types of exceptions and the behaviour when they occur is predefined. Ignoring reset, ARM has six types of exception:

- Prefetch abort - a memory fault (e.g. page fault) during an instruction prefetch;
- Data abort - a memory fault during a read or write transfer;
- Illegal instruction - an emulator trap;
- Software interrupt - a system call (not really an exception, but has similar behaviour);
- Interrupt - a normal, level sensitive interrupt;
- Fast interrupt - similar to normal interrupts, but higher priority.

Of these the majority are quite easy to deal with: software interrupts and illegal instructions can be detected at instruction decode time, as can prefetch aborts (when there is no instruction to decode); interrupts are imprecise and therefore can be inserted anywhere; only data aborts have the ability to cause serious problems because they only evidence *after* the instruction has been decoded and started to execute.

Interrupts. In any processor an interrupt is an asynchronous event. In one sense the arrival of an interrupt can be thought of as the insertion of a 'call' instruction in the normal sequence of instruction fetches. At first glance it would seem that simply arbitrating an extra 'instruction' packet into the instruction stream would suffice; however this simplistic view can cause problems.

The chief problem is that there is some interaction between the interrupt 'instruction' and prefetch stream; the interrupt needs a PC value to synthesise its return address, and the interrupting device cannot know what this is. Furthermore the return address must be valid; if an interrupt is accepted just after a branch has been prefetched it could be inserted into code which should not be run.

Amulet3 implements interrupts by 'hijacking' rather than inserting instructions, the whole operation being performed in the prefetch unit. Although the interrupt signals (in this case they are level-sensitive) change asynchronously with respect to the prefetch unit the mutual exclusion element is better thought of as a synchroniser than as a typical asynchronous arbiter. Figure 15.19 shows a method of implementing this. Here any *change* in the interrupt input will retard the normal request flow until the synchronised state has been latched; the synchronised interrupt signal only changes when done is low.

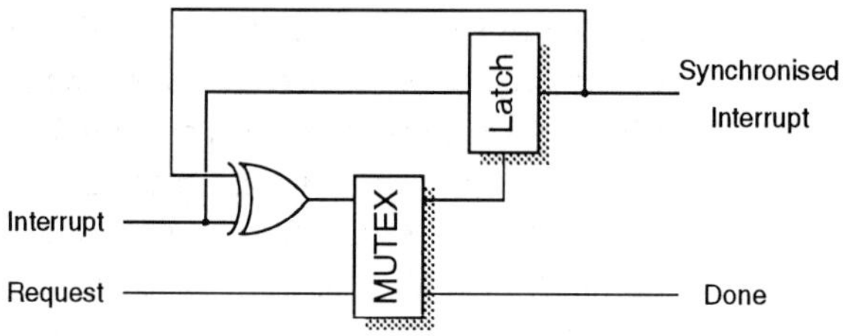

Figure 15.19. Interrupt synchroniser.

When it is known that an interrupt signal has become active the current PC value effectively becomes the address of the interrupt 'instruction' and can be used to form the return address. This can be sent as an instruction but can save time by bypassing the memory. The interrupt can then be disabled to prevent further acknowledgement.

Because this action takes place in the prefetch unit, Amulet3 can treat the interrupt entry as a predicted branch and jump directly to the appropriate interrupt service routine which, in an ARM, are at fixed addresses.

The problem still arises that the interrupt entry may be speculative. If a branch is pending the return address sent to the execution stage may be invalid – in any case it will be wrongly coloured and therefore discarded! However the act of branching updates both the PC and any associated information, including the interrupt enables. As the interrupt has not been serviced the (level-sensitive) request will still be active and another attempt to enter the service routine will be made. This time the branch target address will be saved and there can be no further impediments.

Data aborts. Although it solves the register dependency problems, the reorder buffer was originally introduced to simplify the implementation of data aborts. The ARM architecture specifies that, if an abort occurs, no effects from following instructions will be retained. Earlier Amulet processors did not speculate on memory operations, relying on a fast 'go/no go' decision from any memory management unit. Amulet3 allows for more sophisticated (i.e. slower!) memory management by outputting memory transfer requests and only checking for aborts at the end of the operation. To be of any worth this must allow other, speculative operations to take place in parallel, but these operations cannot be 'retired' until the outcome of the load is known.

The reorder buffer provides a place for speculative results to be held – and forwarded for reuse if necessary – until they can be retired into registers. In the (rare) case of a data abort the speculative results can be discarded, leaving the register bank intact. The discard can be achieved either by using a colouring scheme, tested by the register writeback process, or by marking speculative results as invalid using the same flag as an operation which has failed for other reasons. For certain reasons of implementational expediency Amulet3 uses the latter method, although the asynchronous hazard of invalidating a result whilst a forwarding operation is being attempted must be avoided. (This is achieved by implementing two validity bits and nullifying only one of them; the copy back process uses an AND of these whereas forwarding uses an OR. Forwarding is therefore not disturbed, although the result will be discarded later.)

The reorder buffer accounts only for the register state however; the ARM holds other state bits which also require preservation. Two separate mechanisms are used for these, depending on the frequency of changes.

The first is the current programme status register (CPSR) which holds the processor's flags, operating mode et alia, the whole of which can be represented in ten bits. The flags clearly change frequently during execution and there are many dependences on these bits due, for example, to compare-branch sequences. When attempting a memory operation Amulet3 simply copies the current CPSR state into a history buffer [66] FIFO; successful completion of the transaction discards this entry, but an abort can restore the CPSR state to that at the start of the failed operation.

The other non-register state is a set of five saved programme status registers (SPSRs) which act as a temporary store for the CPSR in various exception entries. These change very rarely and it is uneconomic to enlarge the history buffer to encompass them, although – in theory – they could be changed between a load being issued and an abort being signalled. The solution here was simply to use a semaphore to lock the SPSRs whilst any memory operations are outstanding. This delivers the required functionality very cheaply and the performance penalty is tiny because SPSRs change so rarely.

15.5. Memory – a case study

It seems reasonable that an asynchronous processor should interact with an asynchronous memory system. This implies the need for handshake interfaces on a range of memory systems, including RAM, ROM and caches. This is the subject of the following sections.

An individual memory array is a very regular structure and – under steady voltage, temperature etc. conditions – will produce data in a constant time. At first glance this may suggest that there is not much scope for asynchronous design within a memory system. However each part of the memory will have its own characteristic timing; in some cases even a simple memory will have a variable cycle time. An example is a RAM which will typically take longer to read from than it will to write to.

In fact the memory system is one part of a computer which has extremely variable timing; even a clocked computer will take different times to service a cache hit and a cache miss. An asynchronous system will accommodate such cycle time variation quite naturally and is able to exploit many more subtle variations which would be padded to fill a clock cycle in a synchronous machine.

15.5.1 Sequential accesses

A static RAM (SRAM) stores data in small flip-flops which have only a very weak output drive. To accelerate read access it is normal to use 'sense amplifiers' to detect small (sub-digital) swings in voltage and produce an early digital output. Sense amplifiers, being analogue parts, are quite power-hungry.

Sense amplification is only useful when there has been enough voltage swing for the read bits to be discriminated; it is also only required until the bits can be latched. As this period is certainly less than even half a clock cycle this is an ideal application for a self-timed system. A delay can be inserted to keep the sense amplifiers 'switched off' when the read cycle commences and only switch them on when they may be useful. An extra (known) bit in the RAM may then be discriminated and, when it has been read, used to latch the entire RAM line and disable the sense amplifiers. The same signal can be used to indicate the completion of the read cycle, possibly returning the RAM array to the precharge state in preparation for another cycle (figure 15.20).

When designing such a circuit the RAM completion is easy to detect but the delay before enabling the sense amplifiers is harder to judge. The designer can choose this to be short – to ensure maximum speed – or somewhat longer – to ensure the memory is ready before the amplifiers are enabled thus ensuring minimum power wastage. If the designer errs then either speed or power may be compromised slightly, however functionality is retained.

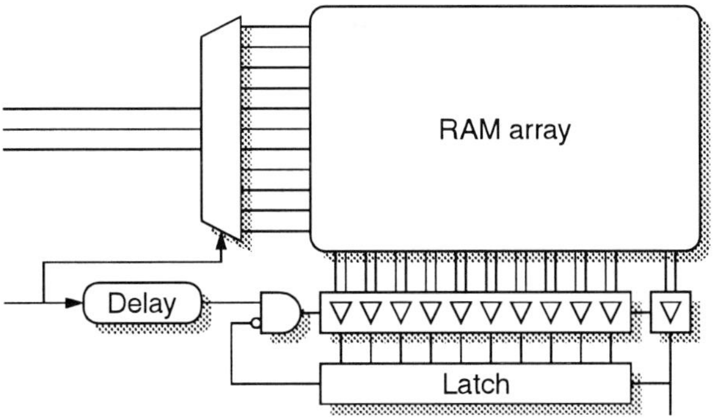

Figure 15.20. **Self-timed sense amplifiers.**

A typical SRAM array is organised to be roughly square; a 1 Kbyte RAM might therefore be organised as (say) 64 × 128 rather than 256 × 32 even though the processor requires only 32 bits on a given cycle. This presents the RAM designer with two choices:

- multiplex 32 sense amplifiers to the required word;
- amplify all 128 bits and ignore the unwanted ones.

The first choice appears better when a read is considered in isolation but cycles are rarely so arranged; typical access patterns (especially code fetches) exhibit considerable sequentiality and this can be exploited in the hardware design.

When using the first option it is possible to delay the RAM precharge and provide a subsequent read operation with a shorter read delay. The Amulet2e cache [49] uses this technique and is therefore able to provide subsequent accesses within a RAM 'line' faster than the first such access. This variation in access time is much less than a whole 'cycle' and therefore would be of no interest to a synchronous designer, but it is exploited automatically in an asynchronous system.

The second option given above can latch the entire RAM line after amplification. It can then service subsequent requests from this latch. This frees the RAM array to be precharged and – possibly – used for other purposes. This technique is exploited in the Amulet3i RAM which is described below.

15.5.2 The Amulet3i RAM

As shown in figure 15.7 on page 283, the Amulet3 processor has a Harvard architecture with separate instruction and data buses. However in the Amulet3i

SoC the memory model is unified; this implies that the buses must 'merge' somewhere outside the processor core.

In practice the local buses are merged in two places: once to get onto the on-chip MARBLE bus (see below) and once for access to the local, high-speed RAM (figure 15.21). In Amulet3i the local RAM is memory-mapped rather than forming a cache, although there is no reason why a cache could not have been implemented here; cache design is discussed later.

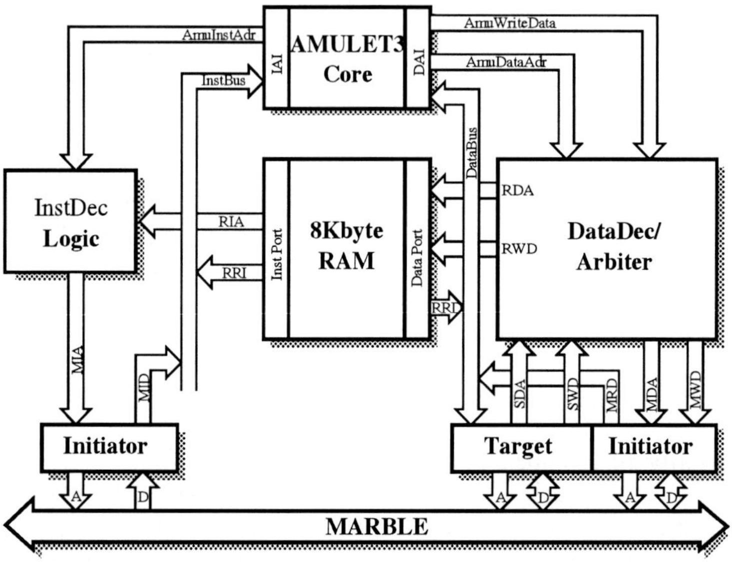

Figure 15.21. Amulet3i local memory.

The local RAM (8 kbytes) is divided into 1 Kbyte blocks; both buses run to each block (figure 15.22). The blocks are interleaved so that – most of the time – there need be no interaction between instruction and data fetches. Only if there is a conflict with both buses requiring data from the same RAM block is there any need for arbitration.

In the case of a conflict there is no clock on which to control an adjudication; access to the block is restricted by a mutual exclusion element ('mutex'), within the Arbiter blocks in figure 15.22, on a 'first-come, first served' basis. Note that, in general, data and instruction accesses are not synchronised and therefore the average wait will be about half the typical RAM access time.

Collisions are further minimised by using the latch at the output of the sense amplifiers (see preceding section) as a form of cache. Here separate latches are provided for instruction and data reads, so sequential accesses rarely need to compete for the arbitrated resource. In practice this gives performance ap-

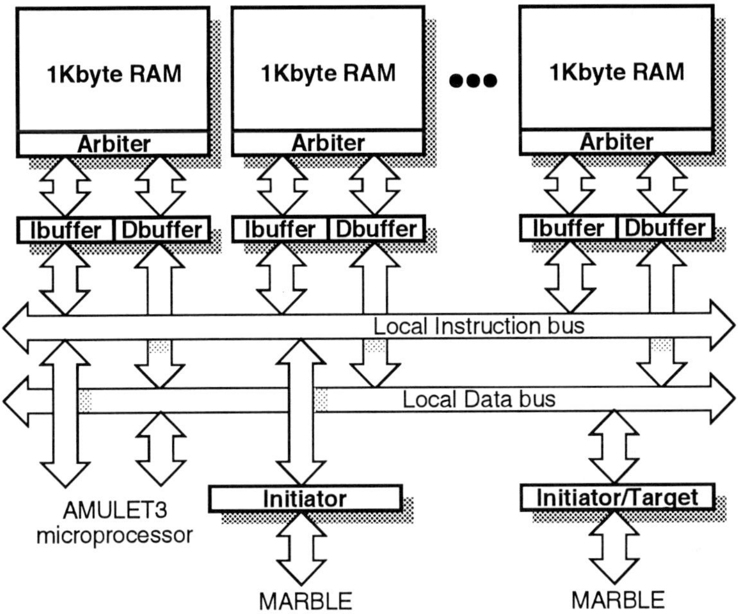

Figure 15.22. **Memory block organisation.**

proaching that of a dual-port RAM despite being implemented with standard SRAM.

The local RAM architecture thus provides memory cycles with two different delays ('random' and 'sequential') with the potential of an added (variable) delay in the rare case of a collision. In Amulet3i this is further complicated because the two local buses cycle in different times; the instruction bus is simplified as a read-only bus and runs noticeably faster than the full-function local data bus, which also permits external bus mastery to allow DMA and test access (fig. 20). The implications of these various timings are absorbed by the asynchronous nature of the system – for instance it is not necessary to slow the instruction fetches down by around 25% to fit a clock period set by the data bus.

The inclusion of arbiters within the memory blocks implies that the access patterns are non-deterministic. Care must therefore be taken to ensure that the system cannot reach a deadlock state. The only possible deadlock that could occur in the memory would occur as follows:

1 A (non-sequential) data transfer needs access to a particular RAM block.

2 This is prevented because an instruction fetch is already using the RAM array.

3 The instruction fetch cannot complete because the instruction decoder is still busy.

4 The processor pipeline is full and is blocked by the data fetch.

5 Deadlock!

To avoid this it is important not to gain access to the shared resource (the RAM array) until it is known that the operation will be able to release it again. A data transfer can always do this but provision has to be made restricting the generation of instruction fetches until they can be guaranteed to release the RAM. In practice the latch following the sense amplifiers forms a convenient, dedicated buffer in which to hold the instruction and allow data accesses to proceed. In Amulet3 the processor throttles its requests so only a single instruction fetch is outstanding at any given time and this must be removed from the 'I buffer' before the next address can be sent (figure 15.23).

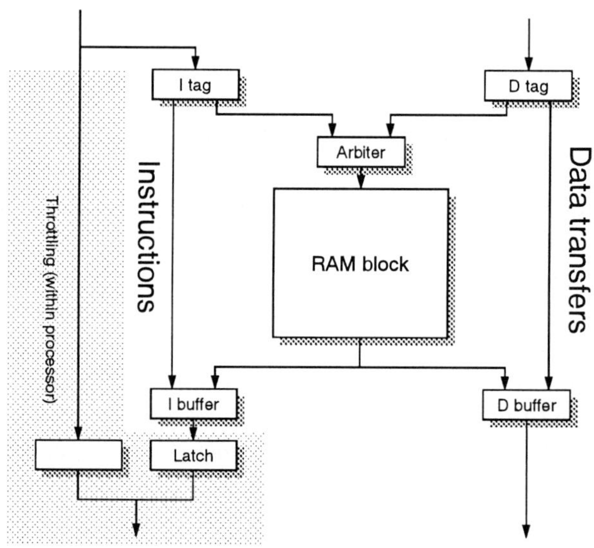

Figure 15.23. Memory block arbitration and throttling.

The need for arbitration is rare and thus the possibility of discovering a deadlock by random simulation even rarer. It is therefore essential to analyse such a non-deterministic system thoroughly to ensure that the opportunities for deadlock are removed.

15.5.3 Cache

An synchronous cache is very similar to an asynchronous RAM; most of the design is a combination of the preceding description of asynchronous RAM and standard cache design techniques. However in order to be efficient there are certain problems, not present in a synchronous cache, which require solution.

The most significant problems are in managing any conflicts between the processor/cache interactions and cache/bus interactions. The first of these to address is the issue of line fetch.

A line fetch generally occurs when a cache miss results from an attempted access to a cacheable location. A cache line comprising the required word and a small number of adjacent words (a cache line) are copied from memory into the cache. The simplest solution to this is to halt the processor, fetch the entire cache line, and allow the processor to proceed as the access is now a cache hit. However this requires a processor stall which is considerably longer than is strictly necessary.

A more efficient scheme is to begin fetching the cache line, forward the required word to the processor as soon as it arrives (it can often be arranged to be the first word fetched) and then allow the processor and line fetch to continue independently. Performance is further enhanced by allowing the processor to use other parts of the cache whilst the line fetch is proceeding ('hit-under-miss') and to use the incoming words as soon as they arrive. Unfortunately in an asynchronous environment this is difficult because the fetched words are arriving with no fixed timing relationship with the processor's cycles.

Initial thoughts may suggest arbitration for the cache. However it is possible to solve this problem without arbitration while maintaining all the desired functions by including a dedicated latch for holding the last fetched line. This latch is called the Line Fetch Latch.

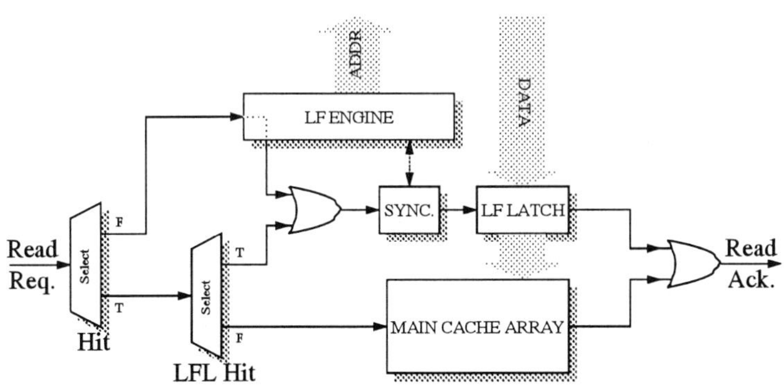

Figure 15.24. **Control circuit request steering.**

The line fetch latch (LFL) (figure 15.24) is actually a set of latches residing just outside the true RAM array. It normally holds the last-fetched cache line. It has its own tag and comparator which allow it to function much like the other cache lines. Note that the LFL holds the only copy of this data. (Incidentally, because the LFL is static and requires no sense amplification when it is read it can provide faster access in an asynchronous system.)

When, as a result of a cache miss, a fetch is needed, a line from the RAM is selected for rejection. For the moment assume that the cache is write-through and therefore the RAM can simply be overwritten. The LFL contents, together with its tag, are then copied into the chosen line and the LFL is marked as 'empty'. This can happen in parallel with the start of the external access.

The processor is then assumed to have a cache hit from within the LFL and attempts to read the appropriate word; this causes a stall at the synchronisation point because the word is empty and – unless the external memory is exceptionally fast – will not have been refilled yet.

As words arrive they are stored in the LFL and individually flagged as 'full'. As soon as the processor can read a word from the LFL (typically after the completion of the first fetch cycle) it can continue. From this time the processor can continue in parallel with the remaining words being fetched.

A subsequent cache cycle could be:

- a cache hit: this can proceed independently and without interaction with the LFL;
- a cache miss: this will cause a stall until the line fetch process is complete and the fetch process can be repeated;
- a LFL hit: this attempts to read the LFL whilst it is being filled. The possibilities are that the required word is already present (the processor continues) or the word is still pending (the processor must stall until it arrives.

Only in the last case is there any interaction between the asynchronous processes. However this interaction is merely a wait which can be implemented with a flip-flop and an AND gate (figure 15.25). The potential wait begins when the LFL is first emptied (caused by, and therefore synchronised with, the processor's action). The wait can end at an arbitrary time but this merely delays a transition; it cannot abort or change an action and can therefore be implemented without arbitration or the risk of metastability. This mechanism was first implemented in the Amulet2e cache system [49] and was also used in TITAC-2 [130].

Both these processors used a simple write-through cache for simplicity. For higher performance a copy-back mechanism is needed. This too can be provided using an extension of the LFL mechanism. In this case the process of

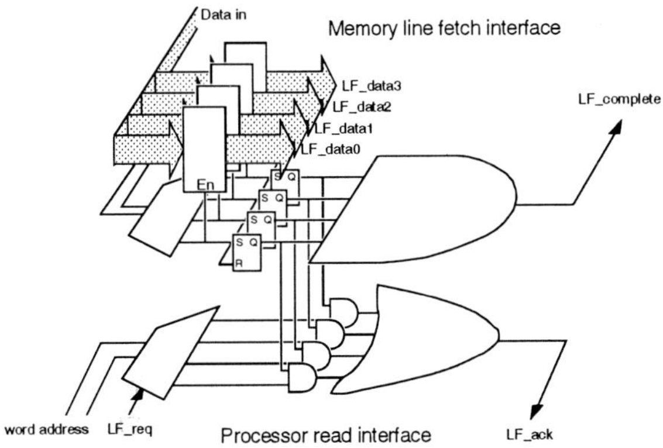

Figure 15.25. **Line-fetch latch read synchronisation.**

line fetching is complicated by the need first to copy the victim line from the cache array before overwriting it. The victim line can be placed in a separate write buffer together with its address as supplied by its tag field. Note that the line fetch is caused by a cache miss, so the rejected line can never be the same as the line being fetched (this becomes more important later). The LFL is then emptied into the RAM array as before and the refilling begins. The writing of the rejected line can be delayed because it is less urgent than satisfying the cache miss.

Each cache line (and the write buffer) also contain a 'dirty' flag which is set if the cache line has been modified. This can be checked and used to determine if the write buffer should be written out (i.e. 'dirty' is true) or is already coherent with the memory; in the latter case the copy-back process can be bypassed.

This process reduces the write traffic but, with a single entry write buffer, does not greatly assist with reducing fetch latency because the write buffer must be emptied before a subsequent fetch. However the write buffer can be extended, albeit at the cost of introducing an arbiter and the potential pitfalls therefrom.

If a second line fetch is needed before an earlier fetch is complete it is theoretically possible for this to overtake any pending write operations. This is also desirable. As the write operation will already be pending – merely waiting for the bus to become free – it is necessary to determine that the new fetch request arrived before the write could begin, which requires arbitration in an asynchronous environment. However, once the decision is made the operations can proceed as before. Two problems remain however:

- if the write buffer becomes full there will be nowhere to evict a line to and the system can deadlock;

- if the required line is one which has recently been evicted the fetch could overtake a pending write and thus lose cache coherency.

The first problem is relatively easy to solve; a simple counter can ensure that only a certain amount of overtaking is allowed and that one space in the write buffer always remains free (i.e. when the last entry is filled the next bus operation must be a write – unless, of course, it is a 'clean' line where the write can be assumed and bypassed). This can be implemented, for example, as a semaphore.

The second problem is harder, but can be solved by forwarding in a similar manner to the forwarding from a processor's reorder buffer. A line fetch checks the addresses of the entries in the write buffer and, if it finds a match, satisfies itself from there instead of requesting the memory bus. In such a case the 'fetch' can be performed with no latency, much more rapidly as it is an internal operation and can copy an entire cache line in a single operation. This is, of course, irrelevant to the functioning of other parts of the system as the whole is self-timed anyway. Such forwarding does not affect the write process (a re-fetched 'dirty' line is returned clean) and can take place regardless of whether the copy-back is pending, in progress, complete or not needed. The write buffer acts, in effect, as a write-through victim cache [56].

There is one caveat to this process; the line fetch occurs in parallel with the eviction of a cache line. It is therefore possible that one entry in the write buffer could be changing *during* the comparison process. As noted above this entry is reserved for the freshly evicted line and can safely be excluded from the comparison, thus averting any possibilities of a false positive due to signals changing.

15.6. Larger asynchronous systems

15.6.1 System-on-Chip (DRACO)

DRACO (DECT Radio Communications Controller) (figure 15.26) is a system on chip based around the Amulet3 processor. In terms of area about half of the (7 mm square) device is an asynchronous 'island' – hence Amulet3i – and the other half comprises synchronous peripherals compiled from VHDL.

The asynchronous subsystem (figure 15.27) is a computer in itself and was developed both with commercial intent and with a view to investigating some new techniques. The processor and RAM have already been discussed. Some other novel asynchronous features are outlined in this section.

15.6.2 Interconnection

Ideally an asynchronous system should be based around an asynchronous bus. Indeed it is arguable that large, fast synchronous systems should also use

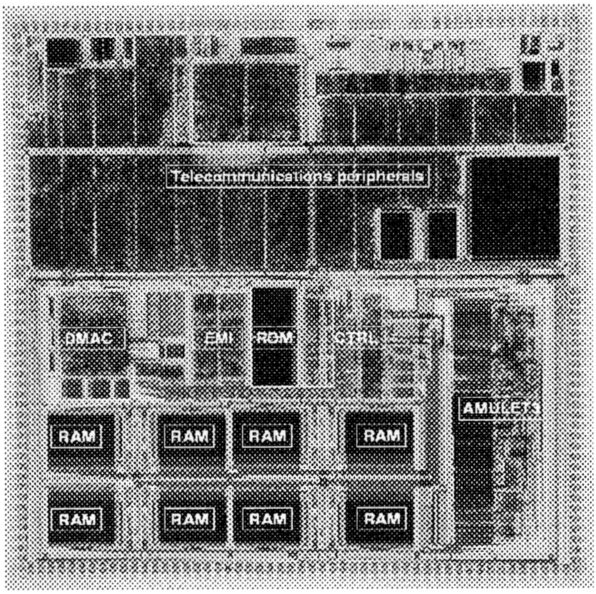

Figure 15.26. DRACO layout.

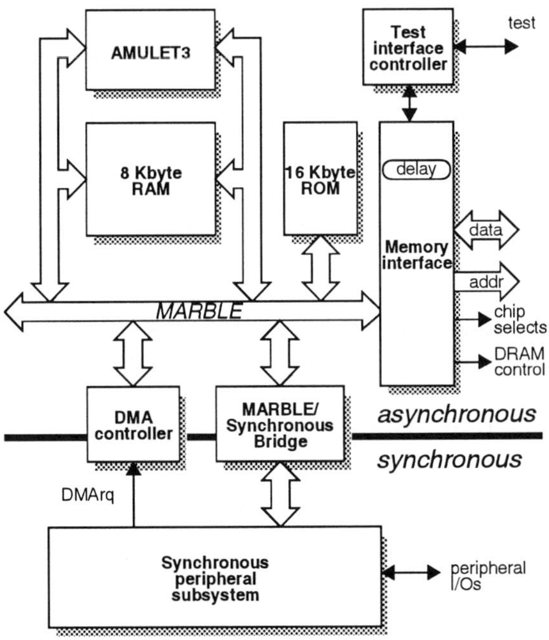

Figure 15.27. Amulet3i asynchronous subsystem.

asynchronous interconnection between their synchronous subsystems to alleviate problems with high-speed clock distribution and clock skew. This model is sometimes referred to as "GALS" (Globally Asynchronous, Locally Synchronous) and may represent an early commercial opportunity for the inclusion of asynchronous circuits in 'conventional' systems.

MARBLE. As a step in developing such an interconnection standard an Amulet3i contains the first implementation of MARBLE [5], a 32-bit, multimaster, on-chip bus which communicates by using handshakes rather than a clock. Apart from this the signal definitions, with 32-bit address and data, look very similar to a conventional bus. MARBLE separates address and data communications, allowing pipelining and interleaving of operations in order to increase the available bandwidth when several devices require global access.

MARBLE is supported by 'initiator' and 'target' interfaces which can be attached to any asynchronous component. These, their address, and the bus wiring provide all that is needed for communication between the various components. In Amulet3i there are four initiators and seven targets. For example the processor's two local buses each terminate in a MARBLE initiator and the local data bus is also a MARBLE target which allows DMA and test data in and out of the RAM from other initiators.

Chain. Chain ('Chip area interconnect') is currently under development as a possible replacement for a conventional bus for on-chip communications. Chain is based around narrow, high-speed, point-to-point links forming a network rather than a bus. The idea is to exploit the potential for fast symbol transmission within an asynchronous system while reducing the number of long distance wires.

By using a delay-insensitive coding scheme Chain relieves the chip designer of the need to ensure timing closure across the whole chip; it also provides tolerance of potential problems such as induced crosstalk on the long interconnection wires. Again the user need only communicate with 'conventional' parallel interfaces.

15.6.3 Balsa and the DMA controller

The DMA controller is a complex, multi-channel unit which was evolved according to an external specification. Whilst the function of the unit is relatively straightforward, even in the asynchronous domain, the unit is notable for being the first real application of Balsa synthesis [11].

The DMA controller comprises a large set of control registers for the many DMA channels and a control unit which chooses an active request and services it. The registers were designed as blocks of custom VLSI to optimise their area. The control logic was written in Balsa, and modified several times as the specification changed. The modifications proved remarkably easy to accommodate in this environment.

Such synthesis is not (yet) suitable for high-performance units, but proved extremely useful in such an application where the performance was limited by other constraints (such as the bus speed, here) and development time predominates. Of course, in an asynchronous environment, it is easy to accommodate devices in any performance range without affecting the overall system functionality.

Part II of this book is an introduction to Balsa and includes a complete source listing of a simpler 4-channel DMA controller.

15.6.4 Calibrated time delays

To be useful in real systems an asynchronous processor must be able to interface with currently available commodity parts. Whilst it is possible – and perhaps even desirable – to have memory devices with handshake interfaces, these are not available 'off the shelf'. Instead commodity memories rely on absolute timing assumptions to guarantee their correct operation and thus a system using them must have an accurate timing reference. This is the one thing lacking in an asynchronous design.

Introducing a clock purely to provide memory timing would negate many of the advantages of the asynchronous system; it is therefore preferable to retain the idea of providing data 'on demand', providing an adequately precise delay can be provided. This delay need not be a clock; a monostable which can be triggered on demand is preferable.

Amulet1 and Amulet2e relied on an externally supplied delay line to provide a bus timing reference. This is a flexible solution in that a short delay can be used repeatedly to provide longer timing intervals, providing a flexible, programmable interface. For example an area of address space could be set up for DRAM devices with (say) 1 delay address set up time, 2 delays RAS address hold, etc. The bus interface can then count delays as appropriate.

This is a reasonable solution but suffers from certain drawbacks:

- the on-chip delay for each delay cycle is not factored;
- delay lines are not particularly precise;
- driving an off-chip delay is power hungry.

A much better solution would be to use an on-chip delay. The chief problem with this is that a fixed delay will be imprecise, varying from chip to chip and also changing with temperature and supply voltage fluctuations. Any on-board delay must therefore be calibratable against an external reference.

Amulet3i uses such a delay. This comprises a chain of delay elements (figure 15.28) which can be 'shorted' at a determined point. This can be calibrated

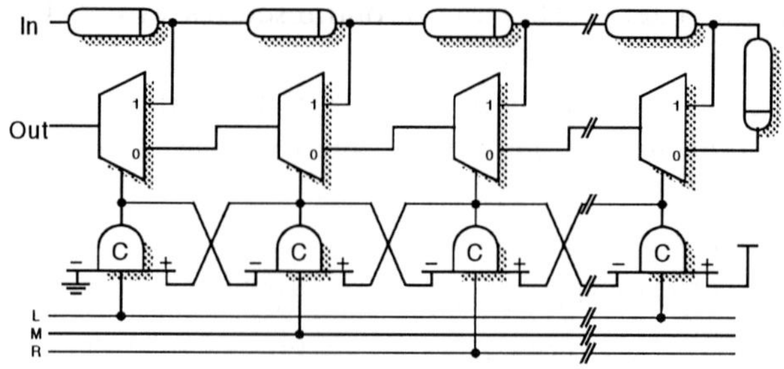

Figure 15.28. Controllable delay circuit.

by counting the number of cycles completed within a known interval and adjusted accordingly using the control wires shown. Once calibrated the delay will only change slowly (e.g. with temperature drift) unless the external conditions change; calibration can therefore be repeated at infrequent intervals under software control. The external timing reference can be a long delay such as a period of a 32 kHz 'watch' oscillator – hardly power expensive!

15.6.5 Production test

Figure 15.27 on page 311 includes a 'test interface controller' block. The DRACO chip was designed as a commercial product, and must therefore be testable in production. The test approach adopted in the design of DRACO is based upon exploiting the MARBLE bus to access the various on-chip modules and their test features. In normal use the external memory interface is a MARBLE target, but for test purposes it can be reconfigured as a bus initiator, enabling an external tester to control the bus and thereby read and write the on-chip memory. Circuits in the test interface controller make this efficient, with support for automatic sequential address generation, and so on.

All of the production tests for the asynchronous subsystem use the external tester to load a test program into the on-chip RAM and then instruct the Amulet3 processor to execute the program. The test runs without intervention from the tester, which must simply wait a given time before inspecting an on-chip memory location to read a pass/fail code. Inevitably the tests run at full speed; there is no external timing input to control them.

Certain on-chip modules are very difficult to test in purely functional mode, so additional access routes (via test registers accessible from MARBLE) are provided to make the tests more efficient. The calibrated time delay is one such circuit, and the processor branch predictor is another. In the latter case, the branch predictor is taken off line so that the processor (of which it is a part) can manipulate its internal state to run optimised tests on its CAM and RAM components.

Although DRACO is not, at the time of writing, in full-scale production, the tests ran without difficulty on prototype sample parts.

15.7. Summary

Devices such as DRACO have demonstrated that it is feasible to build large, functional asynchronous systems. Like any prototype system the chip has its problems. There are two (known) bugs in the asynchronous half of the system: an electrical drive strength problem within the multiplier (which failed to evidence during simulation) and a logic oversight in the prefetch unit which falsely indicates a cycle is 'sequential' under certain specific conditions (a problem if running code from off-chip DRAM). Neither of these is attributable to the asynchronous nature of the device (indeed, there are slightly more bugs in the synchronous part of the device!) and both are readily fixable. The processor is comparable with an ARM9 manufactured on the same process in terms of physical size, performance and energy efficiency; preliminary measurements suggest that is is significantly 'quieter' in terms of EMI.

This chapter has presented possible solutions (though certainly not the only ones!) to many of the problems facing the designer of complex asynchronous processing and memory systems. The majority of the designs described at the beginning of the chapter have been produced by academic groups and could be classified as "research"; however the complexity of a modern system on chip is such that these designs stretch the capability of even a large university group. It has been demonstrated that large, functional asynchronous designs are not only possible, but can be competitive and have some unique advantages in terms of power management and EMI. Asynchronous interconnection may be the only solution for large devices, even those with local clocks. Asynchronous chip design is ready to move to "development".

Epilogue

Asynchronous technology has existed since the first days of digital electronics – many of the earliest computers did not employ a central clock signal. However, with the development of integrated circuits the need for a straightforward design discipline that could scale up rapidly with the available transistor resource was pressing, and clocked design became the dominant approach. Today, most practising digital designers know very little about asynchronous techniques, and what they do know tends to discourage them from venturing into the territory. But clocked design is beginning to show signs of stress – its ability to scale is waning, and it brings with it growing problems of excessive power dissipation and electromagnetic interference.

During the reign of the clock, a few designers have remained convinced that asynchronous techniques have merit, and new techniques have been developed that are far better suited to the VLSI era than were the approaches employed on early machines. In this book we have tried to illuminate these new techniques in a way that is accessible to any practising digital circuit designer, whether or not they have had prior exposure to asynchronous circuits.

In this account of asynchronous design techniques we have had to be selective in order not to obscure the principal goal with arcane detail. Much work of considerable quality and merit has been omitted, and the reader whose interest has been ignited by this book will find that there is a great deal of published material available that exposes aspects of asynchronous design that have not been touched upon here.

Although there are commercial examples of VLSI devices based on asynchronous techniques (a couple of which have been described in this book), these are exceptions – most asynchronous development is still taking place in research laboratories. If this is to change in the future, where will this change first manifest itself?

The impending demise of clocked design has been forecast for many years and still has not happened. If it does happen, it will be for some compelling reason, since designers will not lightly cast aside their years of experience in one design style in favour of another style that is less proven and less well supported by automated tools.

There are many possible reasons for considering asynchronous design, but no single 'killer application' that makes its use obligatory. Several of the arguments for adopting asynchronous techniques mentioned at the start of this book – low power, low electromagnetic interference, modularity, etc. – are applicable in their own niches, but only the modularity argument has the potential to gain universal adoption. Here a promising approach that will support heterogeneous timing environments is GALS (Globally Asynhronous Locally Synchronous) system design. An asynchronous on-chip interconnect – a 'chip area network' such as Chain (described on page 312) – is used to connect clocked modules. The modules themselves can be kept small enough for clock skew to be well-contained so that straightforward synchronous design techniques work well, and different modules can employ different clocks or the same clock with different phases. Once this framework is in place, it is then clearly straightforward to make individual modules asynchronous on a case-by-case basis.

Here, perhaps unsurprisingly, we see the need to merge asynchronous technology with established synchronous design techniques, so most of the functional design can be performed using well-understood tools and approaches. This evolutionary approach contrasts with the revolutionary attacks described in Part III of this book, and represents the most likely scenario for the widespread adoption of the techniques described in this book in the medium-term future.

In the shorter term, however, the application niches that can benefit from asynchronous technology are important and viable. It is our hope in writing this book that more designers will come to understand the principles of asynchronous design and its potential to offer new solutions to old and new problems. Clocks are useful but they can become straitjackets. Don't be afraid to think outside the box!

For further information on asynchronous design see the bibliography at the end of this book, the *Asynchronous Bibliography* on the Internet [111], and the general information on asynchronous design available at the *Asynchronous Logic Homepage*, also on the Internet [47].

References

[1] G. Abouyannis et al. Project PREST EP25242. *European Low Power Initiative for Electronic System Design (ESDLPD) Third International Workshop*, pages 5–49, 2000.

[2] A. Abrial, J. Bouvier, M. Renaudin, P. Senn, and P. Vivet. A new contactless smartcard IC using an on-chip antenna and an asynchronous micro-controller. *IEEE Journal of Solid-State Circuits*, 36(7):1101–1107, July 2001.

[3] T. Agerwala. Putting Petri nets to work. *IEEE Computer*, 12(12):85–94, December 1979.

[4] W.J. Bainbridge and S.B. Furber. Asynchronous macrocell interconnect using MARBLE. In *Proc. International Symposium on Advanced Research in Asynchronous Circuits and Systems (Async'98)*, pages 122–132. IEEE Computer Society Press, April 1998.

[5] W.J. Bainbridge and S.B. Furber. MARBLE: An asynchronous on-chip macrocell bus. *Microprocessors and Microsystems*, 24(4):213–222, April 2000.

[6] T.S. Balraj and M.J. Foster. Miss Manners: A specialized silicon compiler for synchronizers. In Charles E. Leierson, editor, *Advanced Research in VLSI*, pages 3–20. MIT Press, April 1986.

[7] A. Bardsley. The Balsa web pages. `http://www.cs.man.ac.uk/amulet/balsa/projects/balsa`.

[8] A. Bardsley. *Implementing Balsa Handshake Circuits.* PhD thesis, Department of Computer Science, University of Manchester, 2000.

[9] A. Bardsley and D.A. Edwards. Compiling the language Balsa to delay-insensitive hardware. In C. D. Kloos and E. Cerny, editors, *Hardware Description Languages and their Applications (CHDL)*, pages 89–91, April 1997.

[10] A. Bardsley and D.A. Edwards. The Balsa asynchronous circuit synthesis system. In *Forum on Design Languages*, September 2000.

[11] A. Bardsley and D.A. Edwards. Synthesising an asynchronous DMA controller with Balsa. *Journal of Systems Architecture*, 46:1309–1319, 2000.

[12] P.A. Beerel, C.J. Myers, and T.H.-Y. Meng. Automatic synthesis of gate-level speed-independent circuits. Technical Report CSL-TR-94-648, Stanford University, November 1994.

[13] P.A. Beerel, C.J. Myers, and T.H.-Y. Meng. Covering conditions and algorithms for the synthesis of speed-independent circuits. *IEEE Transactions on Computer-Aided Design*, March 1998.

[14] G. Birtwistle and A. Davis, editors. *Proceedings of the Banff VIII Workshop: Asynchronous Digital Circuit Design, Banff, Alberta, Canada, August 28–September 3, 1993*. Springer Verlag, Workshops in Computing Science, 1995. Contributions from: S.B. Furber, "Computing without Clocks: Micropipelining the ARM Processor," A. Davis, "Practical Asynchronous Circuit Design: Methods and Tools," C.H. van Berkel, "VLSI Programming of Asynchronous Circuits for Low Power," J. Ebergen, "Parallel Program and Asynchronous Circuit Design," A. Davis and S. Nowick, "Introductory Survey".

[15] I. Bogdan, M. Munteau, P.A. Ivey, N.L. Seed, and N. Powell. Power reduction techniques for a Viterbi decoder implementation. *European Low Power Initiative for Electronic System Design (ESDLPD) Third International Workshop*, pages 28–48, 2000.

[16] E. Brinksma and T. Bolognesi. Introduction to the ISO specification language LOTOS. *Computer Networks and ISDN Systems*, 14(1), 1987.

[17] E. Brunvand and R.F. Sproull. Translating concurrent programs into delay-insensitive circuits. In *Proc. International Conf. Computer-Aided Design (ICCAD)*, pages 262–265. IEEE Computer Society Press, November 1989.

[18] J.A. Brzozowsky and C.-J.H. Seager. *Asynchronous Circuits*. Springer Verlag, Monographs in Computer Science, 1994. ISBN: 0-387-94420-6.

[19] S.M. Burns. *Performance Analysis and Optimization of Asynchronous Circuits*. PhD thesis, Computer Science Department, California Institute of Technology, 1991. Caltech-CS-TR-91-01.

[20] S.M. Burns. General condition for the decomposition of state holding elements. In *Proc. International Symposium on Advanced Research*

in Asynchronous Circuits and Systems. IEEE Computer Society Press, March 1996.

[21] S.M. Burns and A.J. Martin. Syntax-directed translation of concurrent programs into self-timed circuits. In J. Allen and F. Leighton, editors, *Advanced Research in VLSI*, pages 35–50. MIT Press, 1988.

[22] D.M. Chapiro. *Globally-Asynchronous Locally-Synchronous Systems*. PhD thesis, Stanford University, October 1984.

[23] K.T. Christensen, P. Jensen, P. Korger, and J. Sparsø. The design of an asynchronous TinyRISC TR4101 microprocessor core. In *Proc. International Symposium on Advanced Research in Asynchronous Circuits and Systems*, pages 108–119. IEEE Computer Society Press, 1998.

[24] T.-A. Chu. *Synthesis of Self-Timed VLSI Circuits from Graph-Theoretic Specifications*. PhD thesis, MIT Laboratory for Computer Science, June 1987.

[25] T.-A. Chu and R.K. Roy (editors). Special issue on asynchronous circuits and systems. *IEEE Design & Test*, 11(2), 1994.

[26] T.-A. Chu and L.A. Glasser. Synthesis of self-timed control circuits from graphs: An example. In *Proc. International Conf. Computer Design (ICCD)*, pages 565–571. IEEE Computer Society Press, 1986.

[27] B. Coates, A. Davis, and K. Stevens. The Post Office experience: Designing a large asynchronous chip. *Integration, the VLSI journal*, 15(3):341–366, October 1993.

[28] F. Commoner, A.W. Holt, S. Even, and A. Pnueli. Marked directed graphs. *J. Comput. System Sci.*, 5(1):511–523, October 1971.

[29] J. Cortadella, M. Kishinevsky, A. Kondratyev, L. Lavagno, and A. Yakovlev. Petrify: a tool for manipulating concurrent specifications and synthesis of asynchronous controllers. In *XI Conference on Design of Integrated Circuits and Systems*, Barcelona, November 1996.

[30] J. Cortadella, M. Kishinevsky, A. Kondratyev, L. Lavagno, and A. Yakovlev. Petrify: a tool for manipulating concurrent specifications and synthesis of asynchronous controllers. *IEICE Transactions on Information and Systems*, E80-D(3):315–325, March 1997.

[31] U. Cummings, A. Lines, and A. Martin. An asynchronous pipelined lattice structure filter. In *Proc. International Symposium on Advanced Research in Asynchronous Circuits and Systems*, pages 126–133, November 1994.

[32] A. Davis. A data-driven machine architecture suitable for VLSI implementation. In *Proceedings of the First Caltech Conference on VLSI*, pages 479–494, Pasadena, CA, January 1979.

[33] A. Davis and S.M. Nowick. Asynchronous circuit design: Motivation, background, and methods. In G. Birtwistle and A. Davis, editors, *Asynchronous Digital Circuit Design*, Workshops in Computing, pages 1–49. Springer-Verlag, 1995.

[34] A. Davis and S.M. Nowick. An introduction to asynchronous circuit design. Technical Report UUCS-97-013, Department of Computer Science, University of Utah, September 1997.

[35] A. Davis and S.M. Nowick. An introduction to asynchronous circuit design. In A. Kent and J. G. Williams, editors, *The Encyclopedia of Computer Science and Technology*, volume 38. Marcel Dekker, New York, February 1998.

[36] J.B. Dennis. Data Flow Computation. In *Control Flow and Data Flow — Concepts of Distributed Programming, International Summer School*, pages 343–398, Marktoberdorf, West Germany, July 31 – August 12, 1984. Springer, Berlin.

[37] J.C. Ebergen and R. Berks. Response time properties of linear asynchronous pipelines. *Proceedings of the IEEE*, 87(2):308–318, February 1999.

[38] P.B. Endecott and S.B. Furber. Modelling and simulation of asynchronous systems using the LARD hardware description language. In *Proceedings of the 12th European Simulation Multiconference, Manchester, Society for Computer Simulation International*, pages 39–43, June 1994. ISBN 1-56555-148-6.

[39] K.M. Fant and S.A. Brandt. Null Conventional Logic: A complete and consistent logic for asynchronous digital circuit synthesis. In *International Conference on Application-specific Systems, Architectures, and Processors*, pages 261–273, 1996.

[40] R.M. Fuhrer, S.M. Nowick, M. Theobald, N.K. Jha, B. Lin, and L. Plana. Minimalist: An environment for the synthesis, verification and testability of burst-mode asynchronous machines. Technical Report TR CUCS-020-99, Columbia University, NY, July 1999. http://www.cs.columbia.edu/~nowick/minimalist.pdf.

[41] S.B. Furber and P. Day. Four-phase micropipeline latch control circuits. *IEEE Transactions on VLSI Systems*, 4(2):247–253, June 1996.

[42] S.B. Furber, P. Day, J.D. Garside, N.C. Paver, S. Temple, and J.V. Woods. The design and evaluation of an asynchronous microprocessor. In *Proc. Int'l. Conf. Computer Design*, pages 217–220, October 1994.

[43] S.B. Furber, D.A. Edwards, and J.D. Garside. AMULET3: a 100 MIPS asynchronous embedded processor. In *Proc. International Conf. Computer Design (ICCD)*, September 2000.

[44] S.B. Furber, J.D. Garside, P. Riocreux, S. Temple, P. Day, J. Liu, and N.C. Paver. AMULET2e: An asynchronous embedded controller. *Proceedings of the IEEE*, 87(2):243–256, February 1999.

[45] S.B. Furber, J.D. Garside, S. Temple, J. Liu, P. Day, and N.C. Paver. AMULET2e: An asynchronous embedded controller. In *Proc. International Symposium on Advanced Research in Asynchronous Circuits and Systems*, pages 290–299. IEEE Computer Society Press, 1997.

[46] EU IST-1999-13515, G3Card – generation 3 smartcard, January 2000.

[47] J.D. Garside. The Asynchronous Logic Homepages. `http://www.cs.man.ac.uk/async/`.

[48] J.D. Garside, W.J. Bainbridge, A. Bardsley, D.A. Edwards, S.B. Furber, J. Liu, D.W. Lloyd, S. Mohammadi, J.S. Pepper, O. Petlin, S. Temple, and J.V. Woods. AMULET3i – an asynchronous system-on-chip. In *Proc. International Symposium on Advanced Research in Asynchronous Circuits and Systems*, pages 162–175. IEEE Computer Society Press, April 2000.

[49] J.D. Garside, S. Temple, and R. Mehra. The AMULET2e cache system. In *Proc. International Symposium on Advanced Research in Asynchronous Circuits and Systems (Async'96)*, pages 208–217. IEEE Computer Society Press, March 1996.

[50] D.A. Gilbert. *Dependency and Exception Handling in an Asynchronous Microprocessor*. PhD thesis, Department of Computer Science, University of Manchester, 1997.

[51] D.A. Gilbert and J.D. Garside. A result forwarding mechanism for asynchronous pipelined systems. In *Proc. International Symposium on Advanced Research in Asynchronous Circuits and Systems (Async'97)*, pages 2–11. IEEE Computer Society Press, April 1997.

[52] B. Gilchrist, J.H. Pomerene, and S.Y. Wong. Fast carry logic for digital computers. *IRE Transactions on Electronic Computers*, EC-4(4):133–136, December 1955.

[53] L.A. Glasser and D.W. Dobberpuhl. *The Design and Analysis of VLSI Circuits.* Addison-Wesley, 1985.

[54] S. Hauck. Asynchronous design methodologies: An overview. *Proceedings of the IEEE*, 83(1):69–93, January 1995.

[55] L.G. Heller, W.R. Griffin, J.W. Davis, and N.G. Thoma. Cascode voltage switch logic: A differential CMOS logic family. *Proc. International Solid State Circuits Conference*, pages 16–17, February 1984.

[56] J.L. Hennessy and D.A. Patterson. *Computer Architecture: A Quantitative Approach (2nd edition).* Morgan Kaufmann, 1996.

[57] C.A.R. Hoare. Communicating sequential processes. *Communications of the ACM*, 21(8):666–677, August 1978.

[58] C.A.R. Hoare. *Communicating Sequential Processes.* Prentice-Hall, Englewood Cliffs, 1985.

[59] D.A. Huffman. The synthesis of sequential switching circuits. *J. Franklin Inst.*, pages 161–190, 275–303, March/April 1954.

[60] D.A. Huffman. The synthesis of sequential switching circuits. In E. F. Moore, editor, *Sequential Machines: Selected Papers.* Addison-Wesley, 1964.

[61] H. Hulgaard, S.M. Burns, and G. Borriello. Testing asynchronous circuits: A survey. *Integration, the VLSI journal*, 19(3):111–131, November 1995.

[62] K. Hwang. *Computer Arithmetic: Principles, Architecture, and Design.* John Wiley & Sons, 1979.

[63] ISO/IEC. Mifare identification cards - contactless integrated circuit(s) cards - proximity cards. Standard ISO/IEC Standard 14443 Type A.

[64] H. Jacobson, E. Brunvand, G. Gopalakrishnan, and P. Kudva. High-level asynchronous system design using the ACK framework. In *Proc. International Symposium on Advanced Research in Asynchronous Circuits and Systems*, pages 93–103. IEEE Computer Society Press, April 2000.

[65] D. Jaggar. *Advanced RISC Machines Architecture Reference Manual.* Prentice Hall, 1996.

[66] M. Johnson. *Superscalar Microprocessor Design.* Series in Innovative Technology. Prentice Hall, 1991.

[67] S.C. Johnson and S. Mazor. Silicon compiler lets system makers design their own VLSI chips. *Electronic Design*, 32(20):167–181, 1984.

[68] G. Jones. *Programming in OCCAM*. Prentice-Hall international, 87.

[69] M.B. Josephs, S.M. Nowick, and C.H. van Berkel. Modeling and design of asynchronous circuits. *Proceedings of the IEEE*, 87(2):234–242, February 1999.

[70] G.C. Clark Jr. and J.B. Cain. *Error correcting coding for digital communication*. Plenum, 1981.

[71] G.D. Forney Jr. The Viterbi algorithm. *Proc. IEEE*, 13(3):268–278, 1973.

[72] G. Kane and J. Heinrich. *MIPS RISC Achitecture*. Prentice Hall, 1992.

[73] J. Kessels, T. Kramer, G. den Besten, A. Peeters, and V. Timm. Applying asynchronous circuits in contactless smart cards. In *Proc. International Symposium on Advanced Research in Asynchronous Circuits and Systems*, pages 36–44. IEEE Computer Society Press, April 2000.

[74] J. Kessels, T. Kramer, A. Peeters, and V. Timm. DESCALE: a design experiment for a smart card application consuming low energy. In R. van Leuken, R. Nouta, and A. de Graaf, editors, *European Low Power Initiative for Electronic System Design*, pages 247–262. Delft Institute of Microelectronics and Submicron Technology, July 2000.

[75] Z. Kohavi. *Switching and Finite Automata Theory*. McGraw-Hill, 1978.

[76] A. Kondratyev, J. Cortadella, M. Kishinevsky, L. Lavagno, and A. Yakovlev. Logic decomposition of speed-independent circuits. *Proceedings of the IEEE*, 87(2):347–362, February 1999.

[77] J. Liu. *Arithmetic and control components for an asynchronous microprocessor*. PhD thesis, Department of Computer Science, University of Manchester, 1997.

[78] D.W. Lloyd. VHDL models of asychronous handshaking. (Personal communication, August 1998).

[79] A.J. Martin. The probe: An addition to communication primitives. *Information Processing Letters*, 20(3):125–130, 1985. Erratum: IPL 21(2):107, 1985.

[80] A.J. Martin. Compiling communicating processes into delay-insensitive VLSI circuits. *Distributed Computing*, 1(4):226–234, 1986.

[81] A.J. Martin. Formal program transformations for VLSI circuit synthesis. In E.W. Dijkstra, editor, *Formal Development of Programs and Proofs*, UT Year of Programming Series, pages 59–80. Addison-Wesley, 1989.

[82] A.J. Martin. The limitations to delay-insensitivity in asynchronous circuits. In W.J. Dally, editor, *Advanced Research in VLSI: Proceedings of the Sixth MIT Conference*, pages 263–278. MIT Press, 1990.

[83] A.J. Martin. Programming in VLSI: From communicating processes to delay-insensitive circuits. In C.A.R. Hoare, editor, *Developments in Concurrency and Communication*, UT Year of Programming Series, pages 1–64. Addison-Wesley, 1990.

[84] A.J. Martin. Synthesis of asynchronous VLSI circuits, 1991.

[85] A.J. Martin. Asynchronous datapaths and the design of an asynchronous adder. *Formal Methods in System Design*, 1(1):119–137, July 1992.

[86] A.J. Martin, S.M. Burns, T. K. Lee, D. Borkovic, and P.J. Hazewindus. The design of an asynchronous microprocessor. In Charles L. Seitz, editor, *Advanced Research in VLSI*, pages 351–373. MIT Press, 1989.

[87] A.J. Martin, S.M. Burns, T.K. Lee, D. Borkovic, and P.J. Hazewindus. The first asynchronous microprocessor: The test results. *Computer Architecture News*, 17(4):95–98, 1989.

[88] A.J. Martin, A. Lines, R. Manohar, M. Nyström, P. Penzes, R. Southworth, U.V. Cummings, and T.-K. Lee. The design of an asynchronous MIPS R3000. In *Proceedings of the 17th Conference on Advanced Research in VLSI*, pages 164–181. MIT Press, September 1997.

[89] R. Milner. *Communication and Concurrency*. Prentice-Hall, 1989.

[90] C.E. Molnar, I.W. Jones, W.S. Coates, and J.K. Lexau. A FIFO ring oscillator performance experiment. In *Proc. International Symposium on Advanced Research in Asynchronous Circuits and Systems*, pages 279–289. IEEE Computer Society Press, April 1997.

[91] C.E. Molnar, I.W. Jones, W.S. Coates, J.K. Lexau, S.M. Fairbanks, and I.E. Sutherland. Two FIFO ring performance experiments. *Proceedings of the IEEE*, 87(2):297–307, February 1999.

[92] D.E. Muller. Asynchronous logics and application to information processing. In H. Aiken and W. F. Main, editors, *Proc. Symp. on Application of Switching Theory in Space Technology*, pages 289–297. Stanford University Press, 1963.

[93] D.E. Muller and W.S. Bartky. A theory of asynchronous circuits. In *Proceedings of an International Symposium on the Theory of Switching, Cambridge, April 1957, Part I*, pages 204–243. Harvard University Press, 1959. The annals of the computation laboratory of Harvard University, Volume XXIX.

[94] T. Murata. Petri Nets: Properties, Analysis and Applications. *Proceedings of the IEEE*, 77(4):541–580, April 1989.

[95] C.J. Myers. *Asynchronous Circuit Design.* John Wiley & Sons, July 2001. ISBN: 0-471-41543-X.

[96] National Bureau of Standards. Data encryption standard, January 1997. Federal Information Processing Standards Publication 46.

[97] C.D. Nielsen. Evaluation of function blocks for asynchronous design. In *Proc. European Design Automation Conference (EURO-DAC)*, pages 454–459. IEEE Computer Society Press, September 1994.

[98] C.D. Nielsen, J. Staunstrup, and S.R. Jones. Potential performance advantages of delay-insensitivity. In M. Sami and J. Calzadilla-Daguerre, editors, *Proceedings of IFIP workshop on Silicon Architectures for Neural Nets, StPaul-de-Vence, France, November 1990.* North-Holland, Amsterdam, 1991.

[99] L.S. Nielsen. *Low-power Asynchronous VLSI Design.* PhD thesis, Department of Information Technology, Technical University of Denmark, 1997. IT-TR:1997-12.

[100] L.S. Nielsen, C. Niessen, J. Sparsø, and C.H. van Berkel. Low-power operation using self-timed circuits and adaptive scaling of the supply voltage. *IEEE Transactions on VLSI Systems*, 2(4):391–397, 1994.

[101] L.S. Nielsen and J. Sparsø. A low-power asynchronous data-path for a FIR filter bank. In *Proc. International Symposium on Advanced Research in Asynchronous Circuits and Systems*, pages 197–207. IEEE Computer Society Press, 1996.

[102] L.S. Nielsen and J. Sparsø. An 85 μW asynchronous filter-bank for a digital hearing aid. In *Proc. IEEE International Solid State circuits Conference*, pages 108–109, 1998.

[103] L.S. Nielsen and J. Sparsø. Designing asynchronous circuits for low power: An IFIR filter bank for a digital hearing aid. *Proceedings of the IEEE*, 87(2):268–281, February 1999. Special issue on "Asynchronous Circuits and Systems" (Invited Paper).

[104] D.C. Noice. *A Two-Phase Clocking Dicipline for Digital Integrated Circuits.* PhD thesis, Department of Electrical Engineering, Stanford University, February 1983.

[105] S.M. Nowick. Design of a low-latency asynchronous adder using speculative completion. *IEE Proceedings, Computers and Digital Techniques*, 143(5):301–307, September 1996.

[106] S.M. Nowick, M.B. Josephs, and C.H. van Berkel (editors). Special issue on asynchronous circuits and systems. *Proceedings of the IEEE*, 87(2), February 1999.

[107] S.M. Nowick, K.Y. Yun, and P.A. Beerel. Speculative completion for the design of high-performance asynchronous dynamic adders. In *Proc. International Symposium on Advanced Research in Asynchronous Circuits and Systems*, pages 210–223. IEEE Computer Society Press, April 1997.

[108] International Standards Organization. LOTOS — a formal description technique based on the temporal ordering of observational behaviour. ISO IS 8807, 1989.

[109] N.C. Paver, P. Day, C. Farnsworth, D.L. Jackson, W.A. Lien, and J. Liu. A low-power, low-noise configurable self-timed DSP. In *Proc. International Symposium on Advanced Research in Asynchronous Circuits and Systems*, pages 32–42, 1998.

[110] M. Pedersen. Design of asynchronous circuits using standard CAD tools. Technical Report IT-E 774, Technical University of Denmark, Dept. of Information Technology, 1998. (In Danish).

[111] A.M.G. Peeters. The 'Asynchronous' Bibliography.
`http://www.win.tue.nl/~wsinap/async.html`.
Corresponding e-mail address: `async-bib@win.tue.nl`.

[112] A.M.G. Peeters. *Single-Rail Handshake Circuits.* PhD thesis, Eindhoven University of Technology, June 1996.
`http://www.win.tue.nl/~wsinap/pdf/Peeters96.pdf`.

[113] J.L. Peterson. Petri nets. *Computing Surveys*, 9(3):223–252, September 1977.

[114] Philips Semiconductors. PCA5007 handshake-technology pager IC data sheet. http://www.semiconductors.philips.com/pip/PCA5007H.

[115] J. Rabaey. *Digital Integrated Circuits: A Design Perspective.* Prentice-Hall, 1996.

[116] P. Rakers, L. Connell, T. Collins, and D. Russell. Secure contactless smartcard ASIC with DPA protection. *IEEE Journal of Solid-State Circuits*, 36(3):559–565, March 2001.

[117] M. Renaudin, P. Vivet, and F. Robin. ASPRO-216: A standard-cell QDI 16-bit RISC asynchronous microprocessor. In *Proc. International Symposium on Advanced Research in Asynchronous Circuits and Systems (Async'98)*, pages 22–31. IEEE Computer Society Press, April 1998.

[118] M. Renaudin, P. Vivet, and F. Robin. A design framework for asynchronous/synchronous circuits based on CHP to HDL translation. In *Proc. International Symposium on Advanced Research in Asynchronous Circuits and Systems*, pages 135–144, April 1999.

[119] R. Rivest, A. Shamir, and L. Adleman. A method for obtaining digital signatures and public-key crypto systems, June 1978.

[120] M. Roncken. Defect-oriented testability for asynchronous ICs. *Proceedings of the IEEE*, 87(2):363–375, February 1999.

[121] C.L. Seitz. System timing. In C.A. Mead and L.A. Conway, editors, *Introduction to VLSI Systems*, chapter 7. Addison-Wesley, 1980.

[122] N.P. Singh. A design methodology for self-timed systems. Master's thesis, Laboratory for Computer Science, MIT, 1981. MIT/LCS/TR-258.

[123] J. Sparsø, C.D. Nielsen, L.S. Nielsen, and J. Staunstrup. Design of self-timed multipliers: A comparison. In S. Furber and M. Edwards, editors, *Asynchronous Design Methodologies*, volume A-28 of *IFIP Transactions*, pages 165–180. Elsevier Science Publishers, 1993.

[124] J. Sparsø and J. Staunstrup. Delay-insensitive multi-ring structures. *INTEGRATION, the VLSI Journal*, 15(3):313–340, October 1993.

[125] J. Sparsø, J. Staunstrup, and M. Dantzer-Sørensen. Design of delay insensitive circuits using multi-ring structures. In G. Musgrave, editor, *Proc. of EURO-DAC '92, European Design Automation Conference, Hamburg, Germany, September 7-10, 1992*, pages 15–20. IEEE Computer Society Press, 1992.

[126] R.F. Sproull, I.E. Sutherland, and C.E. Molnar. The counterflow pipeline processor architecture. *IEEE Design & Test of Computers*, 11(3):48–59, 1994.

[127] L. Stok. *Architectural Synthesis and Optimization of Digital Systems.* PhD thesis, Eindhoven University of Technology, 1991.

[128] I.E. Sutherland. Micropipelines. *Communications of the ACM*, 32(6):720–738, June 1989.

[129] Synopsys, Inc. *Synopsys VSS Family Core Programs Manual*, 1997.

[130] A. Takamura, M. Kuwako, M. Imai, T. Fujii, M. Ozawa, I. Fukasaku, Y. Ueno, and T. Nanya. TITAC-2: An asynchronous 32-bit microprocessor based on scalable-delay-insensitive model. In *Proc. International Conf. Computer Design (ICCD'97)*, pages 288–294. MIT Press, October 1997.

[131] H. Terada, S. Miyata, and M. Iwata. DDMPs: Self-timed super-pipelined data-driven multimedia processors. *Proceedings of the IEEE*, 87(2):282–296, February 1999.

[132] M. Theobald and S.M. Nowick. Transformations for the synthesis and optimization of asynchronous distributed control. In *Proc. ACM/IEEE Design Automation Conference*, 2001.

[133] S.H. Unger. *Asynchronous Sequential Switching Circuits*. Wiley-Interscience, John Wiley & Sons, Inc., New York, 1969.

[134] C.H. van Berkel. Beware the isochronic fork. *INTEGRATION, the VLSI journal*, 13(3):103–128, 1992.

[135] C.H. van Berkel. *Handshake Circuits: an Asynchronous Architecture for VLSI Programming*, volume 5 of *International Series on Parallel Computation*. Cambridge University Press, 1993.

[136] C.H. van Berkel, R. Burgess, J. Kessels, A. Peeters, M. Roncken, and F. Schalij. Asynchronous circuits for low power: a DCC error corrector. *IEEE Design & Test*, 11(2):22–32, 1994.

[137] C.H. van Berkel, R. Burgess, J. Kessels, A. Peeters, M. Roncken, and F. Schalij. A fully asynchronous low-power error corrector for the DCC player. In *ISSCC 1994 Digest of Technical Papers*, volume 37, pages 88–89. IEEE, 1994. ISSN 0193-6530.

[138] C.H. van Berkel, R. Burgess, J. Kessels, A. Peeters, M. Roncken, F. Schalij, and R. van de Viel. A single-rail re-implementation of a DCC error detector using a generic standard-cell library. In *2nd Working Conference on Asynchronous Design Methodologies, London, May 30-31, 1995*, pages 72–79, 1995.

[139] C.H. van Berkel, F. Huberts, and A. Peeters. Stretching quasi delay insensitivity by means of extended isochronic forks. In *Asynchronous*

Design Methodologies, pages 99–106. IEEE Computer Society Press, May 1995.

[140] C.H. van Berkel, M.B. Josephs, and S.M. Nowick. Scanning the technology: Applications of asynchronous circuits. *Proceedings of the IEEE*, 87(2):223–233, February 1999.

[141] C.H. van Berkel, J. Kessels, M. Roncken, R. Saeijs, and F. Schalij. The VLSI-programming language Tangram and its translation into handshake circuits. In *Proc. European Conference on Design Automation (EDAC)*, pages 384–389, 1991.

[142] C.H. van Berkel, C. Niessen, M. Rem, and R. Saeijs. VLSI programming and silicon compilation. In *Proc. International Conf. Computer Design (ICCD)*, pages 150–166, Rye Brook, New York, 1988. IEEE Computer Society Press.

[143] H. van Gageldonk. *An Asynchronous Low-Power 80C51 Microcontroller*. PhD thesis, Dept. of Math. and C.S., Eindhoven Univ. of Technology, September 1998.

[144] H. van Gageldonk, D. Baumann, C.H. van Berkel, D. Gloor, A. Peeters, and G. Stegmann. An asynchronous low-power 80c51 microcontroller. In *Proc. International Symposium on Advanced Research in Asynchronous Circuits and Systems*, pages 96–107. IEEE Computer Society Press, April 1998.

[145] P. Vanbekbergen. *Synthesis of Asynchronous Control Circuits from Graph-Theoretic Specifications*. PhD thesis, Catholic University of Leuven, September 1993.

[146] V.I. Varshavsky, M.A. Kishinevsky, V.B. Marakhovsky, V.A. Peschansky, L.Y. Rosenblum, A.R. Taubin, and B.S. Tzirlin. *Self-timed Control of Concurrent Processes*. Kluwer Academic Publisher, 1990. V.I.Varshavsky Ed., (Russian edition: 1986).

[147] T. Verhoeff. Delay-insensitive codes - an overview. *Distributed Computing*, 3(1):1–8, 1988.

[148] A.J. Viterbi. Error bounds for convolutional codes and an asymptotically optimum decoding algorithm. In *IEEE Transactions on Information Theory*, volume 13, pages 260–269, 1967.

[149] P. Viviet and M. Renaudin. CHP2VHDL, a CHP to VHDL translator - towards asynchronous-design simulation. In L. Lavagno and M.B.

Josephs, editors, *Handouts from the ACiD-WG Workshop on Specification models and languages and technology effects of asynchronous design*. Dipartemento di Elettronica, Polytecnico de Torino, Italy, January 1998.

[150] J.F. Wakerly. *Digital Design: Principles and Practices, 3/e*. Prentice-Hall, 2001.

[151] N. Weste and K. Esraghian. *Principles of CMOS VLSI Design – A systems Perspective, 2nd edition*. Addison-Wesley, 1993.

[152] Rik van de Wiel. High-level test evaluation of asynchronous circuits. In *Asynchronous Design Methodologies*, pages 63–71. IEEE Computer Society Press, May 1995.

[153] T.E. Williams. *Self-Timed Rings and their Application to Division*. PhD thesis, Department of Electrical Engineering and Computer Science, Stanford University, 1991. CSL-TR-91-482.

[154] T.E. Williams. Analyzing and improving latency and throughput in self-timed rings and pipelines. In *Tau-92: 1992 Workshop on Timing Issues in the Specification and Synthesis of Digital Systems*. ACM/SIGDA, March 1992.

[155] T.E. Williams. Performance of iterative computation in self-timed rings. *Journal of VLSI Signal Processing*, 6(3), October 1993.

[156] T.E. Williams and M.A. Horowitz. A zero-overhead self-timed 160 ns. 54 bit CMOS divider. *IEEE Journal of Solid State Circuits*, 26(11):1651–1661, 1991.

[157] T.E. Williams, N. Patkar, and G. Shen. SPARC64: A 64-b 64-active-instruction out-of-order-execution MCM processor. *IEEE Journal of Solid-State Circuits*, 30(11):1215–1226, November 1995.

[158] J.V. Woods, P. Day, S.B. Furber, J.D. Garside, N.C. Paver, and S. Temple. AMULET1: An asynchronous ARM processor. *IEEE Transactions on Computers*, 46(4):385–398, April 1997.

[159] C. Ykman-Couvreur, B. Lin, and H. de Man. Assassin: A synthesis system for asynchronous control circuits. Technical report, IMEC, September 1994. User and Tutorial manual.

[160] K.Y. Yun and D.L. Dill. Automatic synthesis of extended burst-mode circuits: Part II (automatic synthesis). *IEEE Transactions on Computer-Aided Design*, 18(2):118–132, February 1999.

Index